The IMA Volumes
in Mathematics
and Its Applications

Volume 17

Series Editors
Avner Friedman   Willard Miller, Jr.

Institute for Mathematics and its Applications
IMA

The **Institute for Mathematics and its Applications** was established by a grant from the National Science Foundation to the University of Minnesota in 1982. The IMA seeks to encourage the development and study of fresh mathematical concepts and questions of concern to the other sciences by bringing together mathematicians and scientists from diverse fields in an atmosphere that will stimulate discussion and collaboration.

The IMA Volumes are intended to involve the broader scientific community in this process.

Avner Friedman, Director
Willard Miller, Jr., Associate Director

* * * * * * * * * *

IMA PROGRAMS

1982-1983	**Statistical and Continuum Approaches to Phase Transition**
1983-1984	**Mathematical Models for the Economics of Decentralized Resource Allocation**
1984-1985	**Continuum Physics and Partial Differential Equations**
1985-1986	**Stochastic Differential Equations and Their Applications**
1986-1987	**Scientific Computation**
1987-1988	**Applied Combinatorics**
1988-1989	**Nonlinear Waves**
1989-1990	**Dynamical Systems and Their Applications**

* * * * * * * * * *

SPRINGER LECTURE NOTES FROM THE IMA:

The Mathematics and Physics of Disordered Media
  Editors: Barry Hughes and Barry Ninham
  (Lecture Notes in Math., Volume 1035, 1983)

Orienting Polymers
  Editor: J.L. Ericksen
  (Lecture Notes in Math., Volume 1063, 1984)

New Perspectives in Thermodynamics
  Editor: James Serrin
  (Springer-Verlag, 1986)

Models of Economic Dynamics
  Editor: Hugo Sonnenschein
  (Lecture Notes in Econ., Volume 264, 1986)

Fred Roberts
Editor

Applications of Combinatorics and Graph Theory to the Biological and Social Sciences

With 213 Illustrations

Springer-Verlag
New York Berlin Heidelberg
London Paris Tokyo Hong Kong

Fred Roberts
Department of Mathematics
Rutgers University
New Brunswick, NJ 08903, USA

Mathematics Subject Classification: 0502, 92-02

Library of Congress Cataloging-in-Publication Data
Roberts, Fred S.
  Applications of combinatorics and graph theory to the biological and social sciences.
  (The IMA volumes in mathematics and its applications; v. 17)
  "Bases on the proceedings of a workshop which was an integral part of the 1987–88 IMA program on applied combinatorics"—Foreword.
  1. Biomathematics—Congresses. 2. Social sciences—Mathematics—Congresses. 3. Combinatorial analysis—Congresses. 4. Graph theory—Congresses. I. Title. II. Series.
  QH323.5.R57   1989   574'.01'51   89-11271

Printed on acid-free paper

© 1989 by Springer-Verlag New York Inc.
All rights reserved. This work may not be translated or copied in whole or in part without the written permission of the publisher (Springer-Verlag, 175 Fifth Avenue, New York, NY 10010, USA), except for brief excerpts in connection with reviews or scholarly analysis. Use in connection with any form of information storage and retrieval, electronic adaptation, computer software, or by similar or dissimilar methodology now known or hereafter developed is forbidden.
The use of general descriptive names, trade names, trademarks, etc. in this publication, even if the former are not especially identified, is not to be taken as a sign that such names, as understood by the Trade Marks and Merchandise Marks Act, may accordingly be used freely by anyone.
Permission to photocopy for internal or personal use, or the internal or personal use of specific clients, is granted by Springer-Verlag New York Inc. for libraries registered with the Copyright Clearance Center (CCC), provided that the base fee of $0.00 per copy, plus $0.20 per page is paid directly to CCC, 21 Congress Street, Salem, MA 01970, USA. Special requests should be addressed directly to Springer-Verlag New York, 175 Fifth Avenue, New York, NY 10010, USA.
ISBN 0-387-97046-0/1989 $0.00 + 0.20

Camera-ready copy prepared by the Institute for Mathematics and Its Applications.
Printed and bound by Edwards Brothers, Ann Arbor, Michigan.
Printed in the United States of America.

9 8 7 6 5 4 3 2 1

ISBN 0-387-97046-0 Springer-Verlag New York Berlin Heidelberg
ISBN 3-540-97046-0 Springer-Verlag Berlin Heidelberg New York

The IMA Volumes
in Mathematics and its Applications

Current Volumes:

Volume 1: Homogenization and Effective Moduli of Materials and Media
   Editors: Jerry Ericksen, David Kinderlehrer, Robert Kohn, J.-L. Lions

Volume 2: Oscillation Theory, Computation, and Methods of Compensated Compactness
   Editors: Constantine Dafermos, Jerry Ericksen,
   David Kinderlehrer, Marshall Slemrod

Volume 3: Metastability and Incompletely Posed Problems
   Editors: Stuart Antman, Jerry Ericksen, David Kinderlehrer, Ingo Müller

Volume 4: Dynamical Problems in Continuum Physics
   Editors: Jerry Bona, Constantine Dafermos, Jerry Ericksen,
   David Kinderlehrer

Volume 5: Theory and Applications of Liquid Crystals
   Editors: Jerry Ericksen and David Kinderlehrer

Volume 6: Amorphous Polymers and Non-Newtonian Fluids
   Editors: Constantine Dafermos, Jerry Ericksen, David Kinderlehrer

Volume 7: Random Media
   Editor: George Papanicolaou

Volume 8: Percolation Theory and Ergodic Theory of Infinite Particle Systems
   Editor: Harry Kesten

Volume 9: Hydrodynamic Behavior and Interacting Particle Systems
   Editor: George Papanicolaou

Volume 10: Stochastic Differential Systems, Stochastic Control Theory and Applications
   Editors: Wendell Fleming and Pierre-Louis Lions

Volume 11: Numerical Simulation in Oil Recovery
   Editor: Mary Fanett Wheeler

Volume 12: Computational Fluid Dynamics and Reacting Gas Flows
   Editors: Bjorn Engquist, M. Luskin, Andrew Majda

Volume 12: Computational Fluid Dynamics and Reacting Gas Flows
Editors: Bjorn Engquist, M. Luskin, Andrew Majda

Volume 13: Numerical Algorithms for Parallel Computer Architectures
Editor: Martin H. Schultz

Volume 14: Mathematical Aspects of Scientific Software
Editor: J.R. Rice

Volume 15: Mathematical Frontiers in Computational Chemical Physics
Editor: D. Truhlar

Volume 16: Mathematics in Industrial Problems
by Avner Friedman

Volume 17: Applications of Combinatorics and Graph Theory to the Biological and Social Sciences
Editor: Fred Roberts

Forthcoming Volumes:

1987-1988: *Applied Combinatorics*

q-Series and Partitions

Invariant Theory and Tableaux

Coding Theory and Applications

Design Theory and Applications

Summer Program 1988: *Signal Processing*

Signal Processing (Volume 1)

Signal Processing (Volume 2)

1988-1989: *Nonlinear Waves*

Solitons in Physics and Mathematics

Solitons in Nonlinear Optics and Plasma Physics

Two Phase Waves in Fluidized Beds, Sedimenation, and Granular Flows

Nonlinear Evolution Equations that Change Type

CONTENTS

Foreword ix

Applications of Combinatorics and Graph Theory to the
Biological and Social Sciences:
Seven Fundamental Ideas 1
   Fred S. Roberts

Social Welfare and Aggregation Procedures: Combinatorial
and Algorithmic Aspects... 39
   Jean-Pierre Barthélemy

Consecutive One's Properties for Matrices and Graphs
Including Variable Diagonal Entries 75
   Margaret B. Cozzens and N.V.R. Mahadev

Probabilistic Knowledge Spaces: A Review 95
   Jean-Claude Falmagne

Uniqueness in Finite Measurement 103
   Peter C. Fishburn and Fred S. Roberts

Conceptual Scaling 139
   Bernhard Ganter and Rudolf Wille

The Micro-Macro Connection: Exact Structure and Process 169
   Eugene C. Johnsen

Sign-Patterns and Stability 203
   Victor Klee

Food Webs, Competition Graphs, Competition-Common
Enemy Graphs and Niche Graphs 221
   J. Richard Lundgren

Qualitatively Stable Matrices and Convergent Matrices 245
   John S. Maybee

Tree Structures in Immunology 259
   J.K. Percus

Meaningless Statements, Matching Experiments,
and Colored Digraphs (Applications of Graph Theory
and Combinatorics to the Theory of Measurement 277
   Fred S. Roberts

Combinatorial Aspects of Enzyme Kinetics 295
   Peter H. Sellers

Spatial Models of Power and Voting Outcomes 315
   Philip D. Straffin, Jr.

Some Mathematics for DNA Restriction Mapping 337
   Michael S. Waterman

FOREWORD

This IMA Volume in Mathematics and its Applications

Applications of Combinatorics and Graph Theory to the Biological and Social Sciences

is based on the proceedings of a workshop which was an integral part of the 1987-88 IMA program on APPLIED COMBINATORICS. We are grateful to the Scientific Committee: Victor Klee (Chairman), Daniel Kleitman, Dijen Ray-Chaudhuri and Dennis Stanton for planning and implementing an exciting and stimulating year-long program. We especially thank the Workshop Organizers, Joel Cohen and Fred Roberts, for organizing a workshop which brought together many of the major figures in a variety of research fields connected with the application of combinatorial ideas to the social and biological sciences.

Avner Friedman

Willard Miller

APPLICATIONS OF COMBINATORICS AND GRAPH THEORY TO THE BIOLOGICAL AND SOCIAL SCIENCES: SEVEN FUNDAMENTAL IDEAS

Fred S. Roberts[*]

Abstract. To set the stage for the other papers in this volume, seven fundamental concepts which arise in the applications of combinatorics and graph theory in the biological and social sciences are described. These ideas are: RNA chains as "words" in a 4 letter alphabet; interval graphs; competition graphs or niche overlap graphs; qualitative stability; balanced signed graphs; social welfare functions; and semiorders. For each idea, some basic results are presented, some recent results are given, and some open problems are mentioned.

Introduction. This paper started out as an overview of the applications of combinatorics and graph theory in the biological and social sciences, or at least of those applications which are emphasized in this volume. I soon realized it was not possible to give such an overview in a reasonable amount of space. I therefore decided to pick out some fundamental ideas which reflect many of the topics in this field. In choosing these ideas, I was naturally prejudiced by ideas which have had a profound influence on my own work. Here is my list of seven ideas, each of which has had remarkable influence on the development of the field:

Seven Fundamental Ideas

(1) RNA chains as "words" in a 4 letter alphabet

(2) Interval graphs

(3) Competition graphs or niche overlap graphs

(4) Qualitative stability

(5) Balanced signed graphs

(6) Social welfare functions

(7) Semiorders.

In this paper, I will define each of these ideas and illustrate it with a simple example. For each idea, I will state some basic results, mention one or two recent results, and describe one or two open problems. However, this paper is by no means a survey, even of the ramifications of the fundamental ideas I will discuss. Rather, I will use these seven ideas to set the stage for the more detailed papers which follow in this volume.

---

[*]Department of Mathematics, Rutgers University, New Brunswick, N.J. 08903
Acknowledgement: The author thanks Garth Isaak and Pey-chun Chen for their helpful comments. He also thanks the Air Force Office of Scientific Research for its support under Grants AFOSR-85-0271 and AFOSR-89-0066 to Rutgers University.

1. RNA Chains as "Words" in a 4-letter Alphabet. Perhaps the combinatorial area in biology that engages the most attention today is the analysis of protein, DNA, and RNA sequences. It is now well-known that information storage within a cell is by means of long nucleic molecules, which can be thought of as long strings of smaller units called nucleotides. For instance, in ribonucleic acids – RNA – each nucleotide for simplicity is one of four "bases" denoted by A (Adenine), C (Cytosine), G (Guanine), and U (Uracil). Sample RNA chains are: $GUGGAAACCUU$, $AACGGUAUCGUU$, $UAUAUGUCCCUCCCUUG$.

The task of determining the sequence of bases in an RNA chain is very important, and not always easy. The first nucleic acid sequence was determined in 1965 by R.W. Holley and his co-workers at Cornell (Holley, et al [1965]). It contained 77 bases (a short chain for nucleic acids). The method used was called the **fragmentation stratagem**. Although this method was used only for a short period of time, it had great historical significance and illustrates an important role for mathematical analysis, and so I describe it briefly here.

Suppose we are given a long unspecified RNA chain. How do we uncover it? Consider

$$CCGAUCGGC.$$

If we know all the bases, but not their order, there are many possible chains. A simple counting argument shows that if there are K_1 A's, K_2 C's, K_3 G's, and K_4 U's, then the number of RNA chains with this makeup is

$$\frac{(K_1 + K_2 + K_3 + K_4)!}{K_1!K_2!K_3!K_4!}$$

Thus, there are

$$\frac{9!}{1!4!3!1!} = 2520$$

RNA chains with the given base makeup.

There are enzymes which break up an RNA chain after each G. After applying such an enzyme, we get

G fragments: CCG, $AUCG$, G, C

We don't know the order of the fragments. However, we can put them back together to find all RNA chains with these G fragments. There are 4! such chains. Actually, we can say more. The fragment C is "abnormal" since it doesn't end in G. It must come last. Thus, there are only $3! = 6$ chains with the given G fragments. (Getting this down to 6 possible chains is certainly much better than getting it down to 2520.)

There are enzymes which break up the chain after each U or C. Applying such an enzyme, we get

U, C fragments: C, C, GAU, C, GGC

The number of RNA chains with this makeup is $5!/3! = 20$. (It is not 5! because C appears three times.) (This is still better than 2520.)

We have not yet made use of both pieces of information. Suppose we list all RNA chains with the given G fragments and see which of them has the proper U, C fragments. The possible chains are

$$CCGAUCGGC$$
$$CCGGAUCGC$$
$$AUCGCCGGC$$
$$AUCGGCCGC$$
$$GCCGAUCGC$$
$$GAUCGCCGC.$$

Of these, only the first has the proper U, C fragments. Thus, we have uncovered the RNA chain uniquely!

This fragmentation stratagem does not always give us the original chain uniquely. (Exercise: Find a smallest "ambiguous" example.)

George Hutchinson showed in [1969] that the fragmentation can be systematized by using graph theory. The G and U, C fragments are used to build a digraph (actually a multidigraph) and then all possible RNA chains with the given fragments correspond to closed eulerian paths in this digraph which begin and end at a designated vertex. For more detailed discussions of the fragmentation stratagem, see Mosimann [1968] or Roberts [1984B].

It should be reiterated that the fragmentation stratagem is not used anymore today, and indeed was used only for a short time before other, more efficient, methods were adopted. I mentioned it for historical reasons and also as an example of how a biological question was solved by the development of new mathematical results. Nowadays, by the use of radioactive marking and high-speed computer analysis, it is possible to sequence long RNA chains rather quickly, and it is becoming feasible to think of sequencing the entire 3-billion base long human genome. The human genome is the total genetic complement of the cell, all the genes on all the chromosomes. There are approximately 100,000 genes distributed in 23 chromosomes. **Mapping** the human genome would require localizing each of its genes; **sequencing** it would requiring determining the exact order of the thousand or more nucleotides which make up each gene. Sequencing the entire human genome would ultimately make it possible to devise ways to treat such genetic disorders as Alzheimer's disease and cystic fibrosis. For more on this topic from a non-technical point of view, see DeLisi [1988] and Pieper [1989].

As for recent work, I wish to mention as an example the work of Waterman and Griggs [1986] on restriction maps of DNA chains. For more on restriction maps, see the paper by Waterman [1989] in this volume and see Goldstein and Waterman [1987]. Mathematically, a DNA chain is like an RNA chain, except THYMINE (T) replaces URACIL (U). In modern molecular biology, one can use **restriction enzymes** to cut the DNA molecule at all occurrences of short specific sequences. The result of such cuts is to leave us with certain sites of two kinds, here denoted A and B, along the linear structure. See Figure 1. Let us call the linear structure with cuts an **A.B restriction map**. It is useful to look at the sub-maps consisting of just A's and just B's. We do not know the ordering of sites on the $A.B$ map or the A sub-map or the B sub-map, but we know something about the (open) intervals between successive sites on these maps. Let us call such an interval an **A-interval**, a **B-interval**, or an **A.B-interval** if it is between two successive sites on, respectively, the A sub-map, the B sub-map, or the $A.B$ map. Again, see Figure 1.

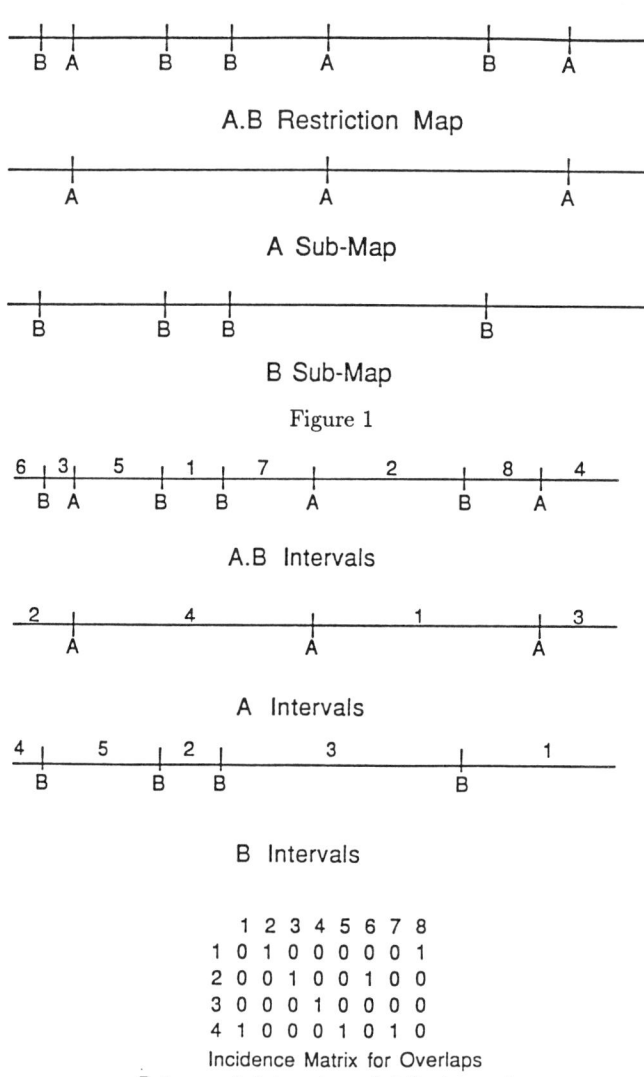

Figure 2

From biochemical data, we know whether or not an A-interval and an $A.B$-interval overlap, and the same for B-intervals and $A.B$-intervals. (Figure 2 shows an arbitrary labelling of the intervals in Figure 1 and the incidence matrix for the overlaps between A subintervals and B subintervals.) The questions now are: Can we determine if the data is consistent? If so, can we determine an ordering of sites?

The solution of Waterman and Griggs is to consider a graph $G(A, B)$ whose vertices are the A-intervals and the B-intervals and which has an edge between an A-interval and a B-interval if and only if these intervals overlap. Waterman and Griggs first observe that we can find $G(A, B)$. Then they observe that the question of consistency is equivalent to the question: Is $G(A, B)$ an interval graph? (See fundamental idea #2.) The latter is easy to check in general (see Section 2), and especially easy in this case since the graph $G(A, B)$ is bipartite. It is also easy to derive the ordering of sites from an assignment of intervals showing that $G(A, B)$ is an interval graph.

I mention one open question posed by Waterman and Griggs: Devise algorithms for computing the minimum number of edge removals needed to convert $G(A, B)$ into a restriction map when it is not initially one.

In closing this section, I should mention that one current area of emphasis is on detecting similarity between two different RNA, DNA, or protein sequences. Such similarities lead to the discovery of shared phenomena. For example, it was discovered that the sequence for platelet derived factor, which causes growth in the body, is 87% identical to the sequence for v-sis, a cancer-causing gene. This led to the discovery that v-sis works by stimulating growth. The quality of a match between two sequences can be determined by a scoring matrix and a charge for introducing gaps in one of the sequences to get a better match. Then a dynamic programming algorithm can be used to determine the largest number of places where two sequences (gaps added) agree. The theory of random graphs can be used to compare two random sequences and predict how good a match one can expect. For results along these lines, and an expansion of the above discussion, see Lander and Waterman [1988].

In recent years, work on determining the similarity between pairs of sequences has expanded to work on detecting matches among a whole cluster of such sequences. In particular, suppose we let a given set of proteins define the vertices of a graph G and call two of these proteins adjacent if their corresponding sequences match sufficiently well (according to some scoring rule). The problem is to identify a set of k vertices in G which generates a subgraph with at least $\alpha\binom{k}{2}$ edges. (Of course, $\alpha = 1$ means we are looking for a clique.) Arratia and Lander [1989] have studied this question, and in particular have computed the expected number of such sets in a random graph and used the results to estimate the size a significant cluster would have to have. There are obviously a variety of relevant graph-theoretical questions which can be formulated here. For instance, are there, at least for certain important classes of graphs arising in practice, good algorithms or heuristics for determining if there is a k-subgraph with at least $\alpha\binom{k}{2}$ edges (and perhaps more structure as well)? The algorithms for clustering vertices of a graph into cliques, which have been developed for social networks and are mentioned briefly in Section 5 below, seem relevant here.

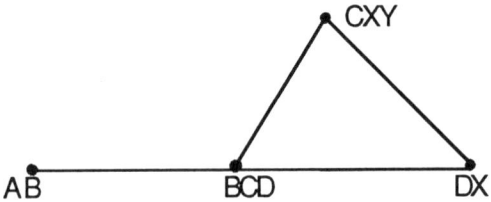

Figure 3

The results described above are only a few of the many known mathematical results about RNA and DNA chains. See the paper in this volume by Waterman [1989] for more about this kind of work.

2. Interval Graphs. Suppose $F = \{S_1, S_2, \ldots, S_N\}$ is a family of sets. The corresponding **intersection graph** of F has vertex set F and an edge between S_i and S_j if and only if $S_i \cap S_j \neq \phi$. For example, suppose

$$S_1 = \{A, B\}, S_2 = \{B, C, D\}, S_3 = \{C, X, Y\}, S_4 = \{D, X\}.$$

Then $F = \{S_1, S_2, S_3, S_4\}$ has the intersection graph shown in Figure 3. A natural question to ask is: What graphs are intersection graphs? The answer was given by Marczewski [1945]: Every graph is the intersection graph of some family of sets.

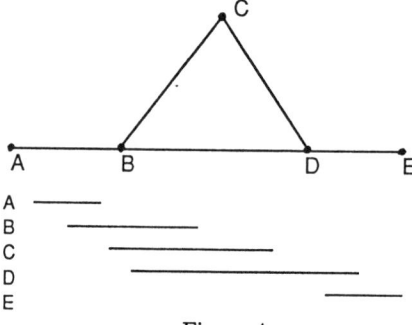

Figure 4

We get "interesting" graphs if we restrict the family of sets. For example, G is an **interval graph** if it is the intersection graph of a family of sets each of which is an interval on the real line. G is a **p-box graph** if it is the intersection graph of a family of "boxes" (generalized rectangles with sides parallel to the coordinate axes) in Euclidean p-space. The graph of Figure 4 is an interval graph, as is easily seen by the family of intervals shown in the figure. The graph of Figure 5 is not an interval graph, as is easy to prove. However, it is a 2-box graph; a family of boxes in 2-space which has this as its intersection graph is shown in Figure 6.

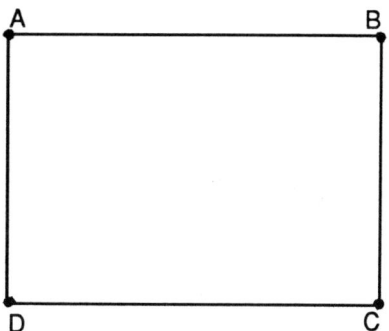

Figure 5

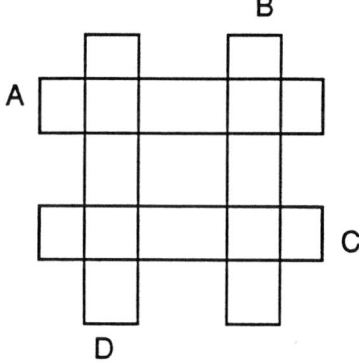

Figure 6

Interval graphs arose from a problem in genetics called Benzer's Problem (Benzer [1959, 1962]) and, independently, from a problem in scheduling due to Hajos [1957]. Benzer's Problem can be formulated as follows: Given information about whether or not two fragments of the gene overlap, is the data consistent with the hypothesis that the gene structure is linear? For instance, the linear structure of Figure 7 shows that the following data is consistent with linearity.

	S_1	S_2	S_3	S_4	S_5	S_6
S_1	1	1	0	0	0	0
S_2	1	1	1	1	0	0
S_3	0	1	1	1	0	0
S_4	0	1	1	1	1	0
S_5	0	0	0	1	1	1
S_6	0	0	0	0	1	1

In this matrix, an i, j entry is 1 if the corresponding fragments overlap, and 0 otherwise. The following data is inconsistent with linearity.

	S_1	S_2	S_3	S_4
S_1	1	1	0	1
S_2	1	1	1	0
S_3	0	1	1	1
S_4	1	0	1	1

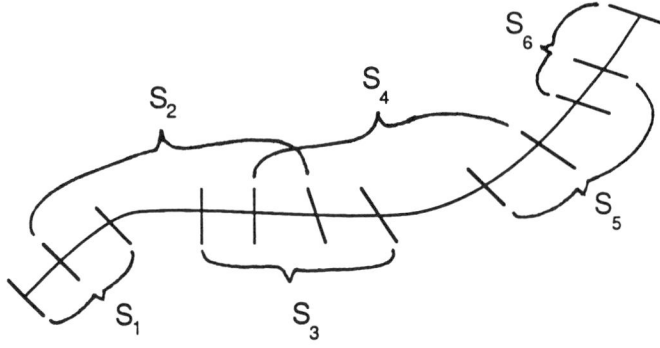

Figure 7

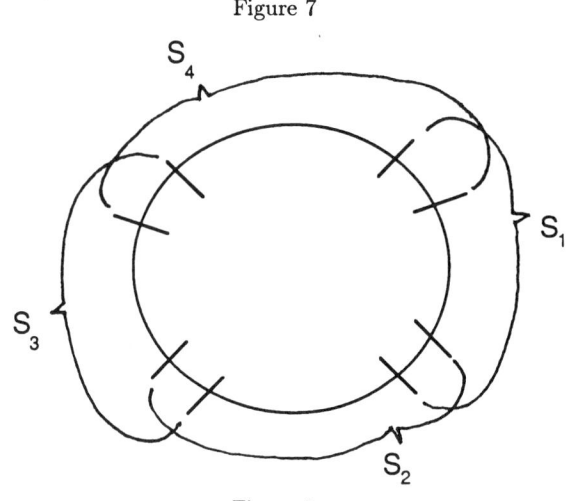

Figure 8

However, this is consistent with a "circular" structure, as shown in Figure 8.

Mathematically, the question of Benzer can be formulated as follows: Define a graph whose vertices are the fragments studied and which has an edge between two fragments if and only if they overlap. Can we tell if the graph is an interval graph?

Benzer's problem was solved by Lekkerkerker and Boland [1962] and by Gilmore and Hoffman [1964], who obtained elegant characterizations of interval graphs. To give a sample result: Gilmore and Hoffman proved that G is an interval graph if and only if G is triangulated (no cycle of length ≥ 4 is a generated subgraph) and the complement of G has an orientation which is transitive.

Another characterization of interval graphs G involves the vertex maximal clique incidence matrix of G. This is a $(0,1)$ matrix M whose i,j entry is 1 if the ith vertex belongs to the jth clique. Fulkerson and Gross [1965] prove that a graph G is an interval graph if and only if the columns of M can be permuted so that the 1's in each row appear consecutively. If this can be done, we say that M has the **consecutive ones property** (for columns). The consecutive ones property has applications to traffic phasing (Stoffers [1968], Roberts [1976], Opsut and Roberts [1981, 1983A,B], Raychaudhuri [1985]) and to information retrieval in computer science (Kambayashi [1981]), to mention two examples. For more on matrices with the consecutive ones property, see the paper by Cozzens and Mahadev [1989] in this volume.

In [1976], Booth and Lueker obtained linear time recognition algorithms for interval graphs. For broad surveys of results about interval graphs, see Golumbic [1980] or Fishburn [1985] or Roberts [1976, 1978B]. For a survey of recent work on interval graphs and other intersection graphs, see Trotter [1988].

The data of Benzer was consistent with the hypothesis of linearity, and played a basic role in establishing the idea that DNA or RNA molecules can be viewed as linear words over a 4-letter alphabet (see fundamental idea #1). Of course there is no longer active interest in Benzer's problem, but there is in variants of it. See for instance the work of Waterman and Griggs on restriction maps for DNA, discussed in Section 1.

Interval graphs and p-box graphs and other intersection graphs, meanwhile, have had a huge number of applications in the biological and social sciences. These include:

Competition and niche overlap in ecology (see fundamental idea #3)

Seriation or sequence dating in archaeology

Seriation in developmental psychology

Measurement and utility theory (see fundamental idea #7)

Traffic phasing

Scheduling of computer systems

Data retrieval

Maintenance problems

Mobile radio telephones

Task assignments.

These applications are summarized in the general references given above; see also Opsut and Roberts [1981].

The study of intersection graphs is a very active area today. The recent survey by Trotter [1988], mentioned earlier, summarizes a variety of recent results and open problems. We mention several recent directions of work and open problems here.

The probability that a random graph is an interval graph is of importance in the study of competition graphs in ecology (see fundamental idea #3). Following some work of Cohen, Komlos, and Mueller [1979] and Cohen [1982], it is known that if an arbitrary graph is chosen at random, the probability that it is an interval graph approaches 0 (as the number of vertices gets large). Recently, Scheinerman [1987A,B], Justicz, Scheinerman, and Winkler [1989], and others have developed a theory of random interval graphs in which an interval graph is generated by randomly choosing n intervals. Some sample results of Scheinerman are the following.

"Almost all" such graphs are hamiltonian. "Almost all" such graphs have chromatic number
$$\frac{n}{2} + o(n).$$

In work on practical problems, there is increasing emphasis on finding solution algorithms which are **on-line** in the sense that one is forced to make choices at the time data becomes available, rather than after having the entire problem spelled out. This is true for example in many applications of graph coloring, for instance to frequency assignment problems (see Cozzens and Roberts [1982B], Hale [1980], Roberts [1986]). When a graph is colored on-line, it is presented one vertex at a time and the color assigned to that vertex must be chosen before a new vertex is presented. The simplest on-line algorithm is the "greedy" algorithm sometimes called "first fit." On-line colorings of interval graphs arise in the study of frequency assignments. They also have applications to dynamic storage problems (see references in Trotter [1988]). Kierstead and Trotter [1981] have shown that for an interval graph, there is an on-line algorithm which always colors the graph in at most three times the chromatic number of colors. Gyarfas and Lehel [1987] have obtained a similar result for the **unit interval graphs** which are intersection graphs of families of closed real intervals of unit length. (The unit interval graphs are equivalent to the indifference graphs discussed in Section 7 below.) Tesman [1989] has extended such results to the generalizations of graph colorings called T-colorings which were introduced by Hale [1980], Cozzens and Roberts [1982B], and others.

As for open questions, the following are several important ones. The first is to characterize the 2-box graphs. Very little is known about these graphs. (There is, however, a relevant literature on the concept of boxicity of a graph; see Section 3.) The characterization problem is also interesting if the boxes in 2-space all have side-length one (then we talk about graphs of cubicity at most 2) and if the boxes all have side-length one and are all contained in an infinite linear strip of fixed height, for example two. This version of the problem probably has application to the frequency assignment problems mentioned above.

The second open problem I want to mention is to characterize the intersection graphs of unit spheres in 2-space. These graphs arise in biochemistry in problems of macro-molecular conformation (see the work of Havel [1982], Havel, Kuntz and Crippen [1983], and Havel, Kuntz, Crippen, and Blaney [1983]) and in communications in the radio and television frequency assignment problem (see references above). These intersection graphs have been studied in a series of papers by Maehara [1984A,B, 1986] and by Fishburn [1983].

A third important open question about interval graphs, which arises in ecology, will be discussed in Section 3. It is the question: Under what conditions on an acyclic digraph is the corresponding competition graph an interval graph?

For more about intersection graphs, see the paper in this volume by Lundgren [1989]. Other graph-theoretical issues are considered in connection with enzyme kinetics in the paper in this volume by Sellers [1989]. Tree structures play a central role in connection with mathematical immunology in the paper in this volume by Percus [1989].

3. Competition Graphs or Niche Overlap Graphs. A species' normal healthy environment is characterized by allowable ranges of different important factors such as temperature, humidity, pH., etc. If there are p factors, if each of these factors is taken to be a dimension in Euclidean p-space, and if we assume that the ranges on the different factors are independent, then the species is represented by a box in p-space. This box is called the species' **ecological niche**.

An old ecological principle says that two species compete if and only if their ecological niches overlap. In [1968], Joel Cohen asked the following questions: If we are given an independent assessment of competition, can we assign to each species in a given ecosystem an ecological niche in Euclidean p-space so that competition between species corresponds to overlap of their niches; and if so, what is the smallest p for which this works?

These questions simply ask the following: Is every graph the intersection graph of boxes in p-space for some p? If so, what is the smallest such p? The answer to the first question is yes. (See Roberts [1969B].) The answer to the second question is called the **boxicity** of the graph.

Unfortunately, computation of the boxicity of a graph is an NP-complete problem - as proved by Cozzens [1981] and then by Yannakakis in [1982]. Thus, the second question is in general hard to answer. Some recent results on computation of boxicity can be found in the papers by Cozzens and Roberts [1983, 1984].

Cohen defined competition as follows. Start with a **food web**, a digraph whose vertices are the species in the ecosystem and which has an arc from species x to species a if x preys on a. D is usually assumed to be acyclic. Figure 9 shows a food web for a Malaysian Rain Forest. This is from data of Harrison [1962], as adapted by Cohen [1978]. Two species x and y are said to **compete** if and only if there is a species a so that (x, a) and (y, a) are arcs in the food web, i.e., if and only if x and y have a common prey. Then the **competition graph** or **niche overlap graph** corresponding to the food web has the same set of vertices and an edge between x and y if and only if they compete. The competition graph for the Malaysian Rain Forest is shown in Figure 10.

We are interested in the boxicity of the competition graph. In [1968], Cohen made the remarkable observation that in a large number of examples of acyclic food webs, the boxicity always turned out to be 1. Since a box in 1-space is an interval, this turns out to be the observation that the competition graph is always an interval graph. In other words, only one ecological dimension suffices to account for competition. (The interpretation of this dimension is unclear.)

Although examples were later found by Cohen and others to show that not every competition graph had boxicity 1, Cohen's original observation and the continued preponderance of examples with boxicity 1 led to a large literature devoted to attempts to explain the observation and to study the properties of competition graphs.

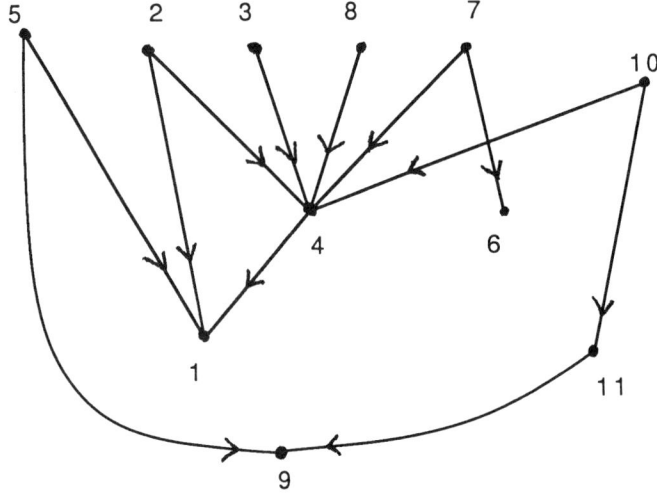

Key

(1) Canopy – leaves, fruits, flowers

(2) Canopy animals – birds, fruit-bats, and other mammals

(3) Upper air animals – birds and bats, insectivorous

(4) Insects

(5) Large ground animals – large mammals and birds

(6) Trunk, fruit, flowers

(7) Middle-zone scansorial animals – mammals in both canopy and ground zones

(8) Middle-zone flying animals – birds and insectivorous bats

(9) Ground – roots, fallen fruit, leaves and truncks

(10) Small ground animals – birds and small mammals

(11) Fungi

Figure 9

The explanations of Cohen's observation are of two types. The first type is statistical, and develops statistical models for randomly generated (acyclic) food webs from which one can show that the corresponding competition graphs are (with high probability) interval graphs. Much of Cohen's [1978] book, **Food Webs and Niche Space**, is devoted to this approach. Recent empirical work on food webs can be found in the papers by Briand and Cohen [1984] and Cohen and Briand [1984]. A new model to account for these observations, the cascade model, is developed by Cohen and Newman [1985], Cohen, Newman, and Briand [1985], Cohen, Briand, and Newman [1986], and Newman and Cohen [1986]. Z. Palko (personal communication)

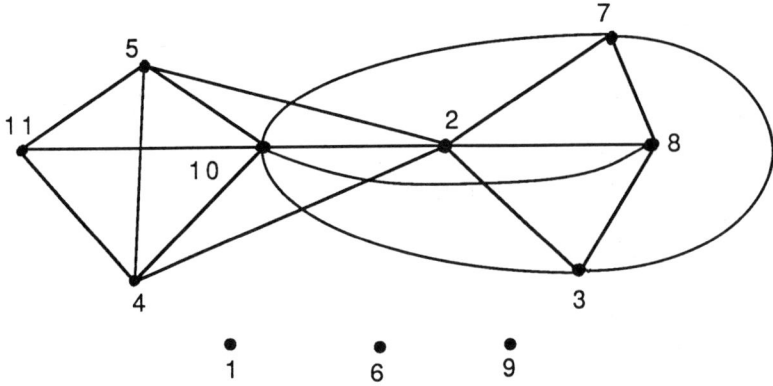

Figure 10

has calculated the asymptotic probability of the competition graph being an interval graph under the cascade model, and showed that it approaches 0.

A second explanation is graph-theoretical. An early example of such an explanation can be found in the paper by Roberts [1978A]. See also Lundgren and Maybee [1983], Dutton and Brigham [1983], and Roberts and Steif [1983] as examples of this kind of explanation. The graph-theoretical explanation (a) involves the analysis of the properties of competition graphs which arise from different kinds of digraphs; and (b) attempts to characterize the acyclic digraphs whose corresponding competition graphs are interval graphs. Quite a bit is known about (a), but (b) remains the **fundamental open problem** in the subject.

As for the fundamental open problem, I should mention early results by Lundgren and Maybee [1984], which essentially say that the competition graph has the Fulkerson-Gross [1965] consecutive ones property. (For more on the consecutive ones property, see Section 2.) I should also mention the observation by Steif [1982] that there can be no forbidden subgraph characterization. Finally, I should mention recent results of Hefner, Kim, Jones, Lundgren and Roberts [1988], which deal with the special case when the degrees of vertices in the digraph are restricted.

Another open problem of some interest concerns the **competition number** $k(G)$ of a graph G, the smallest k so that G together with k isolated vertices is the competition graph of some acyclic digraph. Roberts [1978A] showed that this number is well-defined and that characterization of competition graphs of acyclic digraphs is equivalent to calculation of the competition number of an arbitrary graph. Opsut [1982] showed that computation of $k(G)$ is an NP-complete problem. An attractive open problem is to settle **Opsut's conjecture**: if the open neighborhood of every vertex of G can be covered by two cliques, then $k(G) \leq 2$. For a proof of a variant of this conjecture, see Kim and Roberts [1989].

Other recent directions of research involve variants of the notion of competition graph. One variant is the **common enemy graph**: x and y are adjacent if and only if they have a common predator. These graphs have been studied by Lundgren and Maybee [1984] and by Sugihara [1983]. Another variant is the **competition common enemy graph (CCE graph)**: x and y are adjacent if and only if they have both a common prey and a common predator. These graphs were introduced by Scott [1987]. A third variant is the **niche graph**: x and y are adjacent if and only if they have either a common prey or a common predator. These graphs were introduced by Cable, Jones, Lundgren, and Seager [1987].

Scott [1987] showed that every graph G is a CCE graph of an acyclic digraph if sufficiently many isolated vertices are added to G, and so defined the **double competition number** $dk(G)$ of G as the smallest number of such isolated vertices. She conjectured that $dk(G)$ is always at most 2. Jones, Lundgren, Roberts, and Seager [1987] showed that $dk(G)$ can be arbitrarily large, and asked if there was an interesting infinite family of triangle-free graphs with arbitrarily large $dk(G)$. Seager [1988] settled this question in the affirmative, but the question remains whether this is true for the bipartite graphs. Kim, Roberts, and Seager [1989] show that $dk(G) \leq 2$ for G a bipartite graph in which one class has at most 4 vertices.

As for niche graphs, Cable, Jones, Lundgren, and Seager [1987] show that not every graph has the property that it is a niche graph if sufficiently many isolated vertices are added. They define the **niche number** $n(G)$ of a graph G to be the smallest number of isolated vertices to add to G to make it into a niche graph, if this can be done, and to be ∞ otherwise. A stubborn open problem is the following: Are there graphs G with $2 < n(G) < \infty$? Other recent work on niche numbers can be found in the paper by Bowser and Cable [1988].

While competition graphs were originally introduced with a biological application in mind, they have also found application to a variety of other problems, including communication over noisy channels, frequency assignment problems, and large-scale computer models dealing with energy and economic systems. A sample reference on the latter is the paper by Greenberg, Lundgren, and Maybee [1981]. A summary of a variety of such applications can be found in Raychaudhuri and Roberts [1985].

For other recent work about competition graphs and related concepts, see the papers by Harary, Kim, and Roberts [1989] and Kim, McKee, McMorris, and Roberts [1989] and the paper in this volume by Lundgren [1989].

4. Qualitative Stability. Suppose $A = (a_{ij})$ is an $n \times n$ matrix. We say that A is **semistable** if each eigenvalue has nonpositive real part and **stable** if each eigenvalue has negative real part. These properties of matrices are extremely important in the analysis of the stability properties of dynamical systems. Suppose A is the matrix of a system of homogeneous linear differential equations with constant coefficients. Suppose that for each real vector v, there is a unique positive trajectory x, a mapping from $[0, \infty)$ to $\mathbf{R}^n$, with $x(0) = v$ and $\dot{x} = Ax$. Then A is stable if and only if each positive trajectory $x(t)$ converges to the origin as $t \to \infty$.

In many cases, especially in ecology and economics, the entries in the matrix A are not known exactly. Indeed, sometimes they are known only qualitatively. In the simplest case, they are only known up to sign. In his [1947] book **Foundations of Economic Analysis**, Paul Samuelson introduced in a general way the problem of determining cases in which the stability properties of a matrix A could be deduced if one just knows the signs of the entries. This is called the **sign stability problem**. More precisely, a matrix A is called **sign stable (semistable)** if every matrix with the same pattern of signs is stable (semistable).

As an example, consider the matrix

$$A = \begin{bmatrix} -1 & 1 \\ -1 & -1 \end{bmatrix}.$$

A general matrix with the same sign pattern is given by

$$A' = \begin{bmatrix} -\alpha & \beta \\ -\gamma & -\delta \end{bmatrix},$$

where $\alpha, \beta, \gamma, \delta$ are positive. The characteristic polynomial of A' is given by

$$\lambda^2 + \lambda(\alpha + \delta) + \alpha\delta + \beta\gamma.$$

Using the quadratic formula, one easily verifies that every root of this polynomial has real part negative. Thus, A is sign stable.

A related problem considered by Samuelson is the problem of determining when the signs of a system of simultaneous linear equations determine the signs of a solution – this is called the **sign solvability problem**. If the answer to this problem is positive, we say that the system is **sign solvable**.

As an example, suppose

$$A = \begin{bmatrix} -2 & 0 \\ 1 & -5 \end{bmatrix}, b = (10, 10).$$

The sign pattern is given by the matrix and vector

$$A' = \begin{bmatrix} -\alpha & 0 \\ \beta & -\gamma \end{bmatrix}, b' = (\delta, \varepsilon),$$

where $\alpha, \beta, \gamma, \delta, \varepsilon$ are positive. Then $Ax = b$ is sign solvable. For $A'x' = b'$ has the solution

$$x'_1 = \frac{-\delta}{\alpha}, x'_2 = \frac{-(\beta\delta/\alpha) - \varepsilon}{\gamma}.$$

Both x'_1 and x'_2 are negative independent of the values of $\alpha, \beta, \gamma, \delta$ and ε.

There has been a large body of work on both the sign stability and sign solvability problems and techniques of graph theory and combinatorics have played an important role in what we know about them to date. Surveys of developments on these two problems can be found in the articles by Quirk [1981] and Maybee [1981]. A more recent paper which contains many references on the sign stability problem is the article by Jeffries, Klee, and van den Driessche [1987].

The problem of characterizing the sign solvable matrices was solved by Bassett, Maybee, and Quirk [1968]. We do not present their result here. For recent results, however, see the papers by Manber [1982], Hansen [1983], Klee, Ladner, and Manber [1984], and Klee [1987]. In the Bassett, Maybee, Quirk Theorem, one of the necessary and sufficient conditions for a matrix A to be sign solvable is equivalent to the condition that a certain digraph associated with A have an even cycle. The Bassett, Maybee, Quirk conditions for sign solvability leave one major open problem: Find an efficient algorithm for testing if a digraph has an even cycle.

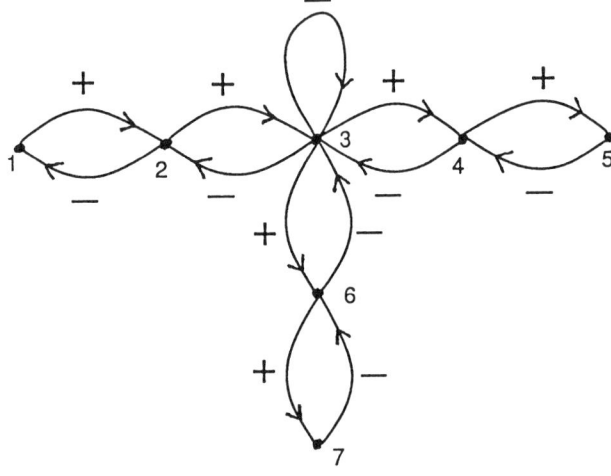

Figure 11

The sign semistability problem was solved by Quirk and Ruppert [1965] and the sign stability problem was solved by Jeffries, Klee, and van den Driessche [1977]. We summarize their results as follows. Given the $n \times n$ matrix A, we associate a digraph $D(A)$ with A as follows. The vertices of A are the elements of $\{1, 2, \ldots, n\}$ and there is an arc from i to j if and only if $a_{ji} \neq 0$. We associate a sign ($+$ or $-$) with each arc of $D(A)$ by using the sign of the corresponding matrix entry. This defines a signed digraph $SD(A)$. This signed digraph is illustrated in Figure 11 if A is the following matrix:

(*)
$$A = \begin{array}{c|ccccccc} & 1 & 2 & 3 & 4 & 5 & 6 & 7 \\ 1 & 0 & -5 & 0 & 0 & 0 & 0 & 0 \\ 2 & 1 & 0 & -1 & 0 & 0 & 0 & 0 \\ 3 & 0 & 4 & -6 & -1 & 0 & -2 & 0 \\ 4 & 0 & 0 & 1 & 0 & -1 & 0 & 0 \\ 5 & 0 & 0 & 0 & 1 & 0 & 0 & 0 \\ 6 & 0 & 0 & 2 & 0 & 0 & 0 & -8 \\ 7 & 0 & 0 & 0 & 0 & 0 & 1 & 0 \end{array}$$

Quirk and Ruppert prove the following:

THEOREM. *A matrix A is sign semi stable if and only if it satisfies the following three conditions:*

(a) *Each 1-cycle in $SD(A)$ is signed* −

(b) *Each 2-cycle in $SD(A)$ has one* + *sign and one* − *sign*

(c) $D(A)$ *has no p-cycle for* $p \geq 3$.

By the Quirk-Ruppert Theorem, the matrix A of (*) is sign semistable.

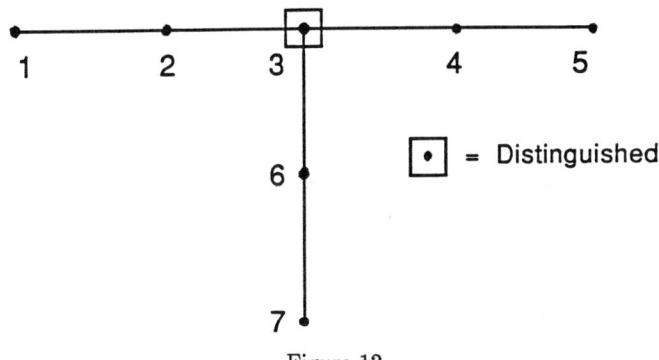

Figure 12

To present the characterization of sign stability, we also associate with A an undirected graph $G(A)$ whose edges are pairs $\{i,j\}$ with both a_{ij} and $a_{ji} \neq 0$. The graph $G(A)$ corresponding to the matrix A of (*) is shown in Figure 12. A vertex i of $G(A)$ is called **distinguished** if $a_{ii} \neq 0$. In Figure 12, the only distinguished vertex is 3. Jeffries, Klee, and van den Driessche study two kinds of colorings of a graph G with a set of distinguished vertices. A δ-**coloring** is a partition of the vertices into two sets, black and white, such that

(1) each distinguished vertex is black

(2) no black vertex has exactly one white neighbor

(3) each white vertex has a white neighbor.

An ε-coloring satisfies conditions (1) and (2) and also

(4) no white vertex has a white neighbor.

Finally, a coloring is called **trivial** if all vertices are black. A δ-coloring for the graph of Figure 12 is obtained by coloring the vertex 3 black and all other vertices white. The only ε-coloring is the trivial one. The coloring conditions are due to Jeffries [1974]; the following theorem first appeared in Jeffries, Klee, and van den Driesche [1977], though we state it in the slightly revised form based on their [1987] paper.

THEOREM. *A matrix A is sign stable if and only if its signed digraph SD(A) satisfies conditions (a) − (c) of the Quirk and Ruppert theorem and its graph G(A) satisfies the following two coloring conditions:*

(d) *Each δ-coloring of G(A) is trivial*

(e) *Each ε-coloring of G(A) is trivial.*

Note that for the matrix of (*), condition (d) is violated; the δ-coloring described above is nontrivial. However, (e) holds.

It should be noted that the results of Jeffries, Klee, and van den Driessche provide efficient algorithms to test sign semistability and sign stability (the algorithms are $O(n + \#$ of nonzero entries of $A))$.

The paper by Bone, Jeffries, and Klee [1988], which has a connection with nonvanishing of species, characterizes the sign patterns that imply (simultaneously) sign stability and positive sign solvability.

Many important open questions remain. Perhaps the most useful is that of generalizing to matrices where we know more than the sign pattern, and in particular if we know whether the entry is large positive, small positive, zero, small negative, or large negative. There are other approaches to incorporating more quantitative information into the theory of qualitative matrices. For one, see Maybee and Wiener [1987].

Another problem which should be mentioned here, and is discussed in the paper in this volume by Maybee [1989], has to do with Hicksian stability. A real $n \times n$ matrix A is called **Hicks stable** if each principal minor A_p of order p satisfies $(-1)^p A_p > 0$. This notion was introduced in a famous book on economic theory by Hicks [1939]. Since then, a considerable body of literature has been devoted to the relationship between Hicksian stability and stability as defined above. In particular, the two notions are known to be equivalent for the so-called Metzlerian matrices and Morishima matrices which arise in economics and, in the latter case, also in the theory of balance (see fundamental notion #5). (For definitions, see Maybee and Quirk [1973], Maybee and Richman [1989], and Roberts [1978B].) Generalizing these matrices, we say that a matrix A is called a **Generalized Metzlerian matrix** or **GM matrix** if its corresponding signed digraph $SD(A)$ has the property that each vertex has a negative loop and whenever I is a negative cycle of length greater than 1 and J is a positive cycle, then either I and J consist of disjoint sets of vertices or the vertices of I form a subset of the vertices of J. Maybee and Quirk [1973] conjecture that among qualitatively specified real matrices A, stability implies Hicksian stability if and only if A is a GM matrix. This conjecture is as yet unsettled. It also remains open to characterize the signed digraphs $SD(A)$ corresponding to GM matrices. See Quirk [1974] and Maybee and Richman [1989].

The papers in this volume by Klee [1989] and by Maybee [1989] deal in more detail with the sign stability and sign solvability problems.

5. Balanced Signed Graphs. Sign patterns play an important role in modeling of social networks. Suppose all the members of a group are represented as the vertices of a signed graph. We study some relationship among the members by letting the edges correspond to the relationship. If the relationship is signed, we can then put a + or − sign on an edge to represent the corresponding sign of the relationship. For instance, if an edge between two individuals means that there is a strong liking or disliking relationship between them, then the sign + can represent liking and the sign − disliking.

Such signed graphs have been used to make precise the notion that a small group of individuals is balanced in the sense of working well together and working without tension. Following experiments of the psychologist Heider in the 1940's (see Heider [1946]), Cartwright and Harary [1956] made the notion of balance precise in the language of the corresponding signed graph: A signed graph is called **balanced** if every circuit has an even number of − signs. (The same notion for signed digraphs was introduced into the economic literature by Morishima [1952] in the language of matrices, and led to the concept now called a Morishima matrix (see Section 4).)

Harary [1954] was able to characterize the balanced signed graphs as follows:

THEOREM. *A signed graph is balanced if and only if the vertices can be partitioned into two classes so that all edges within a class are + and all edges between classes are −.*

Applying this result to the signed graph of Figure 13, we see that this is balanced. The two classes of vertices are $\{a, c, d, f, h\}$ and $\{b, e, g\}$.

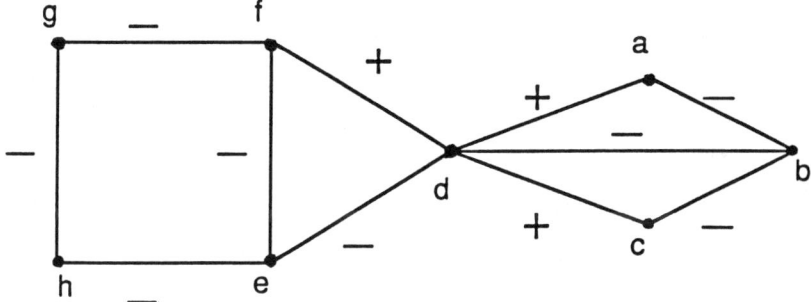

Figure 13

This result has been important in applications in political science. (See some references in Roberts [1979A].) It has also played a role in the linear time algorithm, due to Maybee and Maybee [1983], for recognizing balanced signed graphs. The paper by Hansen [1978] has a similar algorithm, which is only claimed to be quadratic, but which can in fact be made linear (P. Hansen, personal communication). Another balance algorithm appears in the paper by Harary and Kabell [1980].

Ideas related to and similar to balance have played an important role in the theories of clustering, blockmodeling, and structural modeling which have been so important in the social networks literature. For summaries of the network models following on the early balance idea, for instance the models for clustering, ranked clusters of cliques, ranked 2-clusters of cliques, and transitivity models, see the papers by Johnsen [1985, 1986, 1989], the latter appearing in this volume. See also the book by Alexander, Giesen, Munch, and Smelser [1987] and the article by Coleman [1986].

As an open problem, I mention an idea of Batchelder [1988], who asserts that not enough attention is paid in the social networks literature to the dependence of conclusions on the scales of measurement used. He argues that conclusions should be **meaningful** in the sense that their truth or falsity is independent of admissible changes of scale. (See a precise discussion of meaningfulness and other applications of the concept in the papers by Roberts [1985, 1989B], the latter appearing in this volume.) Among other things, Batchelder observes that in the social networks literature, some of the algorithms which cluster vertices into cliques, such as the CONCOR algorithm of Breiger, Boorman and Arabie [1975] and Arabie, Boorman and Levitt [1978], lead to meaningless conclusions even when the scales are as strong as ratio scales. (For definitions of scale types, see Roberts [1979B].) Batchelder's ideas are relevant to the case of balance. Suppose each individual i rates his or her friendship r_{ij} for each other individual j. Suppose a signed graph $S(r)$ is derived from the data (r_{ij}). When is the conclusion that $S(r)$ is balanced a meaningful conclusion? To my knowledge, no work has been done on this problem.

As another open problem on balance, I mention an idea which is closely related to the qualitative stability notion discussed in Section 4. Imagine that the matrix of friendship r_{ij} is calculated as above. This matrix is static, but one can study dynamic models in which the entries r_{ij} change over time according to some rules. A simple example of such a dynamic model is given in the paper by Hubbell, Johnsen, and Marcus [1978]. In such a situation, one can ask what happens to the balance of the corresponding signed graphs over time, and hopefully show that eventually these signed graphs are all balanced or all unbalanced. In this case, one can ask whether the sign pattern of the original matrix determines the final state, balanced or unbalanced. This problem would be interesting to analyze for different dynamic models of change over time.

It is always remarkable when an idea developed for one purpose turns out to be useful for other purposes. Most of the ideas discussed in this paper have this property. Balance is a case in point. To illustrate, I note that many important practical problems can be expressed as problems of maximizing a quadratic polynomial f in $0, 1$ variables. Such an f is called **supermodular** if all quadratic terms have positive coefficients. Now there are good algorithms for maximizing supermodular functions. Some quadratic polynomials in $0, 1$ variables can be brought into supermodular form by a sequence of **switches**, which replace a variable x by a variable $1 - x'$. When can this be accomplished?

To answer this question, associate a signed graph $SG(f)$ with f by taking the vertices to be the variables in f, including an edge between x and y if and only if there is a quadratic term involving x and y, and letting the sign of this edge be the sign of the coefficient of this term. Then Hansen and Simeone [1986] prove that a quadratic polynomial in $0,1$ variables can be brought to supermodular form by a sequence of switches if and only if $SG(f)$ is balanced. (For more general results, see Crama [1987].)

Balance also appears in a series of papers by Greenberg, Lundgren, and Maybee, e.g., [1983,1984A,B]. These are devoted to simplification and analysis of the structure of mathematical models for large, complex systems, for instance those used to analyze economic or energy systems.

For more about balance in particular and social networks in general, see the paper in this volume by Johnsen [1989].

6. Social Welfare Functions. A great deal of work in the mathematics of the social sciences has been concerned with groups making decisions. These can be small groups or large groups such as nations or societies. The classical voting situations are of course an example of a group decisionmaking situation. Each member of the group casts a vote, and then some rule tells us the winner. Much of the literature of group decisionmaking starts from the observation that this method for groups to make decisions omits a great deal of information about the preferences of the individual involved, for instance his or her second choice, his or her strength of preference for one alternative over another, and so on.

One approach to group decisionmaking is to let each individual in a group provide a rank ordering of alternatives in a set A. Ties are often allowed in this ordering. A **profile** of a group is the collection of rankings provided by each member. To give an example, suppose A is the set consisting of the four cities Miami (M), San Francisco (S), Denver (D), and New Orleans (N) and suppose that four individuals rank these cities as places to live. The following rankings are obtained, and these define a profile.

	P_1	P_2	P_3	P_4
$\mathcal{P} =$	M	D	S	M
	D	N	M	D
	N	M	D	$N-S$
	S	S	N	

Here $N - S$ means a tie between N and S.

A fundamental notion in the theory of group decisionmaking is the notion of **social welfare function** or **group consensus function**. This is a rule which tells for each possible profile of the group, what the group's decision will or should be. This decision is either a single winner or, in the more general case, another ranking of the alternatives which in some sense represents the consensus of the individual rankings. In the more general case, two examples of social welfare functions are as follows:

(1) **The Borda count**: Let $B_i(x)$ be the number of alternatives ranked below alternative x by the ith individual. Let $B(x)$ be $\Sigma B_i(x)$. Let the group ranking be determined by the values $B(x)$.

(2) Always use player 1's ranking. (Player 1 is a **dictator**.)

For the profile $\mathcal{P}$ given above, the Borda count is calculated as follows:

$$B_1(M) = 3, B_2(M) = 1, B_3(M) = 2, B_4(M) = 3;$$
$$B(M) = 3 + 1 + 2 + 3 = 9$$
$$B(S) = 0 + 0 + 3 + 0 = 3$$
$$B(D) = 2 + 3 + 1 + 2 = 8$$
$$B(N) = 1 + 2 + 0 + 0 = 3.$$

Thus, the group's ranking is

$$M$$
$$D$$
$$S - N$$

One of the most startling results in the mathematics of the social sciences is the theorem of Kenneth Arrow which says that under some reasonable axioms, there is no possible social welfare function which provides a ranking for every possible profile of rankings. We briefly state Arrow's Impossibility Theorem. The version we present is not necessarily the most modern or the nicest, but it is one of the earlier ones, is important for historical reasons, and will suffice to illustrate the ideas. For alternative versions of impossibility and possibility theorems, see Fishburn [1972]. See also the paper in this volume by Barthelemy [1989].

Arrow's Axioms

Axiom 1. (Positive Association of Social and Individual Values): Suppose the social welfare function provides a ranking in which alternative x is ranked over alternative y. Suppose each individual's rankings are changed so that his or her preferences between alternatives other than x are unchanged and his or her preferences between x and any other alternative are either unchanged or modified in x's favor. Then as a result of the modified profile, the social welfare function again provides a ranking in which x is ranked over y.

To illustrate this axiom, suppose that in the profile $\mathcal{P}$ above, P_2 is changed to

$$P'_2$$
$$\overline{}$$
$$D$$
$$M - N$$
$$S$$

Note that since the Borda count rated M over D and since the only change was to move M up into a tie with N, if the Borda count satisfies Axiom 1, then it must still rate M over D if P'_2 replaces P_2 in the profile $\mathcal{P}$.

Axiom 2. (Independence of Irrelevant Alternatives): Let A_1 be any subset of the set of alternatives A. If a profile is modified in such a way that each individual's ranking among elements in A_1 is unchanged, then the group's ranking resulting from the original and the modified profiles should be the same for alternatives in A_1.

To illustrate this axiom, suppose the profile $\mathcal{P}$ above is changed to the following profile:

$$\mathcal{P}' = \begin{array}{cccc} P_1' & P_2' & P_3' & P_4' \\ \hline M & N & M & M \\ N & M-D & S & N-S \\ S & S & N & D \\ D & & D & \end{array}$$

Since profiles $\mathcal{P}$ and $\mathcal{P}'$ agree on the set $A_1 = \{S, N\}$, if the Borda count satisfies Axiom 2, then it has to rank S and N tied for this new profile since it does for the original one.

Axiom 3 (Citizens' Sovereignty): For each pair of alternatives x and y, there is some profile for which the group ranks x over y.

Axiom 4 (Nondictatorship): There is no individual with the property that, for all alternatives x and y, whenever he ranks x over y, then the group does the same, regardless of the rankings supplied by the other individuals.

ARROW'S IMPOSSIBILITY THEOREM [1951]. *If there are at least two individuals in the group and at least three alternatives in the set of alternatives A, then there is no social welfare function which assigns a ranking to each profile of rankings.*

Obviously the second social welfare function example violates Axiom 4. (Exercise: Which axiom(s) does the first example, the Borda count, violate?) For a discussion of Arrow's Theorem, see for example Fishburn [1972], Roberts [1976], or Sen [1970].

Analogues of Arrow's theorem are important in the mathematical biology literature. For instance, in numerical taxonomy, suppose several phylogenetic trees have been constructed for a set S, where S is a set of biological entities such as a set of species. This situation could arise if different tree construction methods are used on the same data or the same method is used on different sets of data. In this case, one often tries to capture the common elements of the different trees and produce a consensus tree. McMorris [1985] obtained an analogue of Arrow's Impossibility Theorem in this situation, if a phylogenetic tree is interpreted as a graph-theoretic tree with no vertices of degree two and exactly $|S|$ vertices of degree one, each labeled with an element of S.

As an aside, I note that the notion of tree could have been listed as another fundamental idea; however, it did not arise from the biological or social sciences. (It did, however, arise from the organic chemistry research of Cayley.) The notion of tree today has many applications in the biological and social sciences. To name one, I refer to the paper on trees in mathematical immunology by Percus [1989], which appears in this volume.

Among the most important directions of research on social welfare functions after Arrow's Impossibility Theorem, I should mention the idea of Kemeny [1959] and Kemeny and Snell [1962] that we can obtain a "pseudo" social welfare function by first finding a way to measure the distance between any two rankings. Then the social welfare function is given by finding a ranking which minimizes the sum of distances to all the rankings in the profile or the sum of squares of distances to all the rankings in the profile. The former is called a **median** of the profile and the latter a **mean**. We call this a **pseudo social welfare function** because a median and mean are not necessarily unique. (Barthelemy, Flament, and Monjardet [1982] call it a **complete multiprocedure**.)

For example, consider the **symmetric difference distance** d which is defined as follows. Given rankings X and Y, count 2 for each pair of alternatives on which X and Y do the opposite, and 1 for each pair of alternatives for which one of X and Y ranks the pair and the other declares a tie, and sum up the counts to obtain $d(X,Y)$. Now consider the situation where $A = \{M, D, N\}$ and there are three members of the group and consider the following profile:

$$\mathcal{P} = \begin{array}{ccc} P_1 & P_2 & P_3 \\ \hline M & M & D \\ D & D & M \\ N & N & N \end{array}$$

The median of this profile in the space of all rankings is the ranking

$$\begin{array}{c} P \\ \hline M \\ D \\ N \end{array}$$

This is because

$$d(P, P_1) + d(P, P_2) + d(P, P_3) = 0 + 0 + 2 = 2$$

is a minimum. The mean is the ranking

$$\begin{array}{c} Q \\ \hline M - D \\ N \end{array}$$

since

$$d(Q, P_1)^2 + d(Q, P_2)^2 + d(Q, P_3)^2 = 1^2 + 1^2 + 1^2 = 3$$

is minimum.

For recent survey articles on the problem of finding the median, see Barthelemy and Monjardet [1981, 1988].

It has been shown by Wakabayashi [1986] that calculation of the median is an NP-complete problem for the symmetric difference distance, when we consider all rankings (with or without ties). (The result without ties is also proved by Bartholdi, Tovey, and Trick [1987C].) However, Barthelemy, Monjardet, and others observe that medians and means can be calculated efficiently when more structure is introduced into the problem. For instance, let P be a poset and let $d(x, y)$ be the shortest path metric on the diagram of P. A profile of P is a vector of v elements of P and a pseudo social welfare function on P assigns a set of elements of P to each profile of P. Barthelemy and Janowitz [1987] have recently shown that for certain posets P, some simple axioms characterize the pseudo social welfare functions which define the median. In particular, this is true if P is a certain kind of meet semilattice called a median semilattice. Moreover, they show that a polynomial algorithm allows the computation of a median. (However, there can be exponentially many medians, so computation of all of them may be difficult.)

It should be observed that a similar median procedure arises in cluster analysis in the work of Regnier [1965], who tries to find a median equivalence relation among a set of equivalence relations.

It should also be observed that similar poset and lattice-theoretic considerations arise in the literature on concept scaling analysis and knowledge structures, about which one can consult the papers by Ganter and Wille [1989] and by Falmagne [1989] in this volume.

Let me mention some open problems in connection with medians. Suppose we consider a class R of rankings of elements of a set A under the symmetric difference distance. Young and Levenglick [1978] have characterized the median procedure among all pseudo social welfare functions if R is the class of linear orders. However, similar characterizations are not available if R is the set of equivalence relations, preorders, complete preorders, or weak orders (linear orders with ties). Similar characterizations are also needed for the mean.

A number of heuristics for computing medians have been explored, for instance by Marcotorchino and Michaud [1981] and Michaud and Marcotorchino [1979]. However, no approximation algorithms or even approximation heuristics are known which are designed to be within a certain bound of the optimum. Moreover, no good heuristics are known for calculation of the mean.

Three other current directions of work on social welfare functions should be mentioned. One direction involves the calculation of social welfare functions when profiles are derived from spatial representations of both alternatives and group members. The alternatives and individuals are represented as points in a metric space, and some rule based on the metric determines each individual's ranking. Some important results on social welfare functions in this context are obtained by McKelvey [1976] and by Owen and Shapley [1989]. The paper in this volume by Straffin [1989] surveys this direction of work.

Another direction of research has been to apply combinatorial optimization techniques to problems of social welfare functions. Here, I mention the papers by Bartholdi, Tovey, and Trick [1987A,B,C]. In the third paper, it is shown that computing the social welfare function called the Dodgson winner is an NP-complete problem, and it is therefore difficult to determine the winner of an election! The first paper follows on the well-known theorems of Gardenfors [1976], Gibbard [1973], and Satterthwaite [1975] that under certain reasonable conditions, any social welfare function has the property that some individual might be able to achieve a social choice more to his or her liking by misrepresenting his or her preferences. In it, the authors show that in the case of some social welfare functions, "manipulation" is hard because it is NP-complete to find the "right way" to misrepresent one's preferences. For other recent references on social choice and computational complexity, see the papers by Gottinger [1987] and Kelly [1987].

A third direction of research involves the choice of social welfare functions which lead to meaningful statements in the sense already defined in Section 5. For more on this approach, see Aczel, Roberts, and Rosenbaum [1986], Roberts and Rosenbaum [1986], Aczel [1988], Aczel and Roberts [1989], and Roberts [1989C].

For more on social welfare and social choice and voting, see the papers by Barthelemy [1989] and by Straffin [1989] in this volume.

Before closing this section, let me mention that the literature on social welfare functions is closely related to the literature on n-person games. The latter literature has developed the concept of power index which is closely tied to that of social welfare function. Power index is another fundamental idea that might have been fundamental idea #8 in this paper. To see the significance of this idea in the social sciences, it should be noted that just one power index, the Shapley-Shubik index, has had so much influence that the original [1954] Shapley-Shubik paper was recently determined to be one of the ten most cited papers in the political science literature. For more on power indices, see the paper in this volume by Straffin [1989].

7. Semiorders. One of the fundamental necessities in the development of science is the development of our ability to measure things. Historical attempts to understand the nature of measurement go back to the work of Helmholtz in the late 19th century and to Norman Campbell in the 1920's. Modern attempts to put measurement on a firm mathematical foundation have played a major role in helping us to understand what we mean by measurement in biology and the social sciences. These attempts started with the work of Stevens and others in the 1930's in the attempt to measure loudness and led to a fundamental mathematical formulation by Scott and Suppes in [1958] and eventually to the books by Pfanzagl [1968], Krantz, Luce, Suppes, and Tversky [1971], Luce, Krantz, Suppes, and Tversky [1989], Suppes, Krantz, Luce, and Tversky [1989], Roberts [1979B], and Narens [1985].

In studying measurement in the social and biological sciences, some totally new types of scales, unknown in the more well-developed sciences of physics and chemistry, have been uncovered. Here we mention the ideas of additive conjoint measurement developed first by Debreu [1960] and Luce and Tukey [1964] and the notion of semiorder introduced by Luce [1956]. We shall concentrate on the latter idea, referring the reader to any of the books cited above for a discussion of the former and to Fishburn and Roberts [1988] for recent combinatorial results (and open questions) concerning it. We shall also refer the reader to these books and to the papers in this volume by Fishburn and Roberts [1989] and Roberts [1989B] for a discussion of other interesting types of measurement relevant to the biological and social sciences.

Semiorders arise in the measurement of preferences and also in the measurement of comparative loudness, brightness, and so on in psychophysics. Suppose A is a set of alternatives among which we are expressing preferences, and P is a binary relation on A, with xPy interpreted to mean that x is **preferred** to y. For convenience, we say that x is **indifferent** to y, and write xIy, if and only if not xPy and not yPx.

If we "measure" preference, we try to assign numbers to alternatives so that a preferred alternative gets a higher number. That is, we seek a real-valued function f on A so that for all x, y in A,

(1) $$xPy \leftrightarrow f(x) > f(y).$$

If such a function f exists, we call it an **ordinal utility function**. Measurement theorists ask two questions about such a function f: Under what conditions can we find such an f and how unique is it?

If an ordinal utility function exists, what does this say for indifference? It says that for all x, y in A,

(2) $$xIy \leftrightarrow f(x) = f(y).$$

This in turn implies that I is transitive. Unfortunately, indifference is not necessarily transitive. The economist Armstrong [1939] was one of the first to argue that indifference is not transitive. Menger [1951] claims that such arguments go back to Poincaré in the 19th century. One of the most well-known arguments against the transivity of indifference is the coffee-sugar example of Luce [1956]. Luce argues that if you are given a choice between a cup of coffee with no sugar in it and a cup with five spoons of sugar, you are not indifferent between the two cups. However, if you add sugar to the first cup one grain at a time, you will be indifferent between successive cups and hence, if indifference is transitive, between the cup with no sugar and the cup with five spoons of sugar.

If indifference is not transitive, there is no function f satisfying (2) and hence no function f satisfying (1). Based on examples such as the coffee-sugar example, Luce suggests that preference might correspond not to larger than, but to "sufficiently larger than." He makes this precise by introducing a positive constant δ, a **threshold** or **just-noticeable-difference**, and asking for a real-valued function f on A so that for all x and y in A,

(3) $$xPy \leftrightarrow f(x) > f(y) + \delta.$$

The measurement problems become: Under what conditions does there exist such an f and how unique is it?

At least if A is finite, the former question has been answered by Luce's notion of semiorder, which is redefined by Scott and Suppes [1958] as follows: (A, P) is a **semiorder** if it satisfies the following axioms for all x, y, z, w in A:

(1) not xPx

(2) xPy & $zPw \to xPw$ or zPy

(3) xPy & $yPz \to xPw$ or wPz.

Scott and Suppes [1958] prove the following theorem:

THEOREM. *If A is finite and δ is a positive constant, then there is a real-valued function f on A satisfying (3) if and only if (A, P) is a semiorder.*

Semiorders have an indifference analogue called **indifference graphs**. These arise when we can find a real-valued f on A so that for all x, y in A,

(4) $$xIy \leftrightarrow |f(x) - f(y)| \leq \delta.$$

The graph-theoretic interpretation of (4) is as follows: Assign numbers to the vertices of a graph so that two vertices are adjacent if and only if they get numbers which are close. Indifference graphs were introduced by Roberts [1969A] and in many other contexts by many other people.

Semiorders and indifference graphs have fascinating combinatorial properties. They have played an important role in measurement theory and also in economics, psychophysics, seriation in archaeology, radio and television channel assignment, etc. For general references on semiorders and indifference graphs, see Fishburn [1985], Golumbic [1980], and Roberts [1978B, 1979B]. Indifference graphs are closely related to the interval graphs discussed in Section 2; they are intersection graphs of closed intervals of unit length. Semiorders in turn can be thought of as arising from a family of closed unit intervals by the relation "strictly to the right of." If one uses this relation, but allows intervals of arbitrary lengths, one gets the order relations called **interval orders** which were introduced by Fishburn [1970] and have also played an important role in applications. (See Fishburn [1985] and Trotter [1988] for recent results on interval orders.)

The second basic measurement problem, that of specifying how unique a function f satisfying (3) or (4) can be, is a difficult one. Indeed, it remains an open question to specify in a nice closed form the class of "admissible transformations" φ of a 1-1 mapping f satisfying (3) (or (4)) so that $\varphi \circ f$ again satisfies (3) (or (4)).

The notion of meaningfulness introduced in Section 5 is relevant to this uniqueness problem. If f is a scale of measurement, for instance a function satisfying (3) or (4), we might ask when certain conclusions involving f are meaningful. For instance, we might consider the conclusion that $f(a) > f(b)$ or the conclusion that $f(a) = 2f(b)$ or the conclusion that $f(b)$ is between $f(a)$ and $f(c)$. A sample result here is the following from Roberts [1984B]. If f satisfies (4) and $|f(A)| \geq 3$, then the conclusion that $f(b)$ is between $f(a)$ and $f(c)$ is meaningful for all a, b, c in A if and only if the following conditions hold:

(a) For all $a \neq b$ in A, a and b have the same closed neighborhoods in the graph G whose vertex set is A and which has an edge between a and b if and only if aIb; and

(b) The graph G defined in (a) is connected.

Meaningfulness issues lead to a variety of combinatorial and graph-theoretical questions, many of which remain unsolved. Some of these are described in the paper by Roberts [1989B], which appears in this volume.

Another line of work concerning semiorders deals with multiple semiorders. Let $P_1, \ldots, P_n$ be n preference relations on the same set A. When do there exist a real-valued function f on A and positive numbers $\delta_1, \ldots, \delta_n$ so that

$$(5) \qquad xP_jy \leftrightarrow f(x) > f(y) + \delta_j \quad (j=1,...,n).$$

A necessary condition is that each (A, P_j) be a semiorder and the semiorders be nested. This is not sufficient. The case $n = 2$ was solved by Cozzens and Roberts [1982A], who found necessary and sufficient conditions for (5). Necessary and sufficient conditions in the case of arbitrary n were found by Doignon [1987].

It remains an open problem to find a similar solution for the analogous problem for nested families of indifference graphs, where the solution doesn't just reduce it to the semiorder problem.

It should be noted that coloring of nested families of indifference graphs is important in the frequency assignment applications discussed elsewhere in this paper (Section 2). A coloring here means a coloring of each graph simultaneously. While there are good coloring algorithms for indifference graphs, none are known in general for nested families of indifference graphs.

As I have tried to demonstrate, questions of measurement in the biological and social sciences involve a variety of combinatorial and graph-theoretical ideas. Further questions of measurement are discussed in the papers in this volume by Fishburn and Roberts [1989], Roberts [1989B], Cozzens and Mahadev [1989], and Ganter and Wille [1989].

Concluding Remarks. The seven fundamental ideas we have discussed here illustrate the large influence that problems in the biological and social sciences have had on the development of contemporary discrete mathematics. Such problems have led to mathematically challenging and interesting new theories and questions. It seems reasonable to expect such developments to continue.

The new mathematical developments in turn have had some interesting and useful applications in the biological and social sciences. The question of whether such applications have had an important impact on the biological and social sciences is one of the questions this volume is trying to address. I believe that problems of the biological and social sciences are often much more difficult than problems of the physical sciences and that we often do not know enough mathematics to solve them. However, I believe that the breakthroughs of the kind outlined here leave us poised to make important practical contributions in the years ahead.

REFERENCES

Note: papers indicated by * appear in this volume.

[1] ACZÉL, J., *Determining Merged Relative Scores*, mimeographed, Centre for Information Theory and Quantitative Economics, University of Waterloo, Waterloo, Ontario (December 1988).
[2] ACZÉL, J., AND ROBERTS, F.S., *On the Possible Merging Functions*, Math. Soc. Sci. (1989); in press.
[3] ACZÉL, J., ROBERTS, F.S., AND ROSENBAUM, Z., *On Scientific Laws without Dimensional Constants*, J. Math. Anal. & Appl., 119 (1986), pp. 389-416.
[4] ALEXANDER, J.C., GIESEN, B., MUNCH, R., AND SMELSER, N.J. (EDS.), *The Micro-Macro Link*, University of California Press, Berkeley, 1987.
[5] ARABIE, P., BOORMAN, S.A., AND LEVITT, P.R., *Constructing Blockmodels: How and Why*, J. Math. Psychol., 17 (1978), pp. 21-63.
[6] ARMSTRONG, W.E., *The Determinateness of the Utility Function*, Econ. J., 49 (1939), pp. 453-467.
[7] ARRATIA, R., AND LANDER, E.S., *The Distribution of Clusters in Random Graphs*, Adv. Appl. Math. (1989); in press.
[8] ARROW, K., *Social Choice and Individual Values*, Cowles Commission Monograph 12, Wiley, New York, 1951.
[9] *BARTHELEMY, J.P., *Social Welfare and Aggregation Procedures: Combinatorial and Algorithmic Aspects*, in F.S. Roberts (ed.), *Applications of Combinatorics and Graph Theory in the Biological and Social Sciences*, IMA Volumes in Mathematics and its Applications, Springer-Verlag, New York, 1989.
[10] BARTHELEMY, J.P., FLAMENT, C., AND MONJARDET, B., *Ordered Sets and Social Sciences*, in I. Rival (ed.), *Ordered Sets*, D. Reidel, Dordrecht, 1982, pp. 721-758.
[11] BARTHELEMY, J.P., AND JANOWITZ, M.F., *A General Setting for Consensus and the Median Procedure*, mimeographed, Departement Informatique, Ecole Nationale Superieure des Telecommunications, Paris (1987).
[12] BARTHELEMY, J.P., AND MONJARDET, B., *The Median Procedure in Cluster Analysis and Social Choice Theory*, Math. Soc. Sci., 1 (1981), pp. 235-268.
[13] BARTHELEMY, J.P., AND MONJARDET, B., *The Median Procedure in Data Analysis: New Results and Open Problems*, in H.H. Bock (ed.), *Classification and Related Methods of Data Analysis*, North Holland, 1988.
[14] BARTHOLDI, J.J. III, TOVEY, C.A., AND TRICK, M.A., *How Hard is it to Cheat in an Election?*, mimeographed, Institute for Mathematics and its Applications, University of Minnesota, Minneapolis (1987); (A).
[15] BARTHOLDI, J.J. III, TOVEY, C.A., AND TRICK, M.A., *The Computational Difficulty of Manipulating an Election*, mimeographed, School of Industrial and Systems Engineering, Georgia Institute of Technology, Atlanta (1987); to appear in Social Choice and Welfare; (B).
[16] BARTHOLDI, J.J. III, TOVEY, C.A., AND TRICK, M.A., *Voting Schemes for Which it can be Difficult to Tell who Won the Election*, mimeographed, School of Industrial and Systems Engineering, Georgia Institute of Technology, Atlanta (1987); To appear in Social Choice and Welfare; (C).
[17] BASSETT, L, MAYBEE, J., AND QUIRK, J., *Qualitative Economics and the Scope of the Correspondence Principle*, Econometrica, 36 (1968), pp. 544-563.
[18] BATCHELDER, W.H., *Inferring Meaningful Global Network Properties from Individual Actors' Measurement Scales*, in L.C. Freeman, A.K. Romney, and D.R. White (eds.), *Methods in Social Networks Analysis*, George Mason University Press, Fairfax, VA, 1988.
[19] BENZER, S., *On the Topology of the Genetic Fine Structure*, Proc. Nat. Acad. Sci. USA, 45 (1959), pp. 1607-1620.
[20] BENZER, S., *The Fine Structure of the Gene*, Sci. Amer., 206 (1962), pp. 70-84.
[21] BONE, T., JEFFRIES, C., AND KLEE, V., *A Qualitative Analysis of $\dot{x} = Ax + b$*, Discrete Appl. Math., 20 (1988), pp. 9-30.
[22] BOOTH, K.S., AND LUEKER, G.S., *Testing for the Consecutive Ones Property, Interval Graphs, and Graph Planarity Using PQ-tree Algorithms*, J. Comput. Syst. Sci., 13 (1976), pp. 335-379.

[23] BOWSER, S., AND CABLE, C.A., *Some Recent Results on Niche Graphs*, mimeographed, Allegheny College, Meadville, PA (November 1988).

[24] BREIGER, R.L., BOORMAN, S.A., AND ARABIE, P., *An Algorithm for Clustering Relational Data, with Applications to Social Network Analysis and Comparison with Multidimensional Scaling*, J. Math. Psychol., 12 (1975), pp. 328–383.

[25] BRIAND, F., AND COHEN, J.E., *Community Food Webs have Scale-invariant Structure*, Nature, Lond., 307 (1984), pp. 264–266.

[26] CABLE, C., JONES, K., LUNDGREN, J.R., AND SEAGER, S., *Niche Graphs*, mimeographed, Department of Mathematics, University of Colorado, Denver, Colorado (1987); to appear in Discr. Appl. Math.

[27] CARTWRIGHT, D., AND HARARY, F., *Structural Balance: A Generalization of Heider's Theory*, Psych. Rev., 63 (1956), pp. 277–293.

[28] COHEN, J.E., *Interval Graphs and Food Webs: A Finding and a Problem*, RAND Corporation Document 17696-PR, Santa Monica, CA (1968).

[29] COHEN, J.E., *Food Webs and Niche Space*, Princeton University Press, Princeton, N.J., 1978.

[30] COHEN, J.E., *The Asymptotic Probability that a Random Graph is a Unit Interval Graph, Indifference Graph, or Proper Interval Graph*, Discrete Math., 40 (1982), pp. 21–24.

[31] COHEN, J.E., AND BRIAND, F., *Trophic Links of Community Food Webs*, Proc. Nat. Acad. Sci. USA, 81 (1984), pp. 4105–4109.

[32] COHEN, J.E., BRIAND, F., AND NEWMAN, C.M., *A Stochastic Theory of Community Food Webs III. Predicted and Observed Lengths of Food Chains*, Proc. R. Soc. Lond., B 228 (1986), pp. 317-353.

[33] COHEN, J.E., KOMLOS, J., AND MUELLER, T., *The Probability of an Interval Graph, and Why it Matters*, in D.K. Ray-Chaudhuri (ed.), Proc. Symp. on Relations between Combinatorics and other Parts of Mathematics, Amer. Math. Soc., Providence, RI, 1979.

[34] COHEN, J.E., AND NEWMAN, C.M., *A Stochastic Theory of Community Food Webs I. Models and Aggregated Data*, Proc. R. Soc. Lond., B 224 (1985), pp. 421–448.

[35] COHEN, J.E., NEWMAN, C.M., AND BRIAND, F., *A Stochastic Theory of Community Food Webs II. Individual Webs*, Proc. R. Soc. Lond., B 224 (1985), pp. 449–461.

[36] COLEMAN, J.S., *Microfoundations and Macrosocial Theory*, in S. Lindenberg, J.S. Coleman, and S. Nowak (eds.), Approaches to Social Theory, Russell Sage Foundation, New York, 1986, pp. 345–363.

[37] COZZENS, M.B., *Higher and Multi-dimensional Analogues of Interval Graphs*, Ph.D. Thesis, Department of Mathematics, Rutgers University, New Brunswick, N.J. (1981).

[38] *COZZENS, M.B., AND MAHADEV, N.V.R., *Consecutive One's Properties for Matrices and Graphs Including Variable Diagonal Entries*, in F.S. Roberts (ed.), Applications of Combinatorics and Graph Theory in the Biological and Social Sciences, IMA Volumes in Mathematics and its Applications, Springer-Verlag, New York, 1989.

[39] COZZENS, M.B., AND ROBERTS, F.S., *Double Semiorders and Double Indifference Graphs*, SIAM J. Alg. & Discr. Meth., 3 (1982), pp. 566-583; (A).

[40] COZZENS, M.B., AND ROBERTS, F.S., *T-Colorings of Graphs and the Channel Assignment Problem*, Congressus Numerantium, 35 (1982), pp. 191–208; (B).

[41] COZZENS, M.B., AND ROBERTS, F.S., *Computing the Boxicity of a Graph by Covering its Complement by Cointerval Graphs*, Discr. Appl. Math., 6 (1983), pp. 217-228.

[42] COZZENS, M.B., AND ROBERTS, F.S., *On k-suitable Sets of Arrangements and the Boxicity of a Graph*, J. Comb., Info. & Syst. Sci., 9 (1984), pp. 14–24.

[43] CRAMA, Y., *Recognition and Solution of Structured Discrete Optimization Problems*, Ph.D. Thesis, Rutgers Center for Operations Research, Rutgers University, New Brunswick, N.J. (1987).

[44] DEBREU, G., *Topological Methods in Cardinal Utility Theory*, in K.J. Arrow, S. Karlin, and P. Suppes (eds.), Mathematical Methods in the Social Sciences, Stanford University Press, Stanford, CA,, 1960, pp. 16–26.

[45] DELISI, C., *Computers in Molecular Biology: Current Applications and Emerging Trends*, Science, 240 (1988), pp. 47–52.

[46] DOIGNON, J-P., *Threshold Representations of Multiple Semiorders*, SIAM J. Alg. & Discr. Meth., 8 (1987), pp. 77–84.

[47] DUTTON, R.D., AND BRIGHAM, R.C., *A Characterization of Competition Graphs*, Discr. Appl. Math., 6 (1983), pp. 315–317.

[48] *FALMAGNE, J-C., *Probabilistic Knowledge Spaces: A Review*, in F.S. Roberts (ed.), Applications of Combinatorics and Graph Theory in the Biological and Social Sciences, IMA Volumes in Mathematics and its Applications, Springer-Verlag, New York, 1989.
[49] FISHBURN, P.C., *Intransitive Indifference with Unequal Indifference Intervals*, J. Math. Psychol., 7 (1970), pp. 144–149.
[50] FISHBURN, P.C., *The Theory of Social Choice*, Princeton University Press, Princeton, NJ, 1972.
[51] FISHBURN, P.C., *On the Sphericity and Cubicity of Graphs*, J. Comb. Theory, B35 (1983), pp. 309–318.
[52] FISHBURN, P.C., *Interval Graphs and Interval Orders*, Wiley, New York, 1985.
[53] FISHBURN, P.C., AND ROBERTS, F.S., *Unique Finite Conjoint Measurement*, Math. Soc. Sci., 16 (1988), pp. 107–143.
[54] *FISHBURN, P.C., AND ROBERTS, F.S., *Uniqueness in Finite Measurement*, in F.S. Roberts (ed.), Applications of Combinatorics and Graph Theory in the Biological and Social Sciences, IMA Volumes in Mathematics and its Applications, Springer-Verlag, New York, 1989.
[55] FULKERSON, D.R., AND GROSS, O.A., *Incidence Matrices and Interval Graphs*, Pacific J. Math., 15 (1965), pp. 835–855.
[56] *GANTER, B., AND WILLE, R., *Conceptual Scaling*, in F.S. Roberts (ed.), Applications of Combinatorics and Graph Theory in the Biological and Social Sciences, IMA Volumes in Mathematics and its Applications, Springer-Verlag, New York, 1989.
[57] GARDENFORS, P., *Manipulation of Social Choice Functions*, J. Econ. Theory, 13 (1976), pp. 217–228.
[58] GIBBARD, A., *Manipulation of Voting Schemes*, Econometrica, 41 (1973), pp. 587–601.
[59] GILMORE, P.C., AND HOFFMAN, A.J., *A Characterization of Comparability Graphs and of Interval Graphs*, Canad. J. Math., 16 (1964), pp. 539–548.
[60] GOLDSTEIN, L., AND WATERMAN, M.S., *Mapping DNA by Stochastic Relaxation*, Adv. Appl. Math., 8 (1987), pp. 194–207.
[61] GOLUMBIC, M.C., *Algorithmic Graph Theory and Perfect Graphs*, Academic Press, New York, 1980.
[62] GOTTINGER, H.W., *Choice and Complexity*, Math. Soc. Sci., 14 (1987), pp. 1–17.
[63] GREENBERG, H.J., LUNDGREN, J.R., AND MAYBEE, J.S., *Graph-Theoretic Foundations of Computer-Assisted Analysis*, in H.J. Greenberg and J.S. Maybee (eds.), Computer-Assisted Analysis and Model Simplification, Academic Press, New York, 1981.
[64] GREENBERG, H.J., LUNDGREN, J.R., AND MAYBEE, J.S., *Rectangular Matrices and Signed Graphs*, SIAM J. Alg. & Discr. Meth., 4 (1983), pp. 50–61.
[65] GREENBERG, H.J., LUNDGREN, J.R., AND MAYBEE, J.S., *Inverting Graphs of Rectangular Matrices*, Discr. Appl. Math., 8 (1984), pp. 255–265; (A).
[66] GREENBERG, H.J., LUNDGREN, J.R., AND MAYBEE, J.S., *Inverting Signed Graphs*, SIAM J. Alg. & Discr. Meth., 5 (1984), pp. 216–223; (B).
[67] GYARFAS, A., AND LEHEL, J., *On-line and First Fit Colorings of Graphs*, mimeographed, Computer and Automation Institute, Hungarian Academy of Sciences, Budapest, Hungary (1987).
[68] HAJOS, G., *Uber eine Art von Graphen*, Internat. Math. Nachr., 47 (1957), p. 65.
[69] HALE, W.K., *Frequency Assignment: Theory and Applications*, Proc. IEEE, 68 (1980), pp. 1497–1514.
[70] HANSEN, P., *Labelling Algorithms for Balance in Signed Graphs*, in J-C. Bermond, et al. (eds.), Problemes Combinatoires et Theorie des Graphes, Editions du CNRS, Paris, 1978, pp. 215–217.
[71] HANSEN, P., *Recognizing Signsolvable Graphs*, Discr. Appl. Math., 6 (1983), pp. 237–241.
[72] HANSEN, P., AND SIMEONE, B., *Unimodular Functions*, Discrete Appl. Math., 14 (1986), pp. 269–281.
[73] HARARY, F., *On the Notion of Balance of a Signed Graph*, Michigan Math. J., 2 (1954), pp. 143–146.
[74] HARARY, F., AND KABELL, J., *A Simple Algorithm to Detect Balance in Signed Graphs*, Math. Soc. Sci., 1 (1980), pp. 131–136.
[75] HARARY, F., KIM, S., AND ROBERTS, F.S., *Extremal Competition Numbers as a Generalization of Turan's Theorem*, mimeographed, Department of Mathematics, Rutgers University, New Brunswick, NJ (1989).
[76] HARRISON, J.L., *The Distribution of Feeding Habits among Animals in a Tropical Rain Forest*, J. Animal Ecology, 31 (1962), pp. 53–63.

[77] HAVEL, T.F., *The Combinatorial Distance Geometry Approach to the Calculation of Molecular Conformation*, Congressus Numerantium, 35 (1982), pp. 361–371.
[78] HAVEL, T.F., KUNTZ, I.D., AND CRIPPEN, G.M., *The Combinatorial Distance Geometry Approach to the Calculation of Molecular Conformation I. A New Approach to an Old Problem*, J. Theor. Biol., 104 (1983), pp. 359–381.
[79] HAVEL, T.F., KUNTZ, I.D., CRIPPEN, G.M., AND BLANEY, J.M., *The Combinatorial Distance Geometry Approach to the Calculation of Molecular Conformation II. Sample Problems and Computational Statistics*, J. Theor. Biol., 104 (1983), pp. 383–400.
[80] HEFNER, K., JONES, K., KIM, S., LUNDGREN, J.R., AND ROBERTS, F.S., i,j *Competition Graphs*, RUTCOR Research Rep. RRR, # 14-88 Rutgers Center for Operations Research, Rutgers University, New Brunswick, NJ, March 1988.
[81] HEIDER, F., *Attitudes and Cognitive Organization*, J. Psychology, 21 (1946), pp. 107–112.
[82] HICKS, J., *Value and Capital*, Oxford University Press, Oxford, 1939.
[83] HOLLEY, R.W., EVERETT, G.A., MADISON, J.T., MARQUISEE, M., AND ZAMIR, A., *Structure of a Ribonucleic Acid*, Science, 147 (1965), pp. 1462–1465.
[84] HUBBELL, C.H., JOHNSEN, E.C., AND MARCUS, M., *Structural Balance in Group Networks*, in B. Anderson and R.B. Smith (eds.), *Handbook of Social Science Methods*, Irvington Publishers, distributed by Halsted Press, New York, 1978.
[85] HUTCHINSON, G., *Evaluation of Polymer Sequence Fragment Data using Graph Theory*, Bull. Math. Biophys., 31 (1969), pp. 541–562.
[86] JEFFRIES, C., *Qualitative Stability and Digraphs in Model Ecosystems*, Ecology, 55 (1974), pp. 1415–1419.
[87] JEFFRIES, C., KLEE, V., AND VAN DEN DRIESSCHE, P., *When is a Matrix Sign Stable?*, Canad. J. Math., 29 (1977), pp. 315–326.
[88] JEFFRIES, C., KLEE, V., AND VAN DEN DRIESSCHE, P., *Qualitative Stability of Linear Systems*, Linear Alg. & Appl., 87 (1987), pp. 1–48.
[89] JOHNSEN, E.C., *Network Macrostructure Models for the Davis-Leinhardt Set of Empirical Sociomatrices*, Social Networks, 7 (1985), pp. 203–224.
[90] JOHNSEN, E.C., *Structure and Process: Agreement Models for Friendship Formation*, Social Networks, 8 (1986), pp. 257–306.
[91] *JOHNSEN, E.C., *The Micro-Macro Connection: Exact Structure and Process*, in F.S. Roberts (ed.), *Applications of Combinatorics and Graph Theory in the Biological and Social Sciences*, IMA Volumes in Mathematics and its Applications, Springer-Verlag, New York, 1989.
[92] JONES, K., LUNDGREN, J.R., ROBERTS, F.S., AND SEAGER, S., *Some Remarks on the Double Competition Number of a Graph*, Congr. Numerantium, 60 (1987), pp. 17–24.
[93] JUSTICZ, J., SCHEINERMAN, E.R., AND WINKLER, P.M., *Random Intervals*, mimeographed, Department of Mathematics and Computer Science, Emory University, Atlanta, GA (1989).
[94] KAMBAYASHI, Y., *Proceedings of the Conference on Consecutive Retrieval Property*, Institute of Computer Science, Polish Academy of Sciences, Warsaw, Reports 438 and 439, July 1981.
[95] KELLY, J.S., *Social Choice and Computational Complexity*, mimeographed, Syracuse University, Syracuse, NY (1987).
[96] KEMENY, J.G., *Mathematics without Numbers*, Daedalus, 88 (1959), pp. 575–591.
[97] KEMENY, J.G., AND SNELL, J.L., *Mathematical Models in the Social Sciences*, Blaisdell, New York, 1962; (Reprinted by MIT Press, Cambridge, MA, 1972.).
[98] KIERSTEAD, H.A., AND TROTTER, W.T., *An Extremal Problem in Recursive Combinatorics*, Congr. Numerantium, 33 (1981), pp. 143–153.
[99] KIM, S., MCKEE, T., MCMORRIS, F.R., AND ROBERTS, F.S., *p-Competition Graphs*, mimeographed, Department of Mathematics, Rutgers University, New Brunswick, NJ, 1989.
[100] KIM, S., AND ROBERTS, F.S., *On Opsut's Conjecture about the Competition Number*, mimeographed, Department of Mathematics, Rutgers University, New Brunswick, NJ, 1989.
[101] KIM, S., ROBERTS, F.S., AND SEAGER, S., *On 1 0 1-clear $(0,1)$ Matrices and the Double Competition Number of Bipartite Graphs*, mimeographed, Department of Mathematics, Rutgers University, New Brunswick, NJ (1989).
[102] KLEE, V., *Recursive Structure of S∗-Matrices, and an $O(m^2)$ Algorithm for Recognizing Strong Sign-Solvability*, mimeographed, Department of Mathematics, University of Washington, Seattle, (1987); to appear, Linear Alg. & Appl.
[103] *KLEE, V., *Sign-patterns and Stability*, in F.S. Roberts (ed.), *Applications of Combinatorics and Graph Theory in the Biological and Social Sciences*, IMA Volumes in Mathematics and its Applications, Springer-Verlag, New York, 1989.

[104] KLEE, V., LADNER, R., AND MANBER, R., *Signsolvability Revisited*, Lin. Alg. & Appl., 59 (1984), pp. 131–157.

[105] KRANTZ, D.H., LUCE, R.D., SUPPES, P., AND TVERSKY, A., *Foundations of Measurement, Vol. I*, Academic Press, New York, 1971.

[106] LANDER, E.S., AND WATERMAN, M., *Genomic Mapping by Fingerprinting Random Clones: A Mathematical Analysis*, Genomics, 2 (1988), pp. 231–239.

[107] LEKKERKERKER, C.B., AND BOLAND, J. CH., *Representation of a Finite Graph by a Set of Intervals on the Real Line*, Fund. Math., 51 (1962), pp. 45–64.

[108] LUCE, R.D., *Semiorders and a Theory of Utility Discrimination*, Econometrica, 24 (1956), pp. 178–191.

[109] LUCE, R.D., KRANTZ, D.H., SUPPES, P., AND TVERSKY, A., *Foundations of Measurement, Vol. III*, Academic Press, New York, 1989.

[110] LUCE, R.D., AND TUKEY, J.W., *Simultaneous Conjoint Measurement: A New Type of Fundamental Measurement*, J. Math. Psychol., 1 (1964), pp. 1–27.

[111] *LUNDGREN, J.R., *Food Webs, Competition Graphs, Competition-Common Enemy Graphs, and Niche Graphs*, in F.S. Roberts (ed.), Applications of Combinatorics and Graph Theory in the Biological and Social Sciences, IMA Volumes in Mathematics and its Applications, Springer-Verlag, New York, 1989.

[112] LUNDGREN, J.R., AND MAYBEE, J.S., *A Characterization of Graphs with Competition Number m*, Discr. Appl. Math., 6 (1983), pp. 319–322.

[113] LUNDGREN, J.R., AND MAYBEE, J.S., *Food Webs with Interval Competition Graphs*, in Graphs and Applications: Proceedings of the First Colorado Symposium on Graph Theory, Wiley, New York, 1984, pp. 231–244.

[114] MAEHARA, H., *A Digraph Represented by a Family of Boxes or Spheres*, J. Graph Theory, 8 (1984), pp. 431–440; (A).

[115] MAEHARA, H., *Space Graphs and Sphericity*, Discr. Appl. Math., 7 (1984), pp. 55–64; (B).

[116] MAEHARA, H., *Sphericity Exceeds Cubicity for Almost all Complete Bipartite Graphs*, J. Comb. Theory, B40 (1986), pp. 231–235.

[117] MANBER, R., *Graph-theoretical Approach to Qualitative Stability of Linear Systems*, Lin. Alg. Appl., 48 (1982), pp. 457–470.

[118] MARCOTORCHINO, P., AND MICHAUD, P., *Heuristic Approach of the Similarity Aggregation Problem*, Methods of Oper. Res., 43 (1981), pp. 395–404.

[119] MARCZEWSKI, E., *Sur deux Proprietes des Classes d'ensembles*, Fund. Math., 33 (1945), pp. 303–307.

[120] MAYBEE, J.S., *Sign Solvability*, in H. Greenberg and J. Maybee (eds.), Computer-Assisted Analysis and Model Simplification, Academic Press, New York, 1981, pp. 201–257.

[121] *MAYBEE, J.S., *Qualitatively Stable Matrices and Convergent Matrices*, in F.S. Roberts (ed.), Applications of Combinatorics and Graph Theory in the Biological and Social Sciences, IMA Volumes in Mathematics and its Applications, Springer-Verlag, New York, 1989.

[122] MAYBEE, J.S., AND MAYBEE, S.J., *An Algorithm for Identifying Morishima and Anti-Morishima Matrices and Balanced Digraphs*, Math. Soc. Sci., 6 (1983), pp. 99–103.

[123] MAYBEE, J.S., AND QUIRK, J., *The GM-matrix Problem*, Tech. Rep., Dept. of Computer Sci., Univ. of Colorado, Boulder (1973).

[124] MAYBEE, J.S., AND RICHMAN, D.J., *Some Properties of GM-matrices and their Inverses*, Lin. Alg. Appl. (1989) (to appear).

[125] MAYBEE, J.S., AND WIENER, G.M., *From Qualitative Matrices to Quantitative Restrictions*, Linear and Multilinear Alg., 22 (1987).

[126] MCKELVEY, R., *Intransitivities in Multidimensional Voting Models and Some Implications for Agenda Control*, J. Econ. Th., 12 (1976), pp. 472–482.

[127] MCMORRIS, F.R., *Axioms for Consensus Functions on Undirected Phylogenetic Trees*, Math. Biosciences, 74 (1985), pp. 17–21.

[128] MENGER, K., *Probabilistic Theories of Relations*, Proc. Nat. Acad. Sci., 37 (1951), pp. 178–180.

[129] MICHAUD, P., AND MARCOTORCHINO, P., *Modeles d'Optimisation en Analyse des Donnees Relationnelles*, Math. Sci. Hum., 67 (1979), pp. 7–38.

[130] MORISHIMA, M., *On the Laws of Change of the Price System in an Economy which Contains Complementary Commodities*, Osaka Economic Papers, 1 (1952), pp. 101–113.

[131] MOSIMANN, J., *Elementary Probability for the Biological Sciences*, Appleton-Century-Crofts, New York, 1968.

[132] NARENS, L., *Abstract Measurement Theory*, MIT Press, Cambridge, MA, 1985.

[133] NEWMAN, C.M., AND COHEN, J.E., *A Stochastic Theory of Community Food Webs IV. Theory of Food Chain Lengths in Large Webs*, Proc. R. Soc. Lond., B 228 (1986), pp. 355–377.

[134] OPSUT, R.J., *On the Computation of the Competition Number of a Graph*, SIAM J. Alg. & Discr. Meth., 3 (1982), pp. 420–428.

[135] OPSUT, R.J., AND ROBERTS, F.S., *On the Fleet Maintenance, Mobile Radio Frequency, Task Assignment, and Traffic Phasing Problems*, in G. Chartrand, Y. Alavi, D.L. Goldsmith, L. Lesniak-Foster, and D.R. Lick (eds.), *The Theory and Applications of Graphs*, Wiley, New York, 1981, pp. 479–492.

[136] OPSUT, R.J., AND ROBERTS, F.S., *I-Colorings, I-Phasings, and I-Intersection Assignments for Graphs, and their Applications*, Networks, 13 (1983), pp. 327–345; (A).

[137] OPSUT, R.J., AND ROBERTS, F.S., *Optimal I-Intersection Assignments for Graphs: A Linear Programming Approach*, Networks, 13 (1983).

[138] OWEN, G. AND SHAPLEY, L., *The Copeland Winner and the Shapley Value in Spatial Voting Games*, Int. J. Game Theory (1989) (to appear).

[139] *PERCUS, J.K., *Tree Structures in Immunology*, in F.S. Roberts (ed.), *Applications of Combinatorics and Graph Theory in the Biological and Social Sciences*, IMA Volumes in Mathematics and its Applications, Springer-Verlag, New York, 1989.

[140] PFANZAGL, J., *Theory of Measurement*, Wiley, New York, 1968.

[141] PIEPER, G.W., *Computer Scientists Join Biologists in Genome Project*, SIAM News, January 1989, p. 18.

[142] QUIRK, J., *A Class of Generalized Metzlerian Matrices*, in G. Horwich and P. Samuelson (eds.), *Trade, Stability, and Macroeconomics: Essays in Honor of Lloyd A. Metzler*, Academic Press, New York, 1974.

[143] QUIRK, J., *Qualitative Stability of Matrices and Economic Theory: A Survey Article*, in H. Greenberg and J. Maybee (eds.), *Computer-Assisted Analysis and Model Simplification*, Academic Press, New York, 1981, pp. 113–164.

[144] QUIRK, J., AND RUPPERT, R., *Qualitative Economics and the Stability of Equilibrium*, Rev., Economic Studies, 32 (1965), pp. 311–326.

[145] RAYCHAUDHURI, A., *Intersection Assignments, T-Coloring, and Powers of Graphs*, Ph.D. Thesis, Department of Mathematics, Rutgers University, New Brunswick, NJ (1985).

[146] RAYCHAUDHURI, A., AND ROBERTS, F.S., *Generalized Competition Graphs and their Applications*, in P. Brucker and R. Pauly (eds.), *Methods of Operations Research, Vol. 49*, Anton Hain, Konigstein, W. Germany, 1985, pp. 295–311.

[147] REGNIER, S., *Sur Quelques Aspects Mathematiques de la Classification Automatique*, ICC Bull., 4 (1965), pp. 175–191; reprinted, Math. et Sciences Humaines, 82 (1983), 21-30.

[148] ROBERTS, F.S., *Indifference Graphs*, in in F. Harary (ed.), *Proof Techniques in Graph Theory*, Academic Press, New York, 1969, pp. 139–146; (A).

[149] F.S. ROBERTS, *On the Boxicity and Cubicity of a Graph*, in W.T. Tutte (ed.), *Recent Progress in Combinatorics*, Academic Press, New York, 1969, pp. 301–310; (B).

[150] F.S. ROBERTS, *Discrete Mathematical Models, with Applications to Social, Biological, and Environmental Problems*, Prentice-Hall, Englewood Cliffs, N.J, 1976.

[151] F.S. ROBERTS, *Food Webs, Competition Graphs, and the Boxicity of Ecological Phase Space*, in Y. Alavi and D. Lick (eds.), *Theory and Applications of Graphs*, Springer-Verlag, New York, 1978, pp. 477–490; (A).

[152] F.S. ROBERTS, *Graph Theory and its Applications to Problems of Society*, CBMS-NSF Monograph No. 29, SIAM, Philadelphia (1978); (B).

[153] F.S. ROBERTS, *Graph Theory and the Social Sciences*, in R.J. Wilson and L.W. Beinecke (eds.), *Applications of Graph Theory*, Academic Press, London, 1979; (A).

[154] F.S. ROBERTS, *Measurement Theory, with Applications to Decisionmaking, Utility, and the Social Sciences*, Addison Wesley, Reading, MA, 1979; (B).

[155] F.S. ROBERTS, *Applications of the Theory of Meaningfulness to Order and Matching Experiments*, in E. DeGreef and J. Van Buggenhaut (eds.), *Trends in Mathematical Psychology*, North-Holland, Amsterdam, 1984, pp. 283–292; (A).

[156] F.S. ROBERTS, *Applied Combinatorics*, Prentice-Hall, Englewood Cliffs, N.J., 1984; (B) .

[157] F.S. ROBERTS, *Applications of the Theory of Meaningfulness to Psychology*, J. Math. Psychol., 29 (1985), pp. 311-332.

[158] F.S. ROBERTS, *T-Colorings of Graphs: Recent Results and Open Problems*, RUTCOR Research Report RRR #7-86, Rutgers Center for Operations Research, Rutgers University, New Brunswick, N.J., May 1986; To appear in Annals of Discr. Math.

[159] F.S. ROBERTS, *From Garbage to Rainbows: Generalizations of Graph Coloring and their Applications*, in Y. Alavi, G. Chartrand, O.R. Oellermann, and A.J. Schwenk (eds.), Proceedings of the Sixth International Conference on the Theory and Applications of Graphs, Wiley, New York, to appear, 1989; (A).

[160] *ROBERTS, F.S., *Meaningless Statements, Matching Experiments, and Colored Digraphs (Applications of Graph Theory and Combinatorics to the Theory of Measurement)*, in F.S. Roberts (ed.), Applications of Combinatorics and Graph Theory in the Biological and Social Sciences, IMA Volumes in Mathematics and its Applications, Springer-Verlag, New York, 1989; (B).

[161] ROBERTS, F.S., *Merging Normalized Scores*, J. Math. Anal. & Appl. (1989) (to appear); (C).

[162] ROBERTS, F.S., AND ROSENBAUM, Z., *Scale Type, Meaningfulness, and the Possible Psychophysical Laws*, Math. Soc. Sci., 12 (1986), pp. 77-95.

[163] ROBERTS, F.S., AND STEIF, J.E., *A Characterization of Competition Graphs of Arbitrary Digraphs*, Discrete Applied Math., 6 (1983), pp. 323-326.

[164] SAMUELSON, P., *Foundations of Economic Analysis*, Harvard University Press, Cambridge, MA, 1947.

[165] SATTERTHWAITE, M.A., *Strategy-proofness and Arrow's Conditions*, J. Econ. Theory, 10 (1975), pp. 187-217.

[166] SCHEINERMAN, E.R., *An Evolution of Interval Graphs*, mimeographed, Department of Mathematical Sciences, The Johns Hopkins University, Baltimore, MD (1987); (A).

[167] SCHEINERMAN, E.R., *Random Interval Graphs*, mimeographed, Department of Mathematical Sciences, The Johns Hopkins University, Baltimore, MD (1987); to appear in Combinatorica; (B).

[168] SCOTT, D.D., *The Competition-Common Enemy Graph of a Digraph*, Discr. Appl. Math., 17 (1987), pp. 269-280.

[169] SCOTT, D., AND SUPPES, P., *Foundational Aspects of Theories of Measurement*, J. Symb. Logic, 23 (1958), pp. 113-128.

[170] SEAGER, S.M., *The Double Competition Number of Some Triangle-free Graphs*, mimeographed, Department of Mathematics, Mount Saint Vincent University, Halifax, Nova Scotia, Canada (1988).

[171] *SELLERS, P.H., *Combinatorial Aspects of Enzyme Kinetics*, in F.S. Roberts (ed.), Applications of Combinatorics and Graph Theory in the Biological and Social Sciences, IMA Volumes in Mathematics and its Applications, Springer-Verlag, New York, 1989.

[172] SEN, A.K., *Collective Choice and Social Welfare*, Holden-Day, San Francisco, 1970.

[173] SHAPLEY, L.S., AND SHUBIK, M., *A Method for Evaluating the Distribution of Power in a Committee System*, Amer. Polit. Sci. Rev., 48 (1954), pp. 787-792.

[174] STEIF, J.E., *Frame Dimension, Generalized Competition Graphs, and Forbidden Sublist Characterizations*, Henry Rutgers Thesis, Department of Mathematics, Rutgers University, New Brunswick, N.J. (1982).

[175] STOFFERS, K.E., *Scheduling of Traffic Lights - A New Approach*, Transp. Res., 2 (1968), pp. 199-234.

[176] *STRAFFIN, P.D., JR., *Spatial Models of Power and Voting Outcomes*, in F.S. Roberts (ed.), Applications of Combinatorics and Graph Theory in the Biological and Social Sciences, IMA Volumes in Mathematics and its Applications, Springer-Verlag, New York, 1989.

[177] SUGIHARA, G., *Graph Theory, Homology, and Food Webs*, in S.A. Levin (ed.), Population Biology, Proc. Symposia in Applied Mathematics, Vol. 30, American Mathematical Society, Providence, Rhode Island, 1983.

[178] SUPPES, P., KRANTZ, D.H., LUCE, R.D., AND TVERSKY, A., *Foundations of Measurement, Vol. II*, Academic Press, New York, 1989.

[179] TESMAN, B., *Ph.D. Thesis*, Department of Mathematics, Rutgers University, New Brunswick, NJ (1989); in preparation.

[180] TROTTER, W.T., *Interval Graphs, Interval Orders and their Generalizations*, in R.D. Ringeisen and F.S. Roberts (eds.), Applications of Discrete Mathematics, SIAM, Philadelphia, 1988, pp. 45-58.

[181] WAKABAYASHI, Y., *Aggregation of Binary Relations: Algorithmic and Polyhedral Investigations*, thesis, Augsburg (1986).

[182] *WATERMAN, M.S., *Some Mathematics for DNA Restriction Mapping*, in F.S. Roberts (ed.), Applications of Combinatorics and Graph Theory in the Biological and Social Sciences, IMA Volumes in Mathematics and its Applications, Springer-Verlag, New York, 1989.

[183] WATERMAN, M.S., AND GRIGGS, J.R., *Interval Graphs and Maps of DNA*, Bull. Math. Biol., 48 (1986), pp. 189–195.
[184] YANNAKAKIS, M., *The Complexity of the Partial Order Dimension Problem*, SIAM J. Alg. & Discr. Meth., 3 (1982), pp. 351–358.
[185] YOUNG, H.P., AND LEVENGLICK, A., *A Consistent Extension of the Condorcet Election Principle*, SIAM J. Appl. Math., 35 (1978), pp. 285–300.

SOCIAL WELFARE AND AGGREGATION PROCEDURES: COMBINATORIAL AND ALGORITHMIC ASPECTS*

JEAN-PIERRE BARTHÉLEMY†

Abstract. In this paper, we review some aspects about aggregation procedures. First some examples are given: Borda count, Condorcet rule, decisive procedures, Kemeny's medians, Dogson procedure. Then a general definition of an aggregation procedure is proposed and a hierarchy of results (possible/impossible, computable/non computable, easy/hard) is illustrated by several examples. The last part of this paper is devoted to a formal theory of medians and a new possibility result is obtained for social welfare functions.

Key words. Social Welfare, Aggregation, Consensus, Voting, Preferences, Order relations, Medians, Lattices, Computability, NP-Completeness.

1. Introduction. Problems concerning aggregation of preferences or more generally group decision making, can nicely illustrate the connection between combinatorics, algorithmic and the social sciences (Roberts, 1976; Barthélemy, Flament and Monjardet, 1982; Fishburn, 1984; Leclerc, 1988). These problems (or, more precisely, a mathematical point of view on these problems) go back to the late eighteenth century in France with the works of Borda (1784) and Condorcet (1785) on electoral processes: how should an assembly elect the 'best' candidate?

As an alternative to the current *plurality rule*, Borda proposed a method based on ranked votings. Each voter gives the names of the candidates according to an order of preference. Then the sum of the rankings assigned to each candidate is performed and the candidate with the greatest amount of points is the winner. Critizing Borda's procedure, Condorcet proposed a method based on paired comparisons and majority rule. However, as Condorcet himself noticed, this method leads to a drawback: the possibility that there is no winner.

Condorcet, Borda and others, like Lhuillier (cf. Monjardet, 1976), have had some successors. For instance S. Laplace in France, the Reverend Dogson and E.J. Nanson in England. However, despite studies of specific procedures for elections pursued throughout the nineteenth century (cf. Fishburn, 1983), this kind of pursuit became somewhat obsolete in the first part of the twentieth century. Similar preoccupations in a different context (Welfare Economics) appeared in the fifties, leading to a relational formulation of *social welfare functions* (such a function assigns to a profile of *individual preferences* a *social preference*) and to an impossibility theorem (Arrow, 1951). The work of Arrow has had a very numerous lineage and a terrifying amount of literature is devoted to such topics (see, among others: Mukarami, 1968; Sen, 1970; Pattanaik, 1971; Fishburn, 1973; Kelly, 1978; Mirkin, 1979,...)

---

*This paper was written partly during a short visit at the IMA (Minneapolis), partly at the University of Louisville and partly at the E.N.S. Télécom. The author would like to thank F.R. Morris and Ch. Vercken for helpful comments.

†Département Informatique, Ecole Nationale Supèrieure des télécommunications, 46 rue Barrault, 75634 Paris cedex 13. France.

It would have been completely foolish to attempt, in a paper like this, to review the main streams of Social Choice Theory. So, this review will be biased and partial and will omit important topics like, among the most classical:

Variants on Arrow's Conditions (they are available in the books mentioned above).

Strategy theory of voting and manipulation, with the celebrated manipulability theorem of Gibbard and Satterthwaite (Gibbard, 1973; Satterthwaite, 1973, 1975; Moulin, 1983).

Restricted domains of preferences (Guilbaud, 1952; Black, 1958; Inada, 1969; Sen and Pattanaik, 1969; Demange, 1980, ...).

Sophisticated voters, non cooperative voters, cooperating voting (Farquaharson, 1969; Moulin, 1980, 1983; Pattanaik, 1976; Peleg, 1978...).

The uses of heuristic algorithms and of Mathematical Programming to solve difficult aggregation problems (cf. Barthélemy and Monjardet, 1981, 1988, for references).

Besides, this paper emphasizes three points, less classical, but now getting more attention in the current literature, and illustrates them by examples. The first one deals with the *possibility of computing effectively and/or efficiently a social welfare function*.: We propose some considerations about computability in Social Choice theory and review some recent NP-completeness results. The second deals with a *formal theory of aggregation* which involves, rather than the structures of individual and social preferences, structures on sets of preferences. The IMA meeting was devoted to applications of combinatorics and graph theory to Biology and Social Sciences. So, and that is the third point, we will point out connections between Social Welfare and the so-called *Consensus Problem in Taxonomy and/or in Evolution Theory*, each time we will have an opportunity to do this (for more details on this last point, see Barthélemy, Leclerc and Monjardet, 1986).

In the first section, we describe some classical ways to aggregate individual's preferences into a social preference. Then, in the second section, we propose some definitions and state what will be our point of view. The third section illustrates, by examples, different kinds of results: possibility vs. impossibility theorems, computability vs. non computability theorems, easiness vs. difficulty theorems. The last section is devoted to a formal theory of the median procedure.

The reader is assumed to have a basic knowledge on graphs and algorithmic (including NP-completeness, cf. Garey and Johnson, 1978) as well as on order relations as they are used in the mathematics of decision theories (cf. Fishburn, 1972) and on lattice theory (cf. Birkhoff, 1967). Beside the domain of preferences, we shall mention three models that occur in mathematical biology: namely the n-trees, the dendrograms and the phylogenetic trees. It seems appropriate, for the convenience of the reader, to recall here their definitions. Let S be a finite set with $|S| = n$. An n-tree is a set T of non empty subsets of S (called the *clusters* of T) such that $S \in T, \{s\} \in T$ for each $s \in S$ and $A \cap B \in \{A, B, \emptyset\}$ for each $A, B \in T$. The n-trees constitute a combinatorial model for hierarchical clustering. A *dendrogram* is a pair (T, f) with T an n-tree and f a map from T to the set of real numbers such that $f(\{s\}) = 0$ for each singleton $\{s\}$ and $C \subseteq C'$ implies $f(C) \leq f(C')$ for each pair C, C' of clusters of T. A *phylogenetic tree* on S is a pair (P, f), with P a graph theoretic tree and f a map from S to the vertex set V of P such that each vertex in $V - f(S)$ has a degree ≥ 3. The elements of $f(S)$ represent known species, while the elements of $V - f(S)$ represent unknown common ancestors.

Within the relations used to model preferences, that constitute the main topic of this paper, we shall use the following notations that we will not systematically recall:

R will denote the set of all binary relations on S;
L will denote the set of all linear orders on S;
T will denote the set of all tournaments on S;
P will denote the set of all preorders on S;
E will denote the set of all equivalence relations on S;
C will denote the set of all complete preorders on S;
W will denote the set of all weak orders on S;
O will denote the set of all partial orders on S.

Since binary relations R on S are subsets of the cartesian product $S \times S$, we shall use set theoretic notations $R \cap R'$, $R \cup R'$, $R \subseteq R'$, etc. However we shall sometimes write xRy instead of $(x,y) \in R$. In case of models for preferences or choices, this statement has to be understood as: "y is preferred to x" or "y is chosen over x". For a binary relation R, we denote by $R^{-1}, R^c, R^d, I(R)$ and $P(R)$ respectively the inverse of R, the complement of R, the dual of R, the incomparability part of R, the strict part of R. Those relations are defined by:

$xR^{-1}y$ if and only if yRx,

$xR^c y$ if and only if $(x,y) \notin R$,

$xR^d y$ if and only if $(y,x) \notin R$ (so, $R^d = (R^{-1})^c$),

$xI(R)y$ if and only if $(x,y) \notin R$ and $(y,x) \notin R$,

and $xP(R)y$ if and only if xRy and $(y,x) \notin R$.

For a permutation σ of S, we denote by R^σ the binary relation $\{(\sigma(x), \sigma(y)) : (x,y) \in R\}$.

[q] denotes the integer part of the number q.

2. Some examples.

2.1. Plurality rule, Borda count and Condorcet rule. Consider the following situation: 23 voters have to choose one candidate out of three. First let's follow the current plurality rule and assume that the result of the vote is: a is preferred (globally) by 9 voters, b by 6 voters and c by 8 voters. Thus *a is the plurality winner*.

Now assume that each voter choosing c would prefer to see b as a winner rather than a. So, $6 + 8 = 14$ voters (i.e. *majority*) would prefer to see b as a winner rather than a. This drawback led Borda (1784) to propose a somewhat strange solution called the *sum of rankings method* or *Borda count*. Assume that each voter compares the candidates pair by pair and that these comparisons lead to the following profile of linear orders:

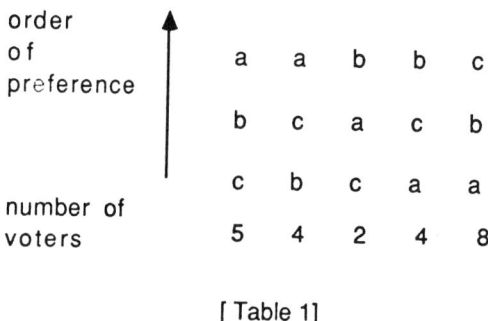

[Table 1]

(notice that this profile is compatible with the previous ballot). Assign 2 points to a first ranked candidate, 1 to a second, 0 to a third. Then the score of a is: $sc(a) = 2(5+4) + 2 = 20$, while $sc(b) = 25$ and $sc(c) = 24$. Thus b *is the Borda winner*. Moreover, Borda count allows us to rank the candidates in such a way that we get as a *social linear order*: $a < c < b$ (notice that generally, candidates may be tied by Borda count).

However, we see on table 1 that 12 voters (hence a majority) would have preferred to see c as a winner rather than b. Criticizing the Borda procedure, Condorcet (1785) proposed a method based on collective paired comparisons. Let $v(x,y)$ denote the number of voters choosing y over x. Then y will be *socially preferred* to x when:

$$v(x,y) \geq v(y,x).$$

In our example, we obtain:
$v(a,b) = 14$, $v(b,c) = 12$ and $v(a,c) = 12$. There is always a majority of voters to prefer c to any other candidate. So, c *is the Condorcet winner*. Moreover we have obtained as a social linear order: $a < b < c$. The valued graph hereunder indicates this order and the strength of the majorities supporting each social preference.

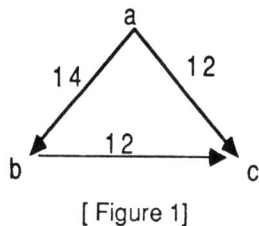

[Figure 1]

However the Condorcet method leads to a drawback. In our example, we have obtained a social linear order; consider now the profile of table 2:

a	b	b	c
c	a	c	b
b	c	a	a
3	1	1	2

[Table 2]

We get $v(b,c) = 5$, $v(a,b) = 4$ and $v(c,a) = 4$. Is a the winner? No, since it is defeated by b. But b is defeated by c and c is defeated by a. In other words, instead of forming a linear order, the social preferences constitute the following tournament:

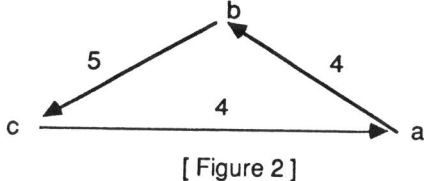

[Figure 2]

This is the well-known *paradox of voting*, called *Effet Condorcet* in the French literature (Guilbaud, 1952).

2.2 Effet Condorcet and decisive procedures. The Condorcet procedure is based on two points:

1°) The social preference on a pair $\{x,y\}$ of candidates depends only on the individual preferences on $\{x,y\}$.

2°) The way to decide between x and y in $\{x,y\}$ is to allow the majority rule.

Indeed point 1 alone suffices to get drawbacks. In order to explain that, let us introduce some notations and one definition. Let S be the finite set of candidates, L the set of all linear orders on S and v a positive integer. Cordorcet intended to construct a function F from L^v to L, assigning to each profile $\pi = (R_1, \ldots, R_v) \in L^v$ of individual preferences, a social preference $F(\pi) \in L$. The requirement of point 1 may be formalized as follows:

(D) If $\pi = (R_1, \ldots R_v)$ and $\pi' = (R'_1, \ldots, R'_v)$ are two profiles of linear orders and if x and y are two candidates such that: $\{i : xR_iy\} = \{i : xR'_iy\}$ (i.e. the voters preferring y to x are the same from the first profile and from the second), then:

$$xF(\pi)y \text{ if and only if } xF(\pi')y.$$

Call any function $L^v \to L$ that satisfies property (D) a *decisive procedure*. Examples of decisive procedures are:
the *absolute dictatorships*: there exists a voter i (called the *dictator*) such that for each profile $\pi = (R_1, \ldots, R_v)$, $F(\pi) = R_i$;

the *persecutions*: there exists a voter i (the *persecuted voter* such that for each profile $\pi = (R_1, \ldots, R_v)$, $F(\pi) = R_i^{-1}$.

The following result (Murakami, 1968; Wilson, 1972; Hansonn, 1972 quoted by Fishburn, 1973, Monjardet, 1978) is a variant of the celebrated Arrow's theorem (cf. 4.1). It shows that these three examples are not extreme and that there exists no reasonable decisive procedure.

THEOREM 1. *Let $F : L^v \to L$ be a decisive procedure such that for each $x, y \in X$, there exists π, $\pi' \in L^v$ such that $xF(\pi)y$ and $yF(\pi')x$. Then F is either a dictatorship, or else a persecution.*

2.3 Kemeny's median. In a classical paper, Kemeny (1959), defines a method to aggregate individual preferences into a social preference. This method is based on the *symmetric difference distance* between linear orders and is justified by giving an axiomatic characterization of this distance. Notice that the same distance has been used implicitly by Kendall (1948) when he defined the correlation coefficient τ between linear orders and that it is used to measure agreement and cohesion between rankings (Monjardet, 1985).

More generally, consider two arbitrary binary relations R and R' on the finite set S. The *symmetric difference distance* $\partial(R, R')$ enumerates the disagreements between R and R' (i.e. $\partial(R, R')$ is the number of ordered pairs (x, y) such that xRy and not $xR'y$, or $xR'y$ and not xRy):

$$\partial(R, R') = |R \Delta R'| = |R \cup R'| - |R \cap R'|.$$

Kemeny defines the *medians* of the profile $\pi(R_1, \ldots, R_v)$ of linear orders as the linear orders R such that the sum:

$$\Sigma_{i=1,2,\ldots v} \partial(R, R_i)$$

is a minimum.

Indeed Kemeny's medians are strongly related with the Condorcet rule. Consider the set T of all the tournaments on S. When v is an odd number, the Condorcet rule assigns to each profile $\pi = (R_1, \ldots R_v)$ of linear orders a unique tournament $C(\pi)$ defined by:

$xC(\pi)y$ if and only if $v(x, y) \geq v(y, x)$ (or equivalently if and only if $v(x, y) > v/2$).

Since each arc of this tournament is supported by a majority of voters, it is quite understandable that it is minimizing the sum of the numbers of disagreements between a tournament R and the orders occuring in π. This point is moreover confirmed by the formula:

(1) $\qquad \Sigma_{i=1,2,\ldots v} \partial(R, R_i) = 2\Sigma_{yRx} v(x, y).$

So, $C(\pi)$ is the unique solution of:

(2) $\qquad \text{Min}_{R \in T} \Sigma_{i=1,2,\ldots v} \partial(R, R_i).$

When v is an even number, we can get several tournament-solutions of (2) by using Condorcet rule and tie breaking. Hence a suitable extension of Condorcet rule is obtained by taking all the linear orders solutions of:

(3) $\qquad\qquad\qquad \text{Min}_{R\in L}\ \Sigma_{i=1,2,\ldots v}\partial(R,R_i).$

When no Effet Cordorcet occurs in π, solutions of (2) and (3) are exactly the same. This connection between Condorcet and Kemeny was first made by Barbut (1967). The Kemeny's median procedure is also related to a problem famous in a combinatorial optimization, the so-called *feedback arc set problem*.

2.4. Medians and feedback: a first glance at the algorithmic complexity of the Kemeny procedure. Dealing with the Condorcet rule, one constructs a complete valued digraph $G(\pi)$ and a matrix $M(\pi)$ associated with each profile π of linear orders. $G(\pi)$ is defined as follows: its vertices are the elements of S; (x,y) is an arc if and only if $v(x,y) \geq v(y,x)$ and the weight of the arc (x,y) is the number:

$$w(x,y) = v(x,y) - v(y,x).$$

The matrix $M(\pi)$ has the numbers $w(x,y)$ as entries. So, it is a square matrix and satisfies the following properties:

1. $M(\pi)$ is an integer matrix;
2. $M(\pi)$ is antisymmetrix $(w(x,y) + w(y,x) = 0)$;
3. All the coefficients of $M(\pi)$ have the same parity $(w(x,y) = 2v(x,y) - v)$.

Call a *Benjamin Franklin matrix*[1] any square matrix fulfilling, 1, 2 and 3 (Benjamin Franklin seems to be the first to have used the matrix $M(\pi)$, cf. Debord, 1987 b). The following lemma is established by Debord (1987 a).

---

[1] (From Debords' Thesis)
London, Sept. 19, 1772
Dear Sir,
In the affair of so much importance to you, where in you ask my advice, I cannot, for want of sufficient premises, advise you what to determine, but if you please I will tell you how. When those difficult cases occur, they are difficult, chiefly because while we have them under consideration, all the reasons pro and con are not present to the mind at the same time; but sometimes one set present themselves, and at other times another, the first being out of sight. Hence the various purposes of inclinations that alternately prevail, and the uncertainty that perplex us.
To get over this, my way is to divide half a sheet of paper by a line into two columns; writing over the Pro and the other the Con. Then, during three or four days consideration, I put down the different heads short hints of the different motives, that at different times occurs to me, for or against the measure. When I have thus got them all together in one view, I endeavor to estimate their weights; and when I find two, one on each side, that seem equal, I strike them both out. If I find a reason pro equal to some two reasons con, I find out the three. If I judge some two reasons con equal to three reasons pro, I strike out the five; and thus proceeding I find at length where the balance lies; and if, after a day or two of further consideration, nothing new that is of importance occurs on either side, I come to a determination accordingly. And, though the weights of reasons cannot be taken with the precision of algebraic quantities, yet when each is thus considered separately and comparatively, and the whole lies before me, I think I can judge better, and am less liable to make a rash step, and in fact I have found great advantage from this kind of equation, in what may be called moral or prudential algebra.
Wishing sincerely that you may determine for the best, I am ever, my dear friend, yours most affectionately,
B. Franklin

LEMMA 1. *Let M be a Benjamin Franklin matrix. Then there exists a profile π of linear orders on S such that $M = M(\pi)$. Moreover, π can be constructed from M in polynomial time.*

Now, let us go gack to the graph $G(\pi)$. Since Kemeny's medians minimize the sum of disagreements between a linear order and a profile π, it is not difficult to become convinced that all the solutions of (3) are obtained by reversing in $G(\pi)$ a set of arcs with minimal sum of weights in order to transform $G(\pi)$ into a transitive digraph. This point is indeed confirmed by formula (4), which comes directly from (3):

(4) $\qquad \Sigma_{i=1,2,\ldots v}\partial(R,R_i) = vn(n-1)/2 - \Sigma_{xRy}w(x,y).$

For instance, applying this principle, from the valued digraph of Figure 2, we obtain two median linear orders $a < b < c$ and $b < c < a$. So problem (3) becomes a variant of the feedback arc set problem. More specifically, consider the following problem:
Name: Feedback arc set (FAS).
Instance: Digraph G, integer k.
Question: Does there exist in G a set A of arcs (called a *feedback*) such that $|A| \leq k$ and every circuit of G contains at least one arc in A?

FAS is known to be NP-complete (Karp, 1972). It is worth noticing that reversing the arcs of a minimum feedback of G transforms G into a circuitless graph (Younger, 1963). This point may be used for the median problem. Consider an instance (G, k) of FAS, complete G by adding between unlinked pairs $\{x, y\}$ the arc (x, y) and the arc (y, x). Let G' be the so-obtained digraph. When, in G', x and y are linked by exactly one arc $((x, y)$ or $(y, x))$, assign the weight 2 to this arc. Assign the weight 0 to each double arc. Completing the weights by antisymmetry, we get a Benjamin Franklin matrix M. So there exits some profile π of linear orders, constructible in polynomial time, such that $G' = G(\pi)$ and we are able to answer YES to the instance (G, k) of FAS if and only if one obtains a linear order by reversing in $G(\pi)$ a set of arcs with sum of weight $\leq 2k$. So, in view of equality (4), we obtain an integer q such that the answer to FAS is YES if and only if the answer to the problem hereunder is YES:

Name: Kemeny's Medians (or LM)
Instance: Set S, profile $\pi = (R_1, \ldots, R_v)$ of linear orders on S, integer q.
Question: does there exist a linear order R such that $\Sigma_{i=1,2,\ldots v}\partial(R,R_i) \leq q$?

So, we have just proved that:

PROPOSITION 1. *Problem LM is NP-complete.*

This feedback arc set interpretation of Kemeny's medians has been used from an algorithmic point of view by Reinelt (1985) and Grötschel, Jünger and Reinelt (1985 *a* and *b*). Many other interpretations of the Condorcet rule and the Kemeny medians have been considered. For example in terms of integer linear programming (cf. among others De Cani, 1969; Bowman and Colantoni, 1973; Marcotorchino and Michaud, 1978; Arditti, 1984...); in terms of quadratic assignment (cf. Hubert and Schulz, 1975). Notice also the polyhedral interpretation in terms of 'permutoèdre' (alias weak Bruhat orders): Guilbaud and Rosensthiel (1963); Feldman-Högaasen (1969); Benzecri (1970); Kreweras (1970); Barbut and Monjardet (1970); Frey, (1971); Marcotorchino and Michaud (1979); Abello and Johnson (1984); Abello (1985). Combinatorial properties of median linear orders may be found in Monjardet (1973). Kemeny's medians constitute an extension of the Condorcet rule, but others are possible and they are studied by Fishburn (1977).

Let us also mention that similar methods occur in cluster analysis. In this case v variables describe the set S of objects. We can associate an equivalence relation on S with each of these variables: x and y are in the same equivalence class if the values of this variable are the same for x and y. So, we obtain v equivalences on $S : R_1, \ldots, R_v$. Regnier (1965), then Mirkin (1974) propose as a good clustering of S, a partition whose associated equivalence relation minimizes the quantity: $\Sigma_{i=1,2,\ldots v} \partial(R, R_i)$ among all the equivalences R in E, ∂ being still defined as the symmetric difference distance between binary relations on S. In the context of mathematical taxonomy the median n-trees associated with a profile $\pi(T_1, \ldots, T_v)$ of n trees have been studied by Margush and McMorris (1981) and by Barthélemy and McMorris (1986). Such a median n-tree minimizes the quantity: $\Sigma_{i=1,2,\ldots v} \partial(T, T_i)$, with the symmetric difference distance ∂ defined as: $\partial(T, T') = |T \Delta T'| = |\{C \subseteq S : C \in T \text{ and } C \notin T', \text{ or } C \notin T \text{ and } C \in T'\}|$. A review of questions concerning the median procedure may be found in Barthélemy and Monjardet (1981, 1988).

2.5 From Condorcet procedure to lattice polynomials. Formula (2) gives a metric point of view on the Condorcet rule. This notion yields also an algebraic approach. In the set $R = 2^{X \times X}$ of all binary relations on S, for each tuple $\pi(R_1, \ldots, R_v) \in R^v$, we denote by $\alpha(\pi)$ the relation:

$$\alpha(\pi) = \bigcup_{W \subset \{1,\ldots,v\};\ |W|=[(v+2)/2]} \bigcap_{i \in W} R_i. \tag{5}$$

The so-obtained function from R^v to R is called the *majority rule*.

We denote by $\beta(\pi)$ the relation:

$$\beta(\pi) = \bigcup_{W \subset \{1,\ldots,v\};\ |W|=[(v+1)/2]} \bigcap_{i \in W} R_i. \tag{6}$$

Obviously, $\alpha(\pi) \subseteq \beta(\pi)$ and if v is an odd number, then $\alpha(\pi) = \beta(\pi)$. In this last case, if π is a profile of linear orders, then $\alpha(\pi)$ is obtained by applying the Condorcet rule.

More generally, define the *algebraic median* of any tuple $\pi = (R_1, \ldots, R_v)$ of binary relations on S as the interval $[\alpha(\pi), \beta(\pi)] = \{R : \alpha(\pi) \subseteq R \subseteq \beta(\pi)\}$. When π is a profile of linear orders, the tournaments obtained by the Condorcet procedure are exactly the tournaments $R \in [\alpha(\pi), \beta(\pi)]$. More generally for a tuple

$\pi = (R_1, \ldots, R_v)$ of binary relations on S the algebraic medians are exactly the relations minimizing the sum: $\sum_{i=1,2,\ldots v} \partial(R, R_i)$. The relation $\alpha(\pi)$ (hence the Condorcet rule, when v is an odd number) is an example of a method to aggregate preferences which uses the lattice laws $\cup$ and $\cap$ on the relations occuring in π.

More generally, consider two sets of binary relations D and M on S, with $D \subseteq M$. A function $F : D^v \to M$ is said to be a *lattice polynomial* if and only if there exists a set W of subsets of $\{1, \ldots, v\}$ such that

(7) $$F_W(\pi) = \bigcup_{W \in \mathcal{W}} \bigcap_{i \in W} R_i$$

for $\pi(R_1, \ldots, R_v)$.

(7) may be interpreted in terms of *decisive sets* (or *coalitions*): there are groups of voters (the so-called *decisive sets* such that y becomes socially preferred to x if and only if y is preferred to x by each member of at least one of these groups. Let us give some usual examples of lattice polynomial functions:

When $|\mathcal{W}| = 1$, F_W is a meet and corresponds to an *absolute oligarchic rule*, that is to say there exists a subset W of $\{1, \ldots, v\}$ such that y is socially preferred to x if and only if y is preferred to x by each member of W. Special cases of oligarchic rules occur when $W = \{1, \ldots, v\}$ or when $W = \{i\}$:

$$F(\pi) = \bigcap_{i \in \{1,\ldots,v\}} R_i \quad \text{(unanimity rule)};$$
$$F(\pi) = R_i \quad \text{(absolute dictatorship)}.$$

Obviously (from common sense as well as from Theorem 1), absolute dictatorships are the only lattice polynomial functions from L^v to L.

When unanimity is replaced by agreement between a given number w of voters (i.e. $\mathcal{W} = \{W : |W| \geq w\}$), we get the *quota rule* F_w:

$$F_w(\pi) = \bigcup_{W \subseteq \{1,\ldots,v\};\ |W|=w} \bigcap_{i \in W} R_i.$$

(the case where $w[(v+2)/2]$ corresponds to the majority rule).

Lattice polynomial functions have been studied and characterized by Monjardet in the concrete framework of tournaments aggregation (1978) and in the abstract framework of lattice theory (1988). In the context of mathematical taxonomy, absolute oligarchic rules on equivalence relations have been characterized by Mirkin (1975). Mirkin's result has been improved and generalized to dendrograms by Leclerc (1984) and rediscovered by Fishburn and Rubinstein (1986). The "lattice consensus methods" for several models occuring in taxonomy have been studied by Neumann and Norton (1986)..

2.6 The Dogson Procedure. To obviate the Effet Condorcet, Dogson (1976) proposed to choose as a winner a candidate who is near to being the Condorcet winner in the following sense: Imagine that a person is empowered to change the ranking of any voter through pairwise interchange of consecutive candidates in this ranking. Then a *Dogson winner* is a candidate who requires the fewest interchanges to become a Condorcet winner.

This method may be stated with the symmetric difference distance. First note that, if R and R' are linear orders, then $\frac{1}{2}\partial(R, R')$ is the minimal number of pairwise interchanges of consecutive elements of R needed to obtain R'. Thus considering the set K_v of all profiles of v linear orders where no Effet Condorcet occurs, the Dogson procedure first solves the problem:

(8) $$\text{Min}_{\pi \in K_v} \Delta(\pi, \pi')$$

with $\Delta(\pi, \pi') = \Sigma_{i=1,2,...v} \, \partial(R_i, R'_i)$, for $\pi = (R_1, \ldots, R_v)$ and $\pi' = (R'_1, \ldots, R'_v)$.

The solutions of (8) for the profile of Table 2 are indicated on Table 3. Considering (1) and (3), we get c as a winner, considering (2) and (4), we get a as a winner.

c	a	b	b	c		a	a	b	c		a	b	c		a	b	b	c	c
a	c	a	c	b		c	b	c	b		c	c	b		c	a	c	a	b
b	b	c	a	a		b	c	a	a		b	a	a		b	c	a	b	a

1	2	1	1	2		3	1	1	2		3	2	2		3	1	1	1
		(1)					(2)					(3)				(4)		

[Table 3]

3. Aggregation procedures.

3.1. A broad variety of domains and codomains. The examples presented in section 2 all dealt with functions assigning to a profile of individual preferences or choices, a social preference or a social choice. In these examples different models both for individual and social preferences occur. Let us recall them.

In the plurality rule, the domain is the set S^v of all v-tuples of candidates (each voter gives the name of one preferred candidate). It could be also the set of v-tuples of subsets of S (each voter gives the name of several preferred candidates). The codomain is the set of all non empty subsets of S.

In the Borda count, the domain is the set L^v of all v-tuples of linear orders on S and the codomain is the set C of all complete preorders on S. But the domain could be the set C^v as well.

In the Condorcet rule, the domain is the set L^v and for v odd, the codomain is the set T of all tournaments on S; for v even the codomain is the set of all non empty subsets of T. Notice that the Condorcet rule works as well (with the same codomains) if we consider T^v as the domain.

For decisive procedures, the domain is L^v and the codomain is L. But condition (D) may be expressed for any sets of binary relations as domain and codomain.

For medians, the domain was L^v and the codomain was $2^L - \{\emptyset\}$. But we have also mentioned the Régnier approach to cluster analysis where the domain is E^v and the codomain is E. More generally problems like (3) make sense if we take as domain and codomain respectively D^v and M, where D and M are any sets of binary relations on S. In that case we must solve:

$$\text{Min}_{R \in M} \Sigma_{i=1,2,\ldots v} \partial(R, R_i),$$

with $(R_1, \ldots, R_v) \in D^v$.

For the Dogson procedure, the domain is L^v and the codomain is $2^S - \{\emptyset\}$.

Moreover, as we have already mentioned, such aggregation problems occur in fields other than preferences and choices. Beside the references to the consensus problem in taxonomy, note the interesting thema of consensus of probabilities (Lehrer and Wagner, 1981). Another remark is that in most cases we don't need to have the number of voters to be fixed. For instance, Kemeny's medians define a map from L^* to $2^L - \{\emptyset\}$, with:

$$L^* = \bigcup_{v>0} L^v.$$

All the above things lead to the following general setting: consider two sets D (the *data*) and M (the *models*) and a positive integer v and call a *D-profile* any tuple $\pi \in D^* = \bigcup_{v>0} D^v$.

A **(D, M, v)-aggregation) procedure** is a map $F : D^v \to M$.

A **(D, M)-complete procedure** is a map $F : D^* \to M$.

A **(D, M, v)-multiprocedure** is a map $F : D^* \to 2^M - \{\emptyset\}$.

A **(D, M)-complete multiprocedure** is a map $F : D^* \to 2^M - \{es\}$.

We capture the traditional definitions of a social choice function and of a social welfare function by calling a *social choice function* either a (W, S, v)-multiprocedure or a (W, S)-complete multiprocedure and a *social welfare function* either a (W, M, v)-procedure or a (W, M)-complete procedure, such that the strict part of any relation in M is a strict partial order. Replacing "procedure" by "multiprocedure", we get the *pseudo social welfare functions* (Roberts, 1988).

3.2 Three approaches to achieve procedures. In Section 2, we state three approaches to define procedures:

1°) **Construction rules.** The way to construct a procedure is explicitly given. The two historical procedures of Borda and Condorcet and some more modern procedures occurring in multicriteria decision making are of this kind.

2°) **Axiomatic constraints.** We retain only procedures satisfying some conditions that arise either from experimental evidence or from ethic considerations. This point of view goes back to Welfare Economics (Arrow, 1951) and it leads essentially to possibility and impossibility theorems. The characterization of (L, L, v)-decisive procedures (2.2) is an illustration of that approach.

3°) **Optimality constraints**. We have at our disposal some criterion measuring the *remoteness* $D(R, \pi)$ of a social preference R from a profile $\pi \in D^*$ and we take as a complete multiprocedure the function assigning to π all the solutions of (12):

$$\text{Min}_{R \in M} D(R, \pi).$$

The median procedure is an example of this approach. This method, advocated by Kemeny (1959), is now attributed to Condorcet himself by Young (1986).

A role for a mathematician in such topics could be -and has already very often been- to build links between these approaches. For instance, if we consider a procedure F defined by optimality constraints, find an axiomatic characterization of F and a method to construct it.

From now on, we shall focus on relationships between point 2°) and points 1°) and 3°), with the additional requirement that the construction of a procedure has to be algorithmic, i.e. it can be done by a computer ... This approach leads to organizing questions and results as in the following hierarchy:

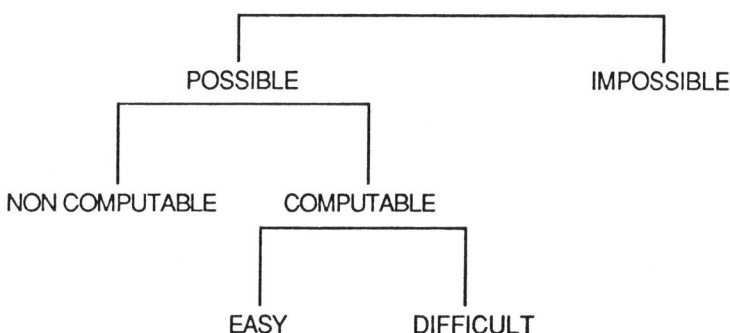

We shall illustrate node of this tree by several examples.

4. A Hierarchy of Problems.

4.1 Possible / impossible. Theorem 1 appears as a possibility theorem: there exist decisive (L, L, v)-procedures (namely, dictatorships, impositions and persecutions). It can be viewed as an impossibility theorem as well, in case we exclude dictatorships, impositions and persecutions: *There exists no (L, L, v)-decisive procedure which is neither imposed, nor persecutive, nor dictatorial.* It may also be used in experimental, or observational situations: A group (or an individual) assigns, in a decision task, to some profile of linear orders a linear order. We observe enough situations to know that the process is neither dictatorial, nor persecuting, nor imposed. So, we have to conclude that if the decision process obeys some rules, then these rules do not apply consistently to each profile or do not lead to a decisive procedure.

As a corollary of theorem 1, we get a special case of Arrow's impossibility theorem: Define a *unanimity* (D, M, v)-procedure as a procedure such that: $R \subseteq F(\pi)$ for each constant profile $\pi = (R, R, \ldots, R)$. So, we get:

COROLLARY 1. *There exists no decisive, unanimity non dictatorial (L, L, v)-procedure.*

Results like this remain essentially if we take as D any set of binary relations, as M any set of transitive relations and if we exclude some undesired solutions. We shall present here a somewhat general result within preferences modeled by order relations. This result is an aggregation of theorems obtained, among others, by Arrow (1951), Mas-Collel and Sonnenschein (1972), Blair and Pollack (1979), Blau (1979), Barthélemy (1981). It has been reported in Barthélemy, Flament and Monjardet (1982).

First, we make some hypotheses on individual and social preferences.

H_0: (1) (transitivity and extended codomain condition). D and M are two sets of binary relations on S with $D \subseteq M \subseteq O$ and:

(2) (stability of the domain). If $R \in D$, then for each permutation σ of S we have $R^\sigma \in D$. For each $R \in D$, we have $R^\sigma \in D$. For each $R \in D$, we have $R^{-1} \in D$.

H_1 (separation). For all distinct $s, t, u \in S$, there exists $R \in D$ such that sRt and tRu.

H_2 ($\wedge$ are allowed, figure 3). For all distinct $s, t, u \in S$, there exists $R \in D$ such that tRs, uRs and $tl(R)u$.

H_3 ($|$. are allowed, figure 3). For all distinct $s, t, u \in S$, there exists $R \in D$ such that sRt and $sl(R)u$ and $tl(R)u$.

Notice that condition H_2 excludes $D = L$, and that condition H_3 excludes $D = W$. Conditions H_0, H_1, H_2 and H_3 are all compatible with the following sets of partial orders: semi-orders, interval orders, orders of dimension 2, semilattices ($n \geq 5$), lattices ($n \geq 6$), distributive, modular etc. lattices ($n \geq 6$), boolean lattices ($n \geq 8$), graded orders, partial orders, ...

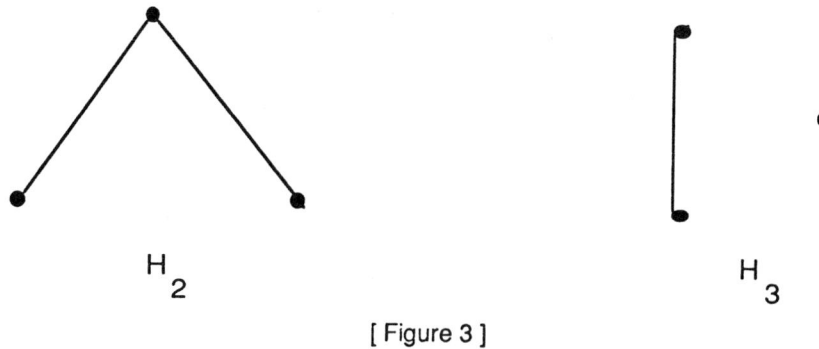

[Figure 3]

Next, we define some special (D, M, v)-procedures.

Let F be a (D, M, v)-procedure. We say that: F is an *oligarchic procedure*, if and only if there exists some subset W of $\{1, \ldots, v\}$ such that, for each D-profile $\pi = (R_1, \ldots, R_v)$

$$\bigcap_{i \in W} R_i \subseteq F(\pi), \quad \text{and} \quad \bigcup_{i \in W} R_i \subseteq (F(\pi))^d.$$

The set W is called an F-*oligarchy*. The second condition means that $xR_j y$ for some member j of the oligarchy W, then we cannot have $yR_j x$: that is each member of W is a *vetoer* (Fishburn, 1973, cf. also Guha, 1972. Also called a *weak dictator*, according to Mas-Collel and Sonnenschein's, 1972, terminology).

When, for an oligarchic procedure F, the set W reduces to one member j, we say that F is a *dictatorship* and that j is a *dictator*. Replacing the inclusion by the equality, we get the *absolute oligarchic procedures* the *absolute dictatorships*, as already defined in 2.5.

We now state the classical notion of *binary independence* for a procedure. This concept can be viewed as an extension of the decisiveness condition as introduced in 2.2 to non complete and non antisymmetric relations: The social preference on a pair of candidates depends only on the individual preferences on that pair.

Let F be a (D, M, v)-procedure. Denote by F_{xy}, the restriction of F to $\{x, y\} \times \{x, y\}$. We say that the F is *independent*, when for each $\pi, \pi' \in D^v$, the equalities:

$$\{i : xR_i y\} = \{i : xR'_i y\} \text{ and } \{i : yR_i x\} = \{i : yR'_i x\}$$

implies the equality

$$F_{xy}(\pi) = F_{xy}(\pi').$$

Define an *arrowian* (D, M, v)- procedure as an independent unanimity procedure.

THEOREM 2. Let F be an arrowian (D, M, v)-procedure.

(i) If H_0, H_1 and H_2 are satisfied, then F is an oligarchic procedure

(ii) If H_0, H_1, H_2 and H_3 are satisfied, then F is an absolute oligarchic procedure.

Theorem 2 admits many dictatorship results as corollaries essentially when M is not "stable" under intersections of elements of D.

COROLLARY 2. Let F be an arrowian (D, M, v)-procedure.

(i) If H_0, H_1 and H_2 are satisfied and if there exists R and R' in D such that $R \cap R'$ is contained in no element of M, then F is a dictatorship.

(ii) If H_0, H_1, H_2 and H_3 are satisfied and if there exists R and R' in D such that $R \cap R' \notin M$, then F is an absolute dictatorship.

Notice that the classical Arrow's theorem is just the case (i) of Corollary 2, with $D = M = W$.

In the case of complete multiprocedures, the Young and Levenglick (1979) theorem provides a characterization of Kemeny's medians.

THEOREM 3. The median procedure is the only (L, L)-complete multiprocedure which is consistent, invariant under permutations of S and Condorcet.

This theorem uses the following definitions:

A (D, M)-complete multiprocedure F is *consistent* if and only if, when π and π' are such that: $F(\pi) \cap F(\pi') \neq \emptyset$, then $F(\pi\pi') = F(\pi) \cap F(\pi')$.

F is *invariant under permutations of S* if and only if for every permutation σ of S and every profile π, we have $F(\pi^\sigma) = \{R^\sigma : R \in F(\pi)\}$.

An (L, L)-complete multiprocedure is *Condorcet* if and only if:

(C_1): For all $x, y \in S$ such that $v(x, y) < v(y, x)$ for the profile π, there is not $R \in F(\pi)$ such that y covers x; and

(C_2): For all $x \in S$ such that $v(x, y) = v(y, x)$ for each $y \in S$, if R' is a linear order obtained from a linear order R by changing the place of x in R, then: $R' \in F(\pi)$ if and only if $R \in F(\pi)$.

Notice that the Condorcet property is equivalent to the fact that if R is a median linear order of the profile π, then the median procedure (on π) restricted to the intervals of R is 'stable' (this stability on interval-solutions has been first noted by Jacquet-Lagrèze, 1969).

To conclude with the possibility theorems, notice (beside the numerous results reported in books like Murakami, 1968; Sen, 1970; Pattanaik, 1971; Fishburn, 1973; Kelly, 1978; Mirking, 1979, ...) a characterization of the Borda rule as a social choice function by Young (1974), a characterization of the majority rule by May (1952), an extension of Theorem 2 to transitively valued relations by Leclerc

(1984), an Arrow-like theorem for equivalence relations by Mirkin (1975) and for phylogenetic trees by McMorris (1985); an axiomatic characterization of the median procedure for n-trees by Barthélemy and McMorris (1986) and other results on consensus in taxonomy (Barthélemy, 1988; Barthélemy, Leclerc and Monjardet, 1986; Margush and McMorris, 1981; McMorris and Neumann, 1983; Neumann and Norton, 1986;...).

4.2 Computable / Non Computable. Recently some works emphasize the property of non computability of some social welfare, or social choice functions (see Lewis, 1985 a,b; Gottinger, 1987; Kelly, 1988). The three first references do not belong to the field of discrete mathematics. They deal with domains involving recursive topologies. The last reference presents some nice examples of non computable social welfare functions that might be stated in the terminology of aggregation procedures. Its main result may be stated as follows: May (1952) characterized the majority rule (cf 2.5) for complete relations with three conditions. Whereas the majority rule is obviously computable, the removal of any of these three conditions leads to non computable solutions.

First, have an informal glance at the notion of computable procedure. Clearly, this notion is not trivial if and only if we deal with complete procedures or complete multiprocedures (procedures and multiprocedures are just functions from a finite set to a finite set). Consider any model of computation (Turing Machine, Random Access Memory, or even programs written in a language like PASCAL, ...). We just need two facts:

1° This model involves programs that can be designed as a finite number of instructions using a discrete set of symbols. Thus the programs form a countable set.

2° The programs in the model compute functions up to a convenient encoding scheme.

Consider two sets D and M of binary relations. Use some encoding scheme Γ both for D-profiles and models (i.e. relations in M) and say that a (D, M)-complete procedure F is *computable* if and only if there exists a program which, for every profile π, starts at the input $\Gamma(\pi)$, runs and stops with $\Gamma(F(\pi))$ as output. So there are countabily many computable complete procedures and, each time a system of axioms leads to an uncountable set of solutions, we are sure to meet such solutions that are non computable. In this situation, it is not difficult with the help of a standard diagonal argument to exhibit, with a "constructive flavor", such non computable procedures. As an example of that, consider the dictatorships. It would be surprising that an (L, L)-unanimity and decisive (by restriction to each D^v) complete procedure might not be computable. But unanimity and decisiveness provide no way to obtain recursively the dictators from one D^v to another. In view of this, the non computability of dictatorships could also seem natural. In any case, let us play with the diagonal and construct non computable dictators.

If $F : L^v \to L$ is unanimity and decisive, then there exists for each integer v an integer $f(v) \leq v$ such that $F(R_1, \ldots, R_v) = F_{f(v)}$. That is to say unanimity and decisive procedures are in one-to-one correspondence with the functions $f : \mathbf{N} \to \mathbf{N}$ such that, for each $n \in \mathbf{N}$, $f(n) \leq n$. So, it is sufficient to construct such a function which is not computable, to get a non computable (L, L)-procedure.

Since all the programs of our model of computation form a countable set, there are countabily many functions $f : \mathbf{N} \to \mathbf{N}$ such that, for each $n \in N$, $f(n) \leq n$. List them: $f_1, f_2, \ldots, f_n, \ldots$. Define a function f by:

$$f(n) = f_{n-1}(n) - 1, \text{ for } n > 1 \text{ and } f_{n-1}(n) > 1;$$
$$f(n) = 2 = f_{n-1}(n) + 1, \text{ for } n > 1 \text{ and } f_{n-1}(n) = 1;$$
$$f(1) = 1.$$

So, we have, for each n, $f(n) \leq n$. But f is not computable: for each $n > 1$, $f(n) \neq f_{n-1}(n)$ and $f(n) \neq f_1(n)$ for some n. So, f cannot belong to the list $f_1, f_2, \ldots, f_n, \ldots$

We could continue to play this game to obtain non computable oligarchies (absolute or not) etc. In particular Theorem 2 restated for complete procedures just indicates the existence of arrowian (D, M)-procedures but does not ensure their computability. But now come back to the scope of Theorem 1. A condition supported by the voters may lead to a recursive way to compute the dictators. As an example, consider the consistency property adapted to procedures: a complete (D, M)-procedure F is consistent if and only if: when π and π' are such that: $F(\pi) = F(\pi')$, then $F(\pi\pi') = F(\pi) = F(\pi')$. It is easy to show that, if F is a consistent (L, L)-dictatorship, then only two functions f and g may be associated to F; namely $f(n) = 1$ and $g(n) = n$. So, a decisive, consistent, unanimity (L, L)-procedure is computable.

4.3 Difficult / Easy. As usual *easy* will mean "solvable in polynomial time". So, an NP- complete problem is deemed to be *difficult*, unless $P = NP$. Notice however that this definition of easiness is a purely technical point of view; from a practical point of view, the NP-hardness of a problem just indicates the way to attempt to solve it: good heuristic and convenient mathematical programming methods. It says neither that there is no way to solve it, nor that such a resolution will necessarily cost an unreasonable amount of time.

We have already proved in Section 2.4 the NP-completeness of the decision version of Kemeny's medians (problem LM). Using the same argument (reduction from Feedback Arc Set) it is not difficult to prove that the following problem is NP-complete also:

Name: $(LM)'$.
Instance: Set S, integer v, v-tuple $(R_1, \ldots, R_v)$ of binary relations on S integer k.

Question: Does there exist a linear order R such that $\Sigma_{1 \leq i \leq v} \partial(R, R_i) \leq k$?

The difference between LM and $(LM)'$ is that in LM the profiles are constituted by linear orders, but the number v of voters is not fixed; while, in $(LM)'$ the number of voters is fixed but the relations occuring in the profiles can be anything. The NP-status of the problem "between" LM and $(LM)'$ is unknown. This problem may be stated as follows:

Name: $(LM)''$.
Instance: Set S, integer v, v-tuple $(R_1, \ldots, R_v)$ of linear orders on S integer k.

Question: Does there exist a linear order R such that $\Sigma_{1 \leq i \leq v} \partial(R, R_i) \leq k$?

One of the difficulties met in an attempt to solve the NP-status of $(LM)''$ is that we do not have any way to recognize, for a given n and a given v, the matrices having as entries the numbers $v(x, y)$ for some profile π of linear orders. This last problem has been studied by Dridi (1980) and Fishburn (1987).

The NP-completeness of problems like (LM) and $(LM)'$ has been extended by Wakabayashi (1986, Theorem 4 below) in the following way: Let M be a set of binary relations on S and consider the following problem:

Name: M-medians $(M\ M)$.
Instance: Set S, integer v, v-tuple $(R_1, \ldots, R_v)$ of binary relations on S integer k.

Question: Does there exist a relation $R \in M$ such that $\Sigma_{1 \leq i \leq v} \partial(R, R_i) \leq k$?

The analog of $LM)'$ is:

Name: $(M\ M)'$
Instance: Set S, v-tuple $(R_1, \ldots, R_v) \in M^*$, integer k.

Question: Does there exist a relation $R \in M$ such that $\Sigma_{1 \leq i \leq v} \partial(R, R_i) \leq k$?

THEOREM 4. *Both MM and (MM)' are NP-complete for* $M = L, C, W, P, O$ *or* E.

In fact, taking different sets as data (D) and models (M), we get many other NP- complete problems (problems DMM and $(DMM)'$). From Wakabayashi's results one deduces about thirty theorems...

Notice also that the NP-completeness of LM (in fact already obtained by Orlin, 1981) has been also established by Bartholdi, Tovey and Trick (1988 a). In the same paper the NP-hardness of the Dogson procedure is shown as well. Moreover these authors (1988 b) have shown that it can be NP- difficult to manipulate an election, even when the winner can be computed in polynomial time.

5. A formal theory of medians.

5.1 Formal aggregation. In Sections 2, 3 and 4 the sets D and M were well specified: They were essentially sets of binary relations on S. But, as we have already noticed, other sets of data and models may also be considered: n-trees, phylogenetic trees, valued relations, probability measures etc. It turns out that many results have a strong likeness, despite the broad variety of sets D and M they involve. From that comes the idea that it is perhaps not extremely useful to state twice, third times and more the same results. A mathematician, in such a situation could also be used to find why those results are, in fact, the same and to establish the reasons that explain them. There are at least two ways to achieve this goal.

1°) Find conditions large enough on data and models that will provide a general theorem, or find general tools for proofs. That was done, for the first point by Mirkin (1979) for binary relations and by Leclerc (1984) for transitively valued relations. In this last reference, a general oligarchic theorem is established and many old (Arrow's Theorem, arrowian Mirkin's theorem for equivalence relations, ...) and new (aggregation of dendrograms, ...) results are derived. For the second point, we can mention the use of ultrafilters to achieve dictatorship results (see Monjardet, 1983).

2°) Forget what are the objects in D and M and consider they are just sets equiped with some "abstract" structure, then look, according to these structures at functions: $D^v \to M$. Proceeding this way, two approaches are already available: the one uses linear algebra (Rubinstein and Fishburn, 1986); the other uses order structures. Up to now, this second approach was mainly concerned with the extension to lattices and semilattices of the lattice polynomial rules as presented in 2.5 and some results are reported in Barthélemy, Leclerc and Monjardet (1986). A deep study of (X, X, v)-procedures, where L is a semilattice, or a lattice, may be found in Monjardet (1988).

Within this second approach, we shall report here results on "abstract median procedure" as a multiprocedure extension of majority rule (Barbut, 1961; Monjardet 1980; Barthélemy, 1981; Bandelt and Barthélemy, 1984; Leclerc, 1988; Barthélemy and Janowitz, 1988). To achieve a formal description of medians, we shall proceed in three steps:

1°) *Medians in metric spaces and graphs.* The minimal structure to give meaning to formulas like (2) (2.3) is that of a metric space. Despite its weakness, this general framework is sufficient to obtain, for medians, the property of consistency (cf. 4.1, Theorem 2). In the cases we considered, medians involve the symmetric difference distance between binary relations, and, as we have noticed, this metric enumerates the number of disagreements between two relations. In 2.6., this remark has been made more precise for linear orders: $\partial(R, R') = 2\times$ (minimum number interchanges of consecutive elements to transform R into R'). Notice that, in the case of tournaments, we get: $\partial(R, R' = 2\times$ (minimum number of arcs to reverse to transform R into R'). So, at least in these two cases, ∂ may be interpreted as the shortest path metric in a graph whose vertices are preferences and edges are admissible transformations between preferences, such as reversion of two consecutive elements for linear orders. In a special type of graph, namely the *median graphs*, we obtain results that are essentially the same as for tournaments:

each odd numbered family admits a unique median. Moreover this graph theoretic approach suggests some extensions of medians, like local medians and external Condorcet rule.

2°) *Medians in posets.* In many cases, the graph we consider to produce medians can be considered as the diagram of some ordered set. As an example, consider the case of partial orders where:

for $R, R' \in O, \partial(R, R') = 1$ if and only if R covers R' or R' covers R

(Recall that R *covers* R' when $R \subset R'$ and there is no partial order R'' such that $R \subset R'' \subset R'$). That could also be the case for tournaments and linear orders, up to a rooting (for tournaments the transformation graph introduced in 1°) is a hypercube so it can be viewed as a boolean lattice; for linear orders we get the "permutoèdre", which can be ordered as a lattice, cf Barbut and Monjardet, 1970). Few results are known in the general case. But the case where the ordered set is a distributive lattice has been intensively studied by Barbut (1961), followed by Monjardet (1980). Since diagrams of distributive lattices are median graphs, all results for median graphs hold. Moreover the majority rule is given by a formula like (5) and the interval $[\alpha(\pi), \beta(\pi)]$ (cf. 2.5) is obtained. Other results have been obtained for (semi)modular (semi)lattices by Barthélemy (1981) then by Leclerc (1988). Semimodularity is important for "concrete" aggregation since the lattices of all the equivalence relations on S and of all the dendrograms on S are semimodular (for the last assertion, cf. Leclerc, 1979, 1981) as well as the semilattice of all partial orders on S.

3°) *Medians in median semilattices.* This is a special case of 2°) which appears as a generalization of the median theory in distributive lattices. *Median semilattices* are those ordered sets whose covering graph is a median graph (Bandelt, 1984). Furthermore, using such a structure, one obtains:

(a) A formula like (5) (cf. 2.5) for the majority rule.

(b) Using join irreducible elements, the interpretation of $\alpha(\pi)$ as an internal "Condorcet object" and the way to get, from $\alpha(\pi)$, all the Condorcet objects.

(c) An axiomatic characterization of medians based on Consistency and the quasi-Condorcet condition (3.1, Theorem 2).

5.2 Medians in metric spaces and graphs. For any finite set X, denote by X^* the set $\bigcup_{n \in \mathbb{N}} X^n$ of all finite sequences of elements of X. An element $\pi \in X^*$ is called a *profile*. Consider a finite metric space (X, d) and put for $x \in X$, $\pi = (x_1, \ldots, x_v) \in X^*$:

$$D(x, \pi) = \sum_{1 \leq i \leq v} d(x, x_i).$$

Any $m \in X$ that minimizes this sum is called a *median* of π:

(9) $$D(m, \pi) = \text{Min}_{x \in X} D(x, \pi)$$

Set $\text{med}(\pi)$ the set of all the medians of π. In this general framework one can get an easy result, namely that the extension of the notion of consistency occurs (Bandelt and Barthélemy, 1984).

PROPOSITION 2. *Let (X, d) be any finite metric space. If π and $\pi' \in X$ are such that $\text{med}(\pi) \cap \text{med}(\pi') \neq \emptyset$, then $\text{med}(\pi\pi') = \text{med}(\pi) \cap \text{med}(\pi')$.*

Let $G = (X, E)$ be a finite, simple, loopless and connected graph with vertex set X and edge set E. The *shortest path metric* d is defined as usual:

$$d(u, v) = \text{length of a shortest path between } u \text{ and } v.$$

For any two vertices u and v, the following set is called an *interval* in G

$$[u, v] = \{t \in X : d(u, v) = d(u, t) + d(t, v)\}.$$

In other words, the interval $[u, v]$ is the set of all vertices on shortest path between u and v.

For a graph $G = (X, E)$, medians with respect to the shortest path metric have some importance in location theory. They have been studied in the case of trees by Jordan (1869), Zelinka (1968) Slater (1978) and others. In the general case, we know from Slater (1980) that any graph is the induced subgraph of some median set of a graph.

For a vertex x of G, let $N(x)$ denotes the set of all vertices adjacent to x. We say that x is a *local median* of the profile $\pi \in X^*$, when

$$D(x, \pi) \leq \text{Min}_{y \in N(x)} D(y, \pi).$$

Let $\text{med}_{\text{loc}}(\pi)$ be the set of local medians of π. For any profile π, we have: $\text{med}(\pi) \subseteq \text{med}_{\text{loc}}(\pi)$. For two vertices x and y of G and a profile $\pi = (x_1, \ldots, x_v)$, denote by $\pi\{x, y\}$ the number $|\{i : d(x, x_i) < d(y, x_i)\}|$ and put:

$$W(x, \pi) = \text{Max}_{y \in N(X)} \pi\{y, x\}.$$

$G = (X, E)$ is a *median graph* if every family of three vertices of G admits just one median. Notice that a graph G is median if and only if for any three vertices u, v, w, the intersection $[u, v] \cap [v, w] \cap [u, w]$ is a singleton (Avann, 1961; notice that this singleton is the unique median of (u, v, w)). Examples of median graphs are trees and hypercubes. Moreover, each median graph is an isometric subgraph, closed under the ternary median operation of some hypercube (Mulder 1980). Proposition 3 may be found in Bandelt and Barthélemy (1984):

PROPOSITION 3. Let $G = (X, E)$ be a median graph. Then for each $\pi \in X^v$, the following conditions are equivalent:

(i) $x \in \text{med}(\pi)$.

(ii) $x \in \text{med}_{\text{loc}}(\pi)$.

(iii) $\pi\{x, y\} \geq v/2$ for all $y \in N(x)$.

(iv) $W(x, \pi) \leq v/2$.

(v) $W(x, \pi)$ is minimal

If G is a tree and π is the family of all vertices of G, then the equivalence between (iv) and (v) gives the classical result of Jordan (1869), while the equivalence of (i) and (iv) gives the theorem of Zelinka (1969). Condition (iii) stipulates that a median in a median graph can be obtained with a local external majority rule. Each vertex x_i in π is a voter and choose, between x and y, the closest vertex. Can this *local* rule (x and y are adjacent) become global? YES, but only in a few cases; since we can expect an Effet Condorcet occurs. Let us make that more precise. Define a *Condorcet vertex* of the profile $\pi = (x_1, \ldots, x_v)$ as a vertex c such that $\pi\{c, y\} - \pi\{y, c\} \geq 0$ for each vertex y. So, c is a Condorcet vertex if and only if $S(c, \pi) \geq 0$, with $S(c, \pi) = \min_{y \in X, y \neq c}(\pi\{c, y\} - \pi\{y, c\})$. Denote by $\text{cond}(\pi)$ the set of all Condorcet vertices of π. When for each $x \in X$, $S(x, \pi) < 0$, we have an (external) Effet Condorcet (i.e. $\text{cond}(\pi) = \emptyset$). Consider for instance the cube of figure 4 and the profile $\pi = (x, y, z, t, t)$. Clearly t is the unique median of π, but $S(x, \pi) = S(y, \pi) = S(z, \pi) = -3$, while $S(v, \pi) = S(s, \pi) = S(u, \pi) = S(w, \pi) = S(t, \pi) = -1$.

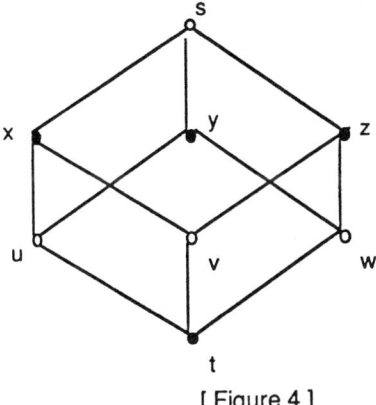

[Figure 4]

Vertices maximizing $S(x, \pi)$ have been introduced by Slater (1975) under the name of *Security Center*. That kind of consideration occurs in the problem of locating a single facility on a network: on the one hand, one can locate the facility at one vertex such that the total distance to the users (the vertices occuring in a profile) is minimized (metric medians, alias Weber points). On the other hand, one can look what would be location of the facility resulting from a voting procedure in which each voter (user) prefers to have a facility as close as possible to her or to him (see Bandelt, 1986; Hansen and Thisse, 1981; Labbé, 1985; Wendell and McKelvey, 1981). In particular, we know from Wendell and McKelvey (1981), Hansen and Thisse (1981) and others that in a tree the medians and the Condorcet vertices of a profile are the same (thus there is no Effet Condorcet). Proposition 4 (Bandelt and Barthélemy, 1984) generalizes this result for median graphs.

PROPOSITION 4. *For a median graph G the following conditions are equivalent*

(i) *G admits no cube as induced subgraph.*
(ii) *For each profile π of vertices, $\text{med}(\pi) = \text{cond}(\pi)$.*
(iii) *For each profile π of vertices, $\text{cond}(\pi) \neq \emptyset$.*

In Proposition 3, the fact that G median $\Rightarrow [(i) \Leftrightarrow (ii)]$ admits a partial converse (Bandelt 1986 a):

PROPOSITION 5. *Let G be a graph without triangle. The two following conditions are equivalent:*

(i) *G is a median graph.*
(ii) *For each profile π of vertices: $\text{med}(\pi) = \text{med}_{\text{loc}}(\pi)$.*

A last result (Bandelt and Barthélemy, 1984) states that, in median graphs, we have the unicity of the median of any odd numbered profile. Moreover, this situation characterizes median graphs:

PROPOSITION 6. *Let $G = (X, E)$ be a graph. The three following conditions are equivalent*

(i) *G is a median graph.*
(ii) *There exists an odd number $v = 2k+1$ such that each profile $\pi \in X^v$ admits a unique median.*
(iii) *For each odd number $v = 2k + 1$, each profile $\pi \in X^v$ admits a unique median.*

5.2 Medians in posets. Let $(X, \leq)$ be a finite poset. An alement v is said to *cover* another element u if $u < v$ and $u < t < v$ for no element t. The *covering graph* $G = (X, E)$ of $(X, \leq)$ is the graph whose vertices are the elements of X and whose edges are those pairs $\{u, v\}$, $u, v \in X$ such that u covers v or v covers u (notice that every bipartite graph occurs as the covering graph of some poset). $(X, \leq)$ is called a *graded poset* if there exists a real valued function h on X such that $h(v) = h(u) + 1$ whenever v covers u. The map h is called a *rank function*. When $(X, \leq)$ satisfies a modularity condition a closed formula, rather than a shortest path algorithm, allows to compute the shortest path metrics on the covering graph (cf. Monjardet, 1981, for a review about metrics on posets). For instance, when $(X, \leq)$

is a semimodular meet semilattice (resp. a semimodular join semilattice, resp. a modular lattice) this distance is given by:

$$d(u,v) = h(u) + h(v) - 2h(u \wedge v)$$

(resp. $d(u,v) = 2h(u \vee v) - h(u) - h(v)$, resp. $d(u,v) = h(u \vee v) - h(u \wedge v)$).

When $(X, \leq)$ is a distributive lattice, then the graph theoretic interval $[u, v]$ is nothing but the order interval $[u \wedge v, u \vee v] = \{t : u \wedge v \leq t \leq u \vee v\}$.

Medians in a poset are considered with respect to the shortest path metric on the covering graph. Some results have been obtained in modular cases (Barthélemy, 1981; Leclerc, 1988):

In a modular lattice, for any profile π, the set $\text{med}(\pi)$ defines a sublattice for the induced order.

In a semimodular semilattice local medians are characterized by a condition like Condition (iii) of Proposition 3.

In a join semimodular lattice medians fulfill the Pareto rule: for $\pi = (x_1, \ldots, x_v) \in X^*$ and $x \in \text{med}(\pi)$, we have: $\wedge_{1 \leq i \leq v} x_i \leq x$. This last result has been broadly extended by Leclerc (1988), in the following way: In any lattice $(X, \leq)$, for each profile $\pi = (x_1, \ldots, x_v) \in X^v$, let $\alpha(\pi)$ denote the *majority rule object*:

(10) $$\alpha(\pi) = \bigvee_{W \in \{1,\ldots,v\}; W = [(v+2)/2]} \bigwedge_{i \in W} X_i .$$

(cf. 2.5, formula 5). Then *a lattice $(X, \leq)$ is upper semimodular if and only if for each profile π and each $x \in \text{med}(\pi)$, we have:* $\alpha(\pi) \leq x$.

For distributive lattices, Barbut (1961) noticed that the Birkhoff and Kiss (1947) equality for ternary medians:

$$(x \vee y) \wedge (y \vee z) \wedge (x \vee z) = (x \wedge y) \vee (y \wedge z) \vee (x \wedge z),$$

extends to any odd numbered profile. More precisely, consider a profile $\pi = (x_1, \ldots, x_v)$ and in addition to $\alpha(\pi)$, define $\beta(\pi)$ as the element:

(11) $$\beta(\pi) = \bigvee_{W \in \{1,\ldots,v\}; W = [(v+1)/2]} \bigwedge_{i \in W} R_i ,$$

then, we have also:

(12) $$\beta(\pi) = \bigwedge_{W \in \{1,\ldots,v\}; W = [(v+2)/2]} \bigvee_{i \in W} x_i .$$

Moreover $\alpha(\pi) = \beta(\pi)$ whenever v is an odd number (in the general case, we have: $\alpha(\pi) \leq \beta(\pi)$). Since the covering graph of a distributive lattice is a median graph, Propositions 3, 4, 5 and 6 hold. Moreover they can be made constructive using Barbut's $\alpha(\pi)$. Using results from Barbut (1961), Monjardet (1980) and Leclerc (1988) one can state:

PROPOSITION 7. *For a finite lattice* $(X, \leq)$, *the following conditions are equivalent:*

(i) $(X, \leq)$ *is a distributive lattice.*
(ii) *There exists an odd integer* $v \geq 3$ *such that* $\mathrm{med}(\pi)$ *is a singleton, for all* $\pi \in X^v$.
(iii) *Each odd numbered profile of elements of* X *admits a unique median.*
(iv) *There exists an odd integer* $v \geq 3$ *such that* $\mathrm{med}(\pi) = \{\alpha(\pi)\}$, *for all* $\pi \in X^v$.
(v) *Each odd numbered profile of elements of* X *admits* $\alpha(\pi)$ *as unique median.*
(vi) *For each integer* v *and each* $x \in X^v$, $\mathrm{med}(\pi) = [a(\pi), \beta(\pi)]$.
(vii) *For each integer* v *and each* $x \in X^v$, *the set* $\mathrm{med}(\pi)$ *is an interval.*

These results have to be compared with what was mentioned in 2.5, about algebraic medians. They show that, within a latticial model for the set of preferences, each time we observe (like in the Condorcet procedure for tournaments) the unicity of median for odd numbered profiles, this enforces the lattice to be distributive and this enforces the median to be obtained, as a lattice polynomial, by the majority rule. Examples of distributive lattices are the set of all complete (resp. antisymmetric) relations on S, the set of all tournaments on S (up to a rooting), the set of all blackien orders (Builbaud, 1952; Black, 1958) and other distributive sublattices of the "permutoèdre" lattice.

4.4 Medians in median semilattices. A *median semilattice* is a meet semilattice such that:

(i) every principal ideal $\{x : x < a\}$ is a distributive lattice and
(ii) any three elements have an upperbound whenever each pair of them does.

In particular, for any three elements $x, y, z \in X$, the upperbound $(x \wedge y) \vee (y \wedge z) \vee (x \wedge z)$ always exists in X. This remark extends easily to each odd numbered profile: for every $\pi \in X^v$, with v as an odd number, $\alpha(\pi)$ exists in X. Moreover, median semilattices are those posets whose covering graph is a median graph (Avann, 1961, in the case of semilattices; Bandelt, 1984, in the general case). Propositions 3, 4, 5, 6 and 7 may be stated mutatis mutandis for median semilattices (Bandelt and Barthélemy, 1984).

PROPOSITION 8. *For a finite semilattice* $(X, \leq)$ *the following conditions are equivalent:*

(i) $(X, \leq)$ *is a median semilattice.*
(ii) *There exists an odd integer* $v \geq 3$ *such that* $\mathrm{med}(\pi)$ *is a singleton, for all* $\pi \in X^v$.
(iii) *Each odd numbered profile of elements of* X *admits a unique median*
(iv) *There exists an odd integer* $v \geq 3$ *such that* $\mathrm{med}(\pi) = \{\alpha(\pi)\}$, *for all* $\pi \in X^v$.
(v) *Each odd numbered profile of elements of* X *admits* $\alpha(\pi)$ *as unique median.*

It is not true, however, that in a median semilattice $(X, \leq)$, for an even numbered profile π, we have $\beta(\pi) \in \text{med}(\pi)$ (because $\beta(\pi)$ does not exist). So the equality $\text{med}(\pi) = [\alpha(\pi), \beta(\pi)]$ does not hold in general. Nevertheless it is easy to construct from $\alpha(\pi)$ the whole set $\text{med}(\pi)$. That is done as follows: Recall that an element s of a poset $(X, \leq)$ is called *join irreducible* if it cannot be expressed as the supremum of elements distincts from s. So, in a finite poset, each element may be stated as the join of (finitely many) join irreducible elements. Denote by J the set of all join irreducible elements of $(X, \leq)$. For $s \in J$ and $\pi = (x_1, \ldots, x_v) \in X^v$, the *index* of s in π is the number:

$$\gamma(s, \pi) = |\{i : s \leq x_i\}|/v.$$

For instance, in the set R of all the binary relations on S, the join irreducible elements are the ordered pairs (x, y) and for such a pair and a profile $\pi \in R^v$, $\gamma((x, y), \pi) = v(x, y)/v$. It can be shown (Barthélemy and Janowitz, 1988) that in a median semilattice, for a profile $\pi = (x_1, \ldots, x_v)$:

$$\alpha(\pi) = \vee\{s \in J : \gamma(s, \pi) > 1/2\}.$$

Moreover for each join irreducible s such that $\gamma(s, \pi) = 1/2$, the element $\alpha(\pi) \vee s$ exists in X and $\text{med}(\pi)$ is the set of all the elements of the form $\alpha(\pi) \vee s_1 \vee \cdots \vee s_p$ such that the indicated supremum exists in X and $\gamma(s_i, \pi) = 1/2$, for $i = 1, 2, \ldots, p$.

These remarks form an abstract version of the Condorcet construction rule as presented in Section 2. They show also that the median decision problem in a median semilattice is "easy": it can be solved in polynomial time with respect to the size of J. Despite this fact, to get the whole set $\text{med}(\pi)$ could be a long task, since $\text{med}(\pi)$ is able to have an exponential cardinality with respect to the size J.

5.5 A Possibility/Easiness Theorem for median semilattices. Let $(X, \leq)$ be any poset with a least element $\underline{0}$ and with J as set of join irreducible elements. Consider an (X, X)-complete multiprocedure F and denote, for $\pi \in X^*$ by $J(F, \pi)$ the set $\{s \in J: \text{there exists } x \in F(\pi) \text{ such that } s \leq x\}$. For instance for J a median semilattice $J(\text{med}, \pi) = \{s : \gamma(s, \pi) \geq 1/2\}$.

F is *efficient* if and only if:

(i) for each constant profile $\pi = (x, \ldots, x)$, we have $F(\pi) = x$, and:

(ii) for each $s \in J$ and each profile $\pi = (x_1, \ldots, x_v)$, such that $x_i = s$ or $x_i = \underline{0}$ we have $F(\pi) \subseteq \{\underline{0}, s\}$.

F is *consistent* if and only if, when π and π' are such that: $F(\pi) \cap F(\pi') \neq \emptyset$, then $F(\pi\pi') = F(\pi) \cap F(\pi')$.

F is *quasi-Condorcet* if and only if for each $s \in J$, each $\pi \in X^*$ such that $\gamma(s, \pi) = 1/2$ and each $x \in X$, $s \vee x$ exists implies that $x \in F(\pi)$ if and only if $x \vee s \in F(\pi)$.

F is *symmetric on join irreducible elements* if and only if for each profile $\pi = (x_1, \ldots, x_v)$ and each permutation σ of $\{1, 2, \ldots, v\}$, we have: $J(F, \pi) = J(F, \sigma(\pi))$, with $\sigma(\pi) = (x_{\sigma(1)}, \ldots, x_{\sigma(v)})$.

F is stable on *join irreducible elements* if and only if for each profiles $\pi = (x_1,\ldots,x_v), \pi' = (x'_1,\ldots,x'_v) \in X^v$:

$$\{i: s \leq x_i\} = \{i: s \leq x'_i\}$$

implies that:

$$s \in J(F,\pi) \text{ if and only if } s \in J(F,\pi').$$

Efficiency is a possible extension of the unanimity condition (cf. 4.1). Consistency and quasi-Condorcet conditions occur in the Young and Levenglick characterization of median linear orders (cf. 4.1). Symmetry asserts that we don't mind the order of the voters. Stability is an extension of decisiveness as introduced in 2.2. It is worth noticing that these two last conditions require only looking at the bricks (i.e. the join irreducible elements) used to build aggregates without regard on the way they are arranged to obtain those aggregates. With all this material, one can now show that (Barthélemy and Janowitz, 1988):

PROPOSITION 9. *Let $(X, \leq)$ be a median semilattice and let F be an (X,X)-complete multiprocedure. The conditions (i) and (ii) are equivalent.*

(i) $F =$ med

(ii) *F is efficient, consistent, quasi-Condorcet, symmetric on join irreducible elements and stable on join irreducible elements.*

The main results for the median procedure in median semilattices may be summarized as a single theorem:

THEOREM 5. *Let $(X, \leq)$ be a median semilattice, with d as the shortest path metric on its covering graph. Let F be an (X,X)-complete multiprocedure. The three following assertions are equivalent:*

(i) *For each profile $\pi = (x_1,\ldots,x_v)$, $F(\pi)$ is the set of all the solutions of:* $\text{Min}_{x \in X} \sum_{1 \leq i \leq v} d(x, x_i)$.

(ii) *$F(\pi)$ is the set of all the elements of the form $\alpha(\pi) \vee s_1 \vee \cdots \vee s_p$ such that the indicated supremum exists in X and $\gamma(s_i, \pi) = 1/2$, for $i = 1, 2, \ldots, p$; with $\alpha(\pi) = \vee\{s \in J: \gamma(s, \pi) > 1/2\}$.*

(iii) *F is efficient, consistent, quasi-Condorcet, symmetric on join irreducible elements and stable on join irreducible elements.*

For the median procedure in median semilattices, this result establishes complete links between the three approaches mentioned in 3.2. Condition (i) is the approach by optimality constraints, condition (ii) is the constructive approach and condition (iii) is the approach by axiomatic constraints. It generalizes the Condorcet procedure as presented in 2.1 and other results like the characterization of the median procedure for n-trees (Barthélemy and McMorris, 1986). It applies to the median semilattice of all the phylogenetic trees on S (Barthélemy, Leclerc and Monjardet, 1986). In the last section, we shall emphasize an application to pseudo social functions.

5.6 A possibility theorem for weak orders. Although we have assumed in the introduction an elementary knowledge about order relations, it seems appropriate to recall here some basic facts about weak orders (for further informations cf. Janowitz, 1984).

A *weak order* on the finite set S is a binary relation R on S which is asymmetric and negatively transitive (xR^cy and yR^cz implies xR^cz). An important point is that R is a weak order if and only if its dual R^d is a complete preorder. So, up to a duality, everything said about complete preorders on S could be said about weak orders as well: The incomparability part $I(R)$ of a weak order R is an equivalence relation. An equivalence class of $I(R)$ is called a *class* of R. The weak order R induces a linear order $L(R)$ on its classes $C_1,\ldots,C_p$. ($C_iL(R)C_j$ if and only if uRv for $u \in C_i$ and $v \in C_j$). The *q-section* S_q of R is the union of the q first classes of R with the convention that: $S_0 = \emptyset$. So, we have $S_i \subseteq S_{i+1}$. Conversely with each chain of subsets of S from $\emptyset$ to S: $S_0 = \emptyset \subseteq S_1 \subseteq \cdots \subseteq S_i \subseteq S_{i+1} \subseteq \cdots \subseteq S$ is associated the weak order R defined by uRv if and only if $i < j$, $u \in S_i$ and $v \in S_j - S_i$. So, we get a bijective map μ from the set W of all the weak orders on S to the set of all chains of subsets of S from $\emptyset$ to S. Moreover for $R, R' \in W$, $R \subseteq R'$ if and only if $\mu(R) \subseteq \mu(R')$ (that is to say $R \subseteq R'$ if and only if each section of R is a section of R'). So, μ is an order isomorphism.

Using this isomorphism it is easy to see that $(W, \subseteq)$ is a median semilattice whose maximal elements are the linear orders, whose minimal elements is the coarse weak order R_0 (i.e. the one-class weak order) and whose join irreducible elements are the weak orders with exactly two classes. In the covering graph of $(W, \subseteq)$, two weak orders R and R' are adjacent if and only if R' is obtained from R either by merging two consecutive classes, or else by breaking a class onto two consecutive ones. So, the shortest path distance between the two weak orders R and R' is equal to the minimal number of consecutive classes to merge or classes to break into two consecutive classes, to transform R into R'. For $A \subset S, A \neq \emptyset$, let R_A denote the weak order with exactly two classes, A and its complement. The R_A are the join irreducible elements of $(W, \subseteq)$. Say that the weak order R and the subset A of X are *compatible* if and only if the set theoretic union $R \cup R_A$ is a weak order. Clearly R and A are compatible if and only if R admits two consecutive sections S_i and S_{i+1} such that: $S_i \subseteq A \subseteq S_{i+1}$. In this case we denote by $R[A]$ the union $R \cup R_A$ and by induction the weak order $R[A_1, A_2, \ldots, A_p]$ when it exists. In particular if $S_0 = \emptyset \subseteq S_1 \subseteq \cdots \subseteq S_i \subseteq S_{i+1} \subseteq \cdots \subseteq S$ is the chain of sections associated with the weak order R, then $R = R_0[S_1, \ldots, S_{p-1}]$.

With these facts we can duplicate for weak orders the definitions of 5.4 and 5.5: For $A \subset S$ and $\pi = (R_1, \ldots, R_v) \in W^v$, the *index* of A in π is the number:

$$\gamma(A, \pi) = |\{i: A \text{ is a section of } R_j\}|/v.$$

Let F be a (W, W)-complete multiprocedure. For $\pi \in X^*$ denote by $J(F, \pi)$ the set $\{A \subset S, A \neq \emptyset$: there exists $R \in F(\pi)$ such that A is a section of $R\}$. F is *efficient* if and only if:

(i) for each constant profile $\pi = (R, \ldots, R)$, we have $F(\pi) = R$, and

(ii) for each $A \subset S$ and each profile π, such that $J(F, \pi) = \{A\}$, we have $F(\pi) \subseteq \{R_0, R_A\}$.

F is *quasi-Condorcet* if and only if for each $A \subset S$, each $\pi \in W^*$, such that $\gamma(A, \pi) = 1/2$ and each $R \in W$, $s \vee x\ R$ and A compatible implies that $R \in F(\pi)$ if and only if $R[A] \in F(\pi)$. F is *symmetric on solution-sections* if and only if for each profile $\pi = (R_1, \ldots, R_v)$ and each permutation σ of $\{1, 2, \ldots, v\}$, we have: $J(F, \pi) = J(F, \sigma(\pi))$. F is *stable on solution sections* if and only if for each profiles $\pi = (R_1, \ldots, R_v), \pi' = (R'_1, \ldots, R'_v) \in W^v$ and each $A \subset S$, $\{i : A$ is a section of $R_i\} = \{i : A$ is a section of $R'_i\}$ implies that: $A \in J(F, \pi)$ if and only if $A \in J(F, \pi')$.

COROLLARY 3. *Let F be a (W, W)-complete multiprocedure. The three following assertions are equivalent:*

(i) *For each profile $\pi = (R_1, \ldots, R_v)$, $F(\pi)$ is the set of all the solutions of $\text{Min}_{R \in W} \sum_{1 \leq i \leq v} d(R, R_i)$, with d as the shortest path metric in the covering graph of $(W, \subseteq)$.*

(ii) *$F(\pi)$ is the set of all the elements of the form $\alpha(\pi)[A_1 \ldots A_p]$ such that this weak order exists and $\gamma(A_i, \pi) = 1/2$, for $i = 1, 2, \ldots, p$; with $\alpha(\pi)$ as the weak order whose sections are all the subsets A of S such that: $\gamma(A, \pi) > 1/2$).*

(iii) *F is efficient consistent, quasi-Condorcet, symmetric on solution-sections and stable on solution-sections.*

This result shows that the decision problem associated to (i) may be solved in polynomial time.

REFERENCES

J.M. ABELLO, *Intrinstic limitations of the majority rule, an algorthmic approach*, SIAM Journal on Algebraic and Discrete Methods 6 (1985), 133–144.

J.M. ABELLO & CH.R. JOHNSON, *How large are transitive simple majority domains?*, SIAM Journal on Algebraic and Discrete Methods 5 (1984), 603–618.

D. ARDITTI, *Un nouvel algorithme de recerche d'un ordre induit par des comparaisons par paires*, in Data Analysis and Informatics III (E. Diday et al. eds), North Holland, Amsterdam (1984), 323–343.

K.J. ARROW, *Social Choice and Individual Values*, Wiley, New York (1951).

S.P. AVANN, *Metric ternary distributive semilattices*, Proceedings of the American Mathematical Society 12 (1961), 407–414.

H.J. BANDELT, *Discrete ordered sets whose covering graphs are median*, Proceedings of the American Mathematical Society 9 (1) (1984), 6–8.

H.J. BANDELT, *Characterizing median graphs*, preprint, to appear in European Journal of Combinatorics (1986 a).

H.J. BANDELT, *Networks with Condorcet Solutions*, preprint, to appear in European Journal of Operations Research (1986 b).

H.J. BANDELT & J.P. BARTHÉLEMY, *Medians in median graphs*, Discrete Applied Mathematics 8 (1984), 131–142.

M. BARBUT (1967), *Médianes, Condorcet et Kendall*, note SEMA, Paris and Mathématiques et Sciences Humaines 69 (1980), 5–13.

M. BARBUT (1961), *Médiane, distributivité, éloignement*, Publications du Centre de Mathématique Sociale EPHE, Paris and Mathématiques et Sciences Humaines 70, (1980) 5–31.

M. BARBUT & B. MONJARDET, *Ordre et Classification, Algébre et Combinatoire*, tomes I et II, Hachette, Paris (1970).

J.P. BARTHÉLEMY, *Trois propriètés de la médiane dans un treillis modulaire*, Mathématiques et Sciences Humaines 75 (1981), 83–91.

J.P. BARTHÉLEMY, *Arrow's theorem: unusual domain and extented codomain*, Mathematical Social Sciences 3 (1982), 79–89. Also published in *Social Choice and Welfare* (P.K. Pattanaik and M. Salles eds), North-Holland, Amsterdam, 19–30.

J.P. BARTHÉLEMY, *Threshold consensus for n-trees*, Journal of Classification (1988), (to appear).

J.P. BARTHÉLEMY, CI. FLAMENT & B. MONJARDET, *Ordered Sets and Social Sciences*, in: *Ordered Sets* (I. Rival, ed.), D. Reidel, Dordrecht (1982), 721–758.

J.P. BARTHÉLEMY & M.F. JANOWITZ, *A formal theory of consensus*, preprint (1988).

J.P. BARTHÉLEMY, B. LECLERC & B. MONJARDET, *On the Use of Ordered Sets in Problems of Comparison and Consensus of Classifications*, Journal of Classification 3 (1986), 187–224.

J.P. BARTHÉLEMY & F.R. MCMORRIS, *The Median Procedure for n-trees*, Journal of Classification 3 (1986), 329–334.

J.P. BARTHÉLEMY & B. MONJARDET, *The Median Procedure in Cluster Analysis and Social Choice Theory*, Mathematical Social Science, 1 (1981), 235–268.

J.P. BARTHÉLEMY & B. MONJARDET, *The Median Procedure in Data Analysis: New Results and Open Problems*, in *Classification and related methods of data analysis*, (H.H. Bock, ed.) North-Holland, Amsterdam (1981) (to appear).

J.J. BARTHOLDI III, C.A. TOVEY & M.A. TRICK, *Voting Schemes for which it can be difficult to tell who won the election*, Social Choice and Welfare (1988) (to appear).

J.J. BARTHOLDI III, C.A. TOVEY & M.A. TRICK, *The computational difficulty of manipulating an election*, Social Choice and Welfare (1988) (to appear).

J.P. BENZECRI ET AL, *L'analyse des données*, tome I: *La taxinomie*, Dunod, Paris (1973).

G. BIRKHOFF, *Lattice Theory*, 3rd. ed, American Mathematical Society, Providence (1967).

G. BIRKHOFF & S.A. KISS, *A ternary operation in distributive lattices*, Bulletin of the American Mathematical Society 53 (1947), 746–752.

D. BLACK, *The Theory of Committees and Elections*, Cambridge University Press, London (1958).

D.H. BLAIR & R.A. POLLAK, *Collective Rationality and Dictatorship: The Scope of Arrow's Theorem*, Journal of Economic Theory 21, (1979), 186–194.

J.H. BLAU, *Semiorders and collective Choice*, Journal of Economic Theory 21 (1979), 195–206.

J.C. BORDA (1784), *Memoire sur les élections au scrutin*, Histoire de l'Académie des Sciences pur 1781, (English translation by A. de Grazia, Isis 44, 1953).

V.J. BOWMAN AND C.S. COLANTONI, *Majority rule under transitivity constraints*, Management Sciences 20 (1973) 1029–1041.

M.J.A. CARITAT, MARQUIS DE CONDORCET, *Essais sur l'application de l'analyse à la probabilité des décisions rendues à la pluralité des voix*, Paris (1785) (reprint, Chelsea publ. 6, New York, 1974).

W.H.E. DAY, F.R. MCMORRIS & D.B. MERONK, *Axioms for consensus based on lower bounds in posets*, Mathematical Social Science, 12 (1986), 185–190.

B. DEBORD, *Caractèrisation des matrices de préférences nettes et méthodes d'agrégation associées*, Mathématiques et Sciences Humaines 97 (1987 a) 5–17.

B. DEBORD, *Axiomatisation de procedures d'agrégation de préférences*, Thesis, Université de Grenoble (1987 b).

J.S. DE CANI, *Maximum likehood paired comparison ranking by linear programming*, Biometrika 3 (1969), 537–545.

G. DEMANGE, *Single peaked order a tree*, Mathematical Social Science 3 (1982), 389–396.

C.L. DOGSON, *A discussion of the various methods of procedure in conducting elections*, Princeton University Library, reprinted in: D. Black (1958), *The Theory of Committees and Elections*, Cambridge University Press, London (1876).

T. DRIDI, *Sur les distributions binaires associées à des distributions ordinales*, Mathématiques et Sciences Humaines 69 (1980), 15–31.

R. FAQUAHARSON, *Theory of Voting*, Yale University Press, Yale (1969).

J. FELDMAN-HÖGAASEN, *Ordres partiels et permutoèdre*, Mathématiques et Sciences Humaines 28 (1969), 28, 27–38.

J.A. FEREJOHN & P.C. FISHBURN, *Representation of Binary Decision Rules by Generalized Decisiveness Structures*, Journal of Economic Theory 21 (1979), 28–45.

P.C. FISHBURN, *Mathematics of Decision Theory*, Mouton, The Hague (1972).

P.C. FISHBURN, *The Theory of Social Choice*, Princeton University Press, Princeton (1973).

P.C. FISHBURN, *Condorcet social choice functions*, SIAM Journal of Applied Mathematics 33, (1977), 469–489.

P.C. FISHBURN, *Research in Decision Theory: a Personal Perspective*, Mathematical Social Sciences 5 (1983), 129–148.

P.C. FISHBURN, *Discrete Mathematics in Voting and Group Choice*, SIAM Journal on Algebraic and Discrete Methods 5 (1984), 263–275.

P.C. FISHBURN, *Decomposing a weighted digraph into Sums of Chains*, Discrete Applied Mathematics 16, (1987), 15–31.

P.C. FISHBURN & A. RUBINSTEIN, *Aggregation of equivalence relations*, Journal of Classification 3 (1986) 61–65.

L. FREY, *Techniques Ordinales en Analyse des Données: Algèbre et Combinatoire*, Hachette, Paris (1971).

M.R. GAREY & D.S. JOHNSON, *Computers and Intractability. A Guide to the Theory of NP-Completeness*, W.H. Freeman and Company, New York (1979).

A. GIBBARD, *Manipulation of voting schemes: a general result*, Econometrica 41 (1973), 587–601.

H.W. GOTTINGER, *Choice and Complexity*, Mathematical Social Sciences 14 (1987) 1–17.

M. GRÖTSCHEL, M. JÜNGER & G. REINELT, *On the acyclic subdigraph problem*, Mathematical programming 33 (1985 a), 28–42.

M. GRÖTSCHEL, M. JÜNGER & G.REINELT, *Facet of the linear ordering polytope*, Mathematical programming 33 (1985 b), 43–60.

A.S. GUHA, *Neutrality, Monotonicity and the Right of Veto*, Econometrica 40 (1972), 821–826.

G.TH. GUILBAUD (1952), *Les théories de l'intérêt général et le problème logique de l'agrégation*, Economie Appliquée, 5 (4), reprinted in Eléments de la Théorie des Jeux, Dunod, Paris (1968). English translation in *Readings in Mathematical Social Sciences*, Science Research Associate, Chicago (1966), 262–307.

G.TH. GUILBAUD & P. ROSENSTHIEL, *Analyse algébrique d'un scrutin*, Mathématiques et Sciences Humaines 4 (1970), 9–33.

P. HANSEN, J.F. THISSE, *Outcomes of Voting and Planning*, Journal of Public Economics 16 (1981), 1–15.

B. HANSSON, *The independence condition in the theory of social choice*, Working paper no. 2, The Mattias Fremling Society, Department of Philosophy, Lund (1972).

L. HUBERT & J. SCHULZ, *Maximum likelihood paired-comparison ranking and quadratic assignment*, Biometrika, 62 (1975), 655–659.

K. INADA, *A note on simple Majority rule*, Econometrica, 37 (1969), 490–506.

E. JACQUET-LAGREZE, *L'agrégation des opinions individuelles*, Informatique et Sciences Humaines 4 (1969), 1–21.

M.F. JANOWITZ, *On the semilattice of weak orders on a set*, Mathematical Social Sciences 8 (1984), 229–239.

C. JORDAN, *Sur les assemblages de lignes*, Journal für die reine und angewandte Mathematik 70 (1869) 185–190.

R.M. KARP, *Reductibility among combinatorial problems*, in R.E. Muller and J.W. Tatcher (eds.), Complexity of Computer Computations, Plenum Press, New York (1972), 85–103.

J.S. KELLY, *Arrow's Impossibility Theorem*, Academic Press, New York (1978).

J.S. KELLY, *Social Choice and Computational Complexity*, preprint.

J.G. KEMENY, *Mathematics without numbers*, Daedalus, 88 (1959), 577–591.

M.G. KENDALL, *Rank Correlation Methods*, Hafner, New-York (1948) (3rd edition 1962).

G. KREWERAS, *Représentation polyèdrique des préordres complets finis*, in: Ordres totaux finis, Gauthiers-Villars, Paris (1971).

M. LABBE, *Essays in Network Location Theory*, Cahiers du entre d'Edudes de Recherche Opérationnelle 27 (1985), 5–130.

B. LECLERC, *Semi-modularité des treillis d'ultramétriques*, Comptes Rendus de l'Académie des Sciences de Paris, A-288 (1979), 575–577.

B. LECLERC, *Description combinatoire des ultramétriques*, Mathématiques et Sciences Humaines 73 (1981), 5–37.

B. LECLERC, *Efficient and binary consensus functions on transitively valued relations*, Mathematical Social Science 8 (1984), 45–61.

B. LECLERC, *Consensus applications in Social Sciences*, in Classification and related methods of data analysis (H.H. Bock, ed.) North-Holland, Amsterdam, (1988) (to appear).

B. LECLERC, *Medians and majorities in semimodular lattices*, Rapport CAMS P031, Paris (1988).

K. LEHRER & C. WAGNER, *Rational Consensus in Science and Society*, D. Reidel, Dordrecht (1981).

A.A. LEWIS, *On effectively computable realizations of choice functions*, Mathematical Social Science, 10 (1985), 43–80.

A.A. LEWIS, *The minimum degree of recursively representable choice functions*, Mathematical Social Science, 10 (1985), 179–188.

A. MAS-COLLEL & H. SONNENSCHEIN, *General possibility Theorem for Group Decisions*, The Review of Economic Studies 39 (1972), 185–192.

J.F. MARCOTORCHINO & P. MICHAUD, *Optimization in Ordinal Data Analysis*, Technical Report, IBM, Paris (1978).

J.F. MARCOTORCHINO & P. MICHAUD, *Optimisation en analyse ordinale des données*, Masson, Paris (1979).

T. MARGUSH & F.R. MCMORRIS, *Consensus n-trees*, Bulletin of Mathematical Biology 43 (1981), 239–244.

K.O. MAY, *A set of independent necessary and sufficient conditions for simple majority decision*, Econometrica 20, (1952), 680–684.

F.R. MCMORRIS, *Axioms for Consensus Functions on Undirected Phylogenetic Trees*, Mathematical Biosciences 74 (1985), 239–244.

F.R. MCMORRIS & D.A. NEUMANN, *Consensus Functions on Trees*, Mathematical Social Sciences 4 (1983), 131–146.

B.G. MIRKIN, *The problems of approximation in space of relations and qualitative data analysis*, Automatika i Telemechanica, translated in information and Remote Control 35 (9) (1974), 1424–1431.

B.G. MIRKIN, *Group Choice*, (P.C. Fishburn ed.), Wiley, New York (1979).

B. MONJARDET, *Tournois et ordres médians pour une opinion*, Mathématiques et Sciences Humaines 43 (1973) 55–70.

B. MONJARDET, *Lhuillier contre Condorcet au pays des paradoxes*, Mathématiques et Sciences Humaines, 54 (1976), 33–43.

B. MONJARDET, *An axiomatic theory of tournament aggregation*, Mathematics of Operation Research 3 (4) (1978), 334–351.

B. MONJARDET, *Théorie et applications de la médiane dans les treillis distributifs finis*, Annals of Discrete Mathematics 9 (1980), 891.

B. MONJARDET, *Metrics on partially ordered sets: a survey*, Discrete Mathematics 35 (1981), 173–184.

B. MONJARDET, *On the use of ultrafilters in social choice theory*, in Social Choice and Welfare (P.K. Pattanaik and M. Salles eds), North-Holland, Amsterdam (1983), 73–78.

B. MONJARDET, *Concordance et consensus d'ordres totaux: les coefficients K et W*, Revue de Statistique Appliquée 33 (1985), 55–85.

B. MONJARDET, *Arrowian characterizations of latticial federation consensus functions*, (1988) submitted.

H. MOULIN, *On strategy-proofness and single peakness*, Public Choice 35 (1980), 437–455.

H. MOULIN, *The Strategy of Social Choice*, North-Holland, Amsterdam (1983).

H.M. MULDER, *n-cubes and Median Graphs*, Journal of Graph Theory 4 (1980), 107–110.

Y. MURAKAMI, *Logic and Social Choice*, New York (1968).

D.A. NEUMANN, *Faithful Consensus Methods for n-trees*, Mathematical Biosciences 63 (1983), 271–287.

D.A. NEUMANN & V.T. NORTON JR., *On Lattice Consensus Methods*, Journal of Classification 3 (1986), 225–255.

J. ORLIN, (1981), unpublished manuscript.

P.K. PATTANAIK, *Voting and Collective Choice*, Cambridge University Press, Cambridge (1971).

P.K. PATTANAIK, *Threats, counter-threats and strategic voting*, Econometrica, 44 (1976), 91–104.

B. PELEG, *Consistent Voting Scheme*, Econometrica 46 (1978), 153–161.

S. REGNIER, *Sur quelques aspects mathématiques des problémes de classification automatique*, ICC Bulletin 4, Rome (1965).

G. REINELT, *The linear ordering Problem: algorithms and applications*, Helderman Verlag, Berlin (1985).

F.S. ROBERTS, *Discrete Mathematical Models, with Applications to Social, Biological and Environmental Problems*, Prentice-Hall, Englewood Cliffs, NJ (1976).

F.S. ROBERTS, *Applications of Combinatorics and Graph Theory to the Biological and Social Sciences. Seven Fundamental Ideas*, (1989), this volume.

A. RUBINSTEIN & P.C. FISHBURN, *Algebraic aggregation theory*, Journal of Economic Theory 38 (1986), 281–297.

M.A. SATTERTHWAITE, *The existence of strategy-proof voting procedure, a topic in Social Choice Theory*, PhD dissertation, University of Wisconsin, Madison (1973).

M.A. SATTERTHWAITE, *Strategy-proofness and Arrow's Conditions: existence and correspondence theorems for voting procedures and social choice functions*, Journal of Economic Theory, 10 (1975) 187–217.

A.K. SEN, *Collective Choice and Social Welfare*, North-Holland Publishing Company, Amsterdam (1970).

A.K. SEN & P.K. PATTANAIK, *Necessary and Sufficient Conditions for rational choice under Majority Decision*, Journal of Economic Theory, 1, (1969), 178–202.

P.J. SLATER, *Maximin facility location*, Journal of Research of the National Bureau of Standard, 79B (1975), 107–105.

P.J. SLATER, *Centers to centroid in graphs*, Journal of Graph Theory 2 (1978), 209–222.

P.J. SLATER, *Median of arbitrary graphs*, Journal of Graph Theory 4 (1980), 389–392.

Y. WAKABAYASHI, *Aggregation of binary relations: algorithmic and polyhedral investigations*, Thesis, Augsburg (1986).

R.E. WENDELL & R.D. MCKELVEY, *New Perspectives in competitive location theory*, European Journal of Operational Research 6 (1981), 174–182.

R. WILSON, *Social Choice without the Pareto Principle*, Journal of Economic Theory, 5 (1972), 89–99.

H.P. YOUNG, *An axiomatization of Borda's Rule*, Journal of Economic Theory, 9 (1974), 43–52.

H.P. YOUNG, *Optimal ranking and choice from pairwise comparisons*, in *Information Pooling and Group Decision Making* (eds. B. Grofman and G. Owen), JAI Press, Greenwich, Conn. (1986), 113–122.

D.H. YOUNGER, *Minimum feedback arc sets for a directed graph*, IEE Trans. Circuit Theory 10 (1963), 238–245.

B. ZELINKA, *Medians and peripherian of trees*, Archivum Math. (Brno) 4 (1968), 87–95.

CONSECUTIVE ONE'S PROPERTIES FOR MATRICES AND GRAPHS INCLUDING VARIABLE DIAGONAL ENTRIES

MARGARET B. COZZENS AND N.V.R. MAHADEV*

Abstract. The general consecutive one's property is applied to matrices with prescribed rows corresponding to sets of elements from a specified set. In applications, the rows and the columns often correspond to vertices of a graph and the entries are determined by the existence of an edge (1) between two vertices or nonexistence of the edge (0) between the two vertices. A 1 on the diagonal implies a loop at the vertex. Similarly, if the matrix represents a relation on a set, all 1's on the diagonal corresponds to a reflexive relation and all 0's on the diagonal corresponds to in irreflexive relation. Often in applications we don't know if the relation is reflexive or irreflexive, nor do we care. We don't care if an element is related to itself or not. What we are interested in is a linear ordering of the vertices (elements) such that the neighbors of an element appear consecutively in the ordering with or without the vertex itself. We present the basics of an algorithm that determines in linear time if such an ordering is possible. This algorithm determines when a diagonal element must be 0 and when it must be 1, and if an ordering is possible, produces the ordering. Roberts showed in 1968 that if all of the diagonal entries are 1 then the corresponding graph (with loops assumed at each vertex) is an indifference graph. In 1970, Tucker provided a characterization of $0-1$ matrices with the consecutive ones property for columns. This characterization can be used for a prescribed set of diagonal entries, but would require 2^n applications to check all possible sets of diagonal entries for the adjacency matrix of an n vertex graph.

This work applies in the social sciences where relationships are "to" a particular individual, country, product, etc.. For example, we might want a linear ordering of countries such that all those countries that are allies of the US appear together in the order, all those allies of Russia appear together in the order, etc., and we don't care if, in the list of allies of the US, the US appears or not. The class of graphs whose corresponding adjacency matrix has the consecutive ones property for some choice of diagonal elements is a class of perfectly orderable (thus perfect) graphs. This class neither contains nor is contained in any of the other known classes of perfectly orderable graphs.

Key words. graph, matrix, consecutive ones, indifference graph, perfectly orderable, pq-trees

AMS(MOS) subject classifications. 05C

1. Background. A matrix whose entries are zeros and ones is said to have the **consecutive ones property for columns** if its rows can be permuted in such a way that 1's in each column occur consecutively. For example $\mathbf{A}_1$ has the consecutive ones property for columns, as shown in Figure 1. $\mathbf{A}_3$ does not

$$\mathbf{A}_1 = \begin{matrix} 1 \\ 2 \\ 3 \\ 4 \end{matrix} \begin{pmatrix} 1 & 0 & 0 & 1 \\ 1 & 1 & 1 & 0 \\ 0 & 1 & 0 & 0 \\ 1 & 0 & 1 & 1 \end{pmatrix} \longrightarrow \mathbf{A}_2 = \begin{matrix} 1 \\ 4 \\ 2 \\ 3 \end{matrix} \begin{pmatrix} 1 & 0 & 0 & 1 \\ 1 & 0 & 1 & 1 \\ 1 & 1 & 1 & 0 \\ 0 & 1 & 0 & 0 \end{pmatrix}$$

---

*Department of Mathematics, Northeastern University, Boston, MA 02115.

$$\mathbf{A}_3 = \begin{pmatrix} 1 & 1 & 0 & 0 \\ 0 & 1 & 0 & 0 \\ 0 & 1 & 1 & 0 \\ 1 & 0 & 1 & 0 \end{pmatrix}$$

Figure 1.

have the consecutive ones property for columns. A comparable definition applies to the consecutive ones property for rows. For symmetric matrices these two properties are equivalent. The consecutive ones property for columns received considerable attention in 1965 when Fulkerson and Gross used this property to characterize interval graphs. An undirected graph G is called an **interval graph** if its vertices can be put into one-to-one correspondence with a set of intervals of real numbers such that two vertices are connected by an edge if and only if their corresponding intervals have nonempty intersections. A maximal clique of a graph is a subset S of the vertices that induces a complete subgraph such that no set properly containing S also induces a complete subgraph. Fulkerson and Gross [10] gave the following characterization of interval graphs.

PROPOSITION 1: (FULKERSON AND GROSS). *An undirected graph G is an interval graph if and only if its clique matrix $\mathbf{M}$ (maximal cliques versus vertices incidence matrix) has the consecutive ones property for columns.*

With this theorem, Fulkerson and Gross developed a polynomial algorithm for testing for the consecutive ones property for columns for an $m \times n$ matrix in $O(mn^2)$ time.

Interval graphs have been used in modeling in many diverse areas such as psychology, archaeology, management science, computer science, and ecology. For instance, interval graphs have been used in seriation using overlap data in [18] and [34], in archeology in [5], in traffic light phasing in [26], [26], [31], and [35], and in ecology in [21]. In each instance Proposition 1 is used to verify that the graph in question is an interval graph, and to provide the specified ordering using the consecutive ones ordering.

In 1970, Alan Tucker provided a graph theoretic characterization of matrices having the consecutive ones property for columns. Given a zero-one $m \times n$ matrix $\mathbf{A}$, define a bipartite graph $G = (V_1, V_2, \mathbf{A})$ with vertex set V consisting of the set of rows V_1 and set of columns V_2 and an edge exists between vertex i in V_1 and j in V_2 if and only if the (i,j)-entry of $\mathbf{A}$ is 1. A V_1-**asteroidal triple** of G is a set of vertices, $\{x, y, z\} \subseteq V_1$, such that between any two of these three vertices there exists a path P such that no vertex on P is adjacent to the third vertex. Figure 2 illustrates a V_1-asteroidal triple $\{x, y, z\}$.

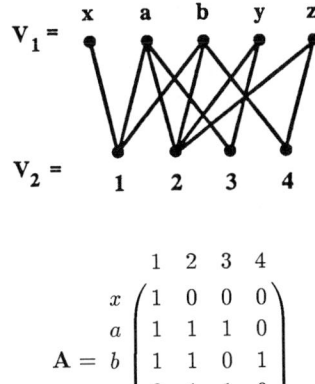

$$A = \begin{pmatrix} & 1 & 2 & 3 & 4 \\ x & 1 & 0 & 0 & 0 \\ a & 1 & 1 & 1 & 0 \\ b & 1 & 1 & 0 & 1 \\ y & 0 & 1 & 1 & 0 \\ z & 0 & 1 & 0 & 1 \end{pmatrix}$$

Figure 2.

PROPOSITION 2: (TUCKER) [36]. *A bipartite graph* $G = (V_1, V_2, \mathbf{A})$ *contains no V_1-asteroidal triple if and only if* $\mathbf{A}$ *has the consecutive ones property for columns.*

With this characterization of the consecutive ones property Tucker [37] was able to provide a forbidden subgraph characterization of those graphs $G = (V_1, V_2, \mathbf{A})$ that contain no V_1 asteroidal triple. However, in terms of recognition, this characterization proved not very useful as the smallest graph in this characterization is the one shown in Figure 2 with 9 vertices, and there are three infinite classes of forbidden graphs, and one other graph. In the 1970's the consecutive ones property, or equivalently the consecutive arrangement property, was frequently applied to a variety of information retrieval problems. Let S represent a set of records or data entries, and I be a collection of inquiries, subsets of S. Can S be placed in sequential storage in such a way that the members of each $J \in I$ are stored in consecutive locations? When this storage layout is possible, the records pertinent to an inquiry can be accessed with two parameters, a starting pointer and a length. Ghosh [11], [12] calls this the **consecutive retrieval property** and as such this property has been studied and applied to various information retrieval problems by Nakama [23],[24], Ghosh [12],[13], Waksman and Green [38], Patrinos and Hakimi [27], Bartholdi et. al. [1], and Gupta [16].

Analyzing how preferences are made enables psychologists and sociologists to predict individual and group behavior. The discipline of utility theory provides mechanisms for quantifying preference. One reasonable measure, due to Luce [20] is the notion of a semiorder. We assign a real number $f(x)$ to each x in a set of alternatives V so that for x and y in V, x is preferred to y if and only if $f(x)$ is sufficiently larger than $f(y)$. Formally we can describe a **semiorder** as a relation P on a set V such that there exists a real valued function f with xPy if and only if $f(x) > f(y) + 1$ for all $x, y \in V$. The complement of a semiorder relation is called an **indifference relation** I, and has the property that xIy if and only if

$|f(x) - f(y)| \leq 1$ for all $x, y \in V$. An **indifference graph** has vertex set V and edge $\{x, y\}$ if and only if xIy. (One is indifferent between alternatives x and y.) Indifference graphs are exactly the same graphs as **unit interval graphs**, interval graphs where the length of each interval is one. Roberts in 1968 [29] proved the following Proposition.

PROPOSITION 3: *A graph is an indifference graph (unit interval graph) if and only if the adjacency matrix of G augmented with ones on the diagonal has the consecutive ones property for columns.*

Indifference graphs have applications in frequency assignment problems as in [6], and in seriation as in [5], [18], and [32]. There are potential applications of indifference graphs in all areas of decision making, whenever there is a need to measure preference, compatibility, etc. and in measuring judgements of similarity or matching in psychology as in [33].

The consecutive ones property for columns applied to indifference graphs provides a linear ordering of the vertices so that the closed neighborhood of each vertex (the vertex and all vertices adjacent to it) appears consecutively in the order. In the information retrieval problem, we wanted all the records pertaining to an inquiry to appear consecutively in storage. We can construct a graph such that each record of S is a vertex and an edge exists between two vertices if and only if they appear in the same subset (inquiry) of I. Now S can be put in sequential linear storage if and only if the closed neighborhoods of each vertex appear consecutively in some linear order if and only if the graph is an indifference graph.

If a graph has an adjacency matrix **A** with all zeros on the diagonal and **A** has the consecutive ones property for columns then there is a linear order of the vertices such that the neighborhoods (without the vertex itself) of each vertex appear consecutively in the order.

Later in this paper we will discuss graphs, and their corresponding adjacency matrices, for which there exists a linear order of the vertex set such that the neighbors of each vertex appear consecutively in the order with or without the vertex itself. For example if V is the set of countries of the world and there is an edge between x and y if x is an ally of y, we want a linear order of countries such that the allies of the United States appear consecutively and we don't care if the United States appears in this group or not, we simply want to identify the allies.

Fast algorithms exist for two classic graph problems for graphs whose adjacency matrix has the consecutive ones property, either with all ones on the diagonal (indifference graphs) or with all zeros. By applying the greedy algorithm to the linear order arising from the consecutive ones property for columns one can find an optimal graph coloring. One can easily determine the size of the largest clique as well, as these graphs are perfect graphs.

Chvàtal [3] defined a class of perfect graphs called perfectly orderable graphs. A graph G is **perfectly orderable** if and only if there exists an acyclic orientation of the edges of G such that for each induced P_4, no substructure of the form •→ —•—•←—• exists. Such a substructure is called an **obstruction**. Many classes of perfectly orderable graphs have been identified, including triangulated (rigid circuit) graphs, comparability graph, Welsh-Powell perfect graphs, brittle graphs, graphs with Dilworth number less than or equal to 3, and graphs with threshold dimension less than or equal to 2. An $O(n\log n)$ algorithm exists for finding optimal colorings and maximum cliques for all perfectly orderable graphs [3].

THEOREM 1: *Let* $\mathbf{A}$ *be a* $(0-1)$ *matrix with the consecutive ones property for columns and let* G *be the graph whose vertices are the rows of* $\mathbf{A}$ *and* $\{i,j\}$ *is an edge if and only if* $i \neq j$ *and the* (i,j)-*entry of* $\mathbf{A}$ *is* 1. *Then* G *is perfectly orderable.*

Proof. Let G be the graph corresponding to the matrix $\mathbf{A}$ with the consecutive ones property for columns and let $v_1, v_2, \ldots, v_n$ be the consecutive ordering of vertices. Define an orientation P on the edges by iPj if and only if v_i precedes v_j in the ordering. We want to show that P is a perfect ordering. Suppose there exists a path a, b, c, d that induces a P_4 such that aPb and dPc. Therefore a precedes b in the ordering and d precedes c in the ordering. Without loss of generality assume bPc, so that b precedes c in the ordering. Three cases exist: d precedes a, d precedes b and not a, d precedes c and not b.

The following matrices represent each case:

$$\begin{array}{c} \quad\;\, d\;\, a\;\, b\;\, c \\ \begin{array}{c} d \\ a \\ b \\ c \end{array}\!\!\left(\begin{array}{cccc} 0 & 0 & 1 & \\ 0 & & 1 & 0 \\ 0 & 1 & & 1 \\ 1 & 0 & 1 & \end{array}\right) \end{array} \qquad \begin{array}{c} \quad\;\, a\;\, d\;\, b\;\, c \\ \begin{array}{c} a \\ d \\ b \\ c \end{array}\!\!\left(\begin{array}{cccc} & 0 & 1 & 0 \\ 0 & & 0 & 1 \\ 1 & 0 & & 1 \\ 0 & 1 & 1 & \end{array}\right) \end{array} \qquad \begin{array}{c} \quad\;\, a\;\, b\;\, d\;\, c \\ \begin{array}{c} a \\ b \\ d \\ c \end{array}\!\!\left(\begin{array}{cccc} & 1 & 0 & 0 \\ 1 & & 0 & 1 \\ 0 & 0 & & 1 \\ 0 & 1 & 1 & \end{array}\right) \end{array}$$

$$\qquad\qquad\text{Case 1} \qquad\qquad\qquad\qquad \text{Case 2} \qquad\qquad\qquad\qquad \text{Case 3}$$

In case 1, column c does not have consecutive ones. In cases 2 and 3 column b does not have consecutive ones. Each case contradicts the consecutive ones property for columns. Therefore no obstructions exists. The ordering from the consecutive ones property is a transitive ordering, therefore the orientation P arising from this order cannot have any directed cycles, for if iPj and jPk then iPk, not kPi. Therefore P is acyclic, and P is a perfect ordering of G. □

Booth and Lueker in 1976 [2] developed a linear time algorithm for determining if a matrix has the consecutive ones property, and they applied this algorithm to recognizing interval graphs in linear time, and to recognizing planar graphs in linear time. They invented a PQ-tree data structure expressly for this purpose. PQ-trees are used to represent all the permutations of a set V subject to the constraints of consecutivity induced by subsets, neighborhoods, etc.. Only a small amount of storage is required for this representation. A more complete description of the PQ-data structure along with various modifications will be presented in the next section.

2. Mixed Diagonal Consecutive Ones. As was indicated in the last section, we are often interested in a linear ordering of the vertices of a graph such that the neighbors of each vertex, with or without the vertex itself, appear consecutively in the order. Equivalently, given a zero-one matrix with blank diagonal entries is there some assignment of zeroes and ones to the diagonal entries so that the resulting matrix has the consecutive ones property for columns?

DEFINITION 2.1. *A mixed diagonal consecutive ones graph (MDC-graph) is a graph whose adjacency matrix with some assignment of zeros and ones to the diagonal elements has the consecutive ones property for columns. (Because of symmetry, the rows have the consecutive ones property as well, and the same permutation and diagonal assignment applies.)*

Proposition 3 implies that every indifference (unit interval) graph is an MDC-graph, since if each diagonal entry is 1 then the adjacency matrix of an indifference graph has the consecutive ones property for columns. MDC-graphs are perfectly orderable by Theorem 1, therefore there exist $O(n \log n)$ algorithms for finding optimal colorings, and the largest clique for MDC-graphs. Indeed, these algorithms are greedy algorithms applied to the vertices in the order presented by the perfect ordering. Hence, recognizing MDC-graphs is of considerable importance in applications. Figure 3 gives some examples of MDC-graphs with their linear order of vertices, and corresponding adjacency matrix.

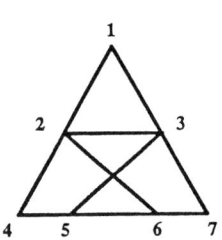
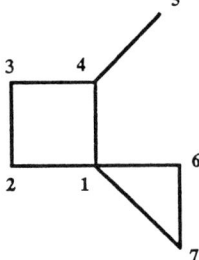

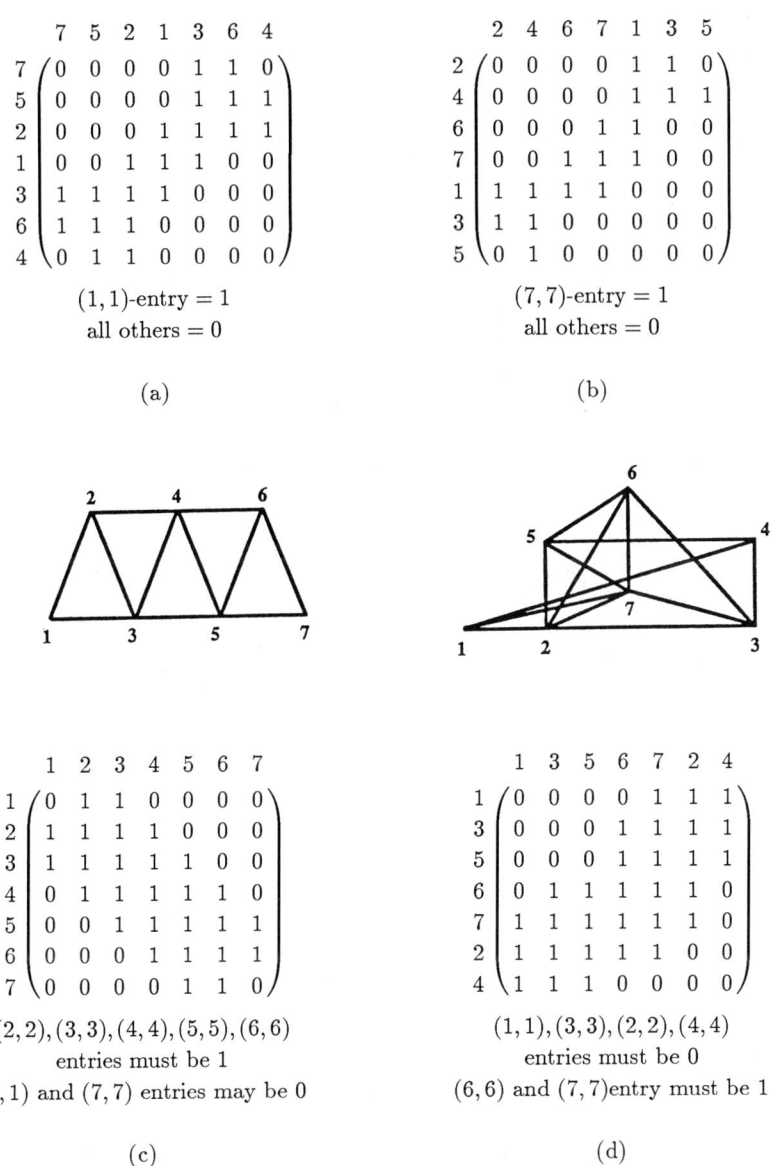

Figure 3

The graph in Figure 3(a) is an MDC-graph, but it is not an interval graph, since it contains a circuit of length 4, as well as an asteroidal triple of vertices $\{1, 4, 7\}$, each of which is forbidden by various characterizations of interval graphs. The graph in Figure 3(d) is an MDC-graph that is not a comparability graph (does not have a transitive orientation) and is not triangulated (has a cycle of length 4), two previously known classes of perfectly orderable graphs. If we extend the graph in Figure 3(c) arbitrarily far to the right in a similar manner, we have a graph with arbitrarily high Dilworth number [9]. and arbitrarily high threshold dimension [7], yet these graphs are MDC-graphs. Thus we have MDC graphs that do not appear in any of the standard classes of perfectly orderable graphs.

Once an ordering of vertices is given such that the adjacency matrix with prescribed diagonal entries has the consecutive ones property for columns we have proven that the given graph is an MDC-graph. This is not always easy to do, but for those graphs shown in Figure 3 we have proved that each is an MDC-graph. Figure 4 illustrates a few graphs that are not MDC-graphs. Since MDC-graphs are perfectly orderable, thus perfect, and C_5 is not perfect, C_5 in Figure 4(a) is not an MDC-graph. It is harder to show that the graph in Figure 4(b) is not an MDC-graph but one can illustrate the algorithm on this graph. Figure 4(b) also gives an example of a perfectly orderable graph that is not an

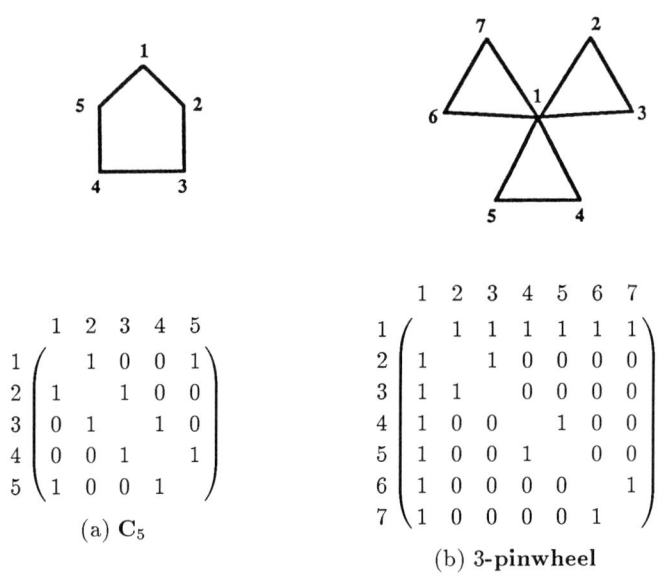

Figure 4

MDC graph. This graph has an acyclic orientation and no P_4's so no obstruction exists. This graph is also a triangulated comparability graph that is not an MDC-graph.

At the moment there does not exist a characterization of MDC-graphs so we must be content with a linear time recognition algorithm for MDC-graphs. This MDC-linear algorithm utilizes the PQ-tree data structures of Booth and Lueker by modifying the implementation to allow for both the possibility of a 1 as a diagonal entry for a vertex or a 0 as a diagonal entry, and maintains the dual possibility as long as a consecutive arrangement is still possible.

A *PQ*-tree T is a rooted tree whose internal nodes are of two types, P nodes and Q nodes. The children of a P node may be permuted arbitrarily while the children of a Q node must appear in the order listed or the reverse order only. We will use circles to represent P nodes and rectangles to represent Q nodes as shown in Figure 5.

Figure 5

The effect of the Q nodes is to restrict the number of permutations by making some of the sibling relationships permanent. The leaves of T are labeled by the vertex set or set to be arranged.

The general *PQ*-tree algorithm developed by Booth and Lueker [2] makes use of a pattern matching routine **REDUCE** which attempts to apply the set of 11 templates shown in Figure 6. Each template consists of a **pattern** to be matched against the current *PQ*-tree and a **replacement** to be substituted for the pattern. The templates are applied from the bottom to the top of the tree.

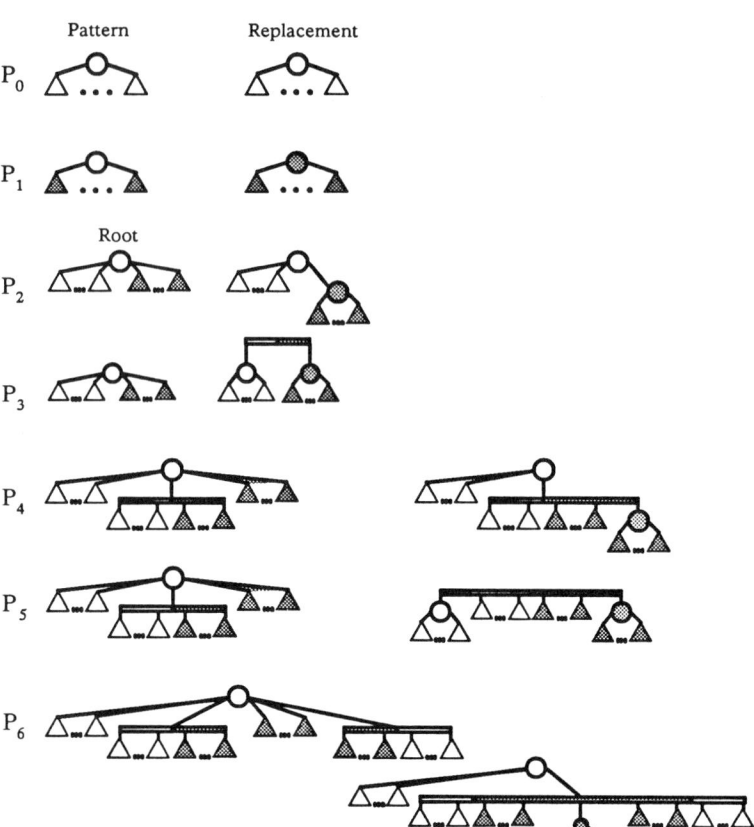

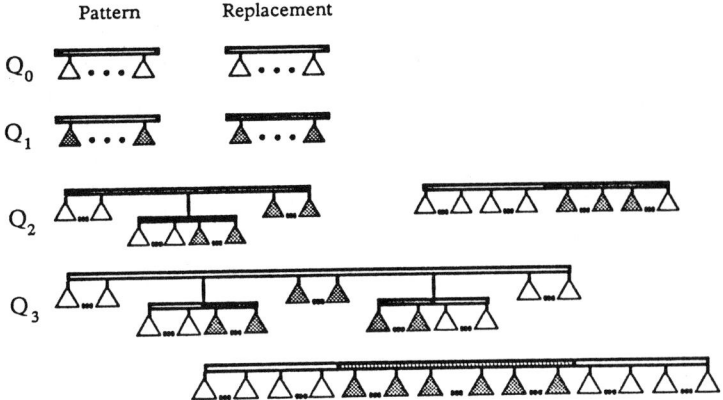

Figure 6

To prepare the tree for the application of the routine **REDUCE**, a routine **BUBBLE** is applied. This routine supplies the information necessary for the matching by marking the nodes in a constraint subset or neighborhood. Marked nodes are shown as dark nodes in Figure 6.

Before we modify the general PQ-tree algorithm in order to recognize MDC-graphs, we will illustrate the basic algorithm on the set $I = \{A, B, C, D\}$ where $\{A, B\}$ must appear together and $\{B, C\}$ must appear together and $\{A, D\}$ must appear together. The initial stage of the algorithm has a root node that is a P node with the elements of I as children, as shown in Figure 7(a).

Figure 7

Figure 7(b) shows the marking process for first processing the subset $\{A, B\}$. Template P_2 is applied with the result shown in Figure 8(a). Figure 8(b) shows the marking process applied to subset $\{B, C\}$.

(a)   (b)

Figure 8

Template P_3 is applied to the subtree containing A and B with the result shown in Figure 9(a). Template P_4 is applied to reduce the tree still further as shown in Figure 9(b). Figure 9(c) marks the new tree with the subset $\{A, D\}$.

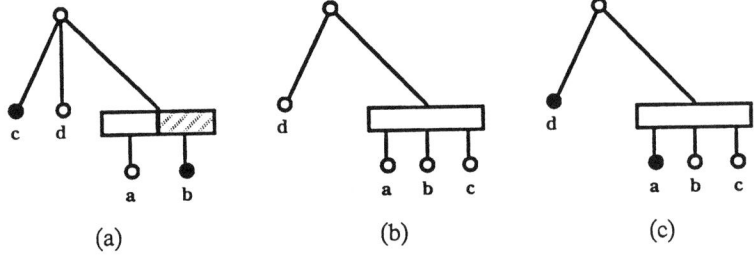

(a)   (b)   (c)

Figure 9

Applying P_4 again yields the final tree as shown in Figure 10. Two orders are possible: $D - A - B - C$ and $C - B - A - D$.

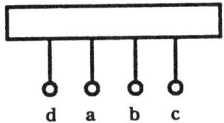

Figure 10

In testing a graph for the MDC-property we are trying to find a linear ordering of vertices such that the neighbors of a vertex appear consecutively in the order, with or without the vertex. We process each vertex neighborhood with 2 or more vertices, one vertex at a time. To allow for the possibility that a vertex may appear with its neighbors we give additional rules for marking, implemented in the **BUBBLE** routine. As a node x is processed its neighbors are marked. If the parent of x is a P node then:

(a) mark x if all of its siblings are marked and unbarred;
(b) don't mark x if all of its siblings are unmarked and unbarred;
(c) otherwise, double x and mark one of the nodes, placing bars under each copy of x as shown below:

If the parent of x is a Q-node, then double x as shown above unless x is between two marked vertices, at least one of which is unbarred in which case x is marked, or x is between two unmarked vertices at least one of which is unbarred in which case x is unmarked. The first instance is illustrated in Figure 11.

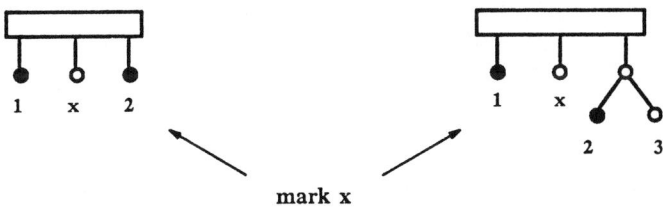

Figure 11

The validity of these rules for marking, as well as the validity of other aspects of the modified PQ-tree algorithm is detailed in [8]. Analogous to the first marking rule, we can give criteria requiring 1 for a diagonal entry.

LEMMA 1: *If graph G contains $K_4 - \{e\}$ (shown in Figure 12) as an induced subgraph and G is an MDC-graph then the $(1,1)$ or $(3,3)$ entries of the adjacency matrix must be 1, and both may be 1.*

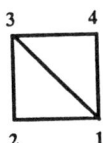

Figure 12

Proof. The neighbors of 2 are 1 and 3 and not 4, and the neighbors of 4 are 1 and 3 and not 2. Therefore $\{1,3\}$ must appear consecutively in the order without 2 and 4 between them. Therefore the consecutive ordering induced on $1,2,3,4$ must be one of the following or its reversal:

(i) 2134: 1 and 3 appear with its neighbors.

(ii) 4132: 1 and 3 appear with its neighbors.

(iii) 1324: 3 appears with its neighbors.

(iv) 1342: 3 appears with its neighbors.

(v) 3124: 1 appears with its neighbors.

(vi) 3142: 1 appears with its neighbors.

In all cases either 1 or 3 appears in the middle of its neighborhood, hence the $(1,1)$ entry or the $(3,3)$ entry is 1. In cases (iii)-(vi) either the $(1,1)$ entry is 1 or the $(3,3)$ entry is 1 and both could be 1. □

Situations also occur that prescribe 0 as a diagonal entry of the adjacency matrix.

LEMMA 2: *If graph G contains C_4 as an induced subgraph and G is an MDC-graph then there exists an adjacent pair of vertices x and y of C_4 such that the (x,x) and (y,y) entries of the adjacency matrix must be 0.*

Proof. Let G be a graph and the cycle $1-2-3-4-1$ be an induced 4-cycle in G. $\{2,4\}$ must appear together without 1 in the middle as neighbors of 3 and without 3 in the middle as neighbors of 1. Similarly $\{1,3\}$ must appear together without 2 in the middle as neighbors of 4 and without 4 in the middle as neighbors of 2. Therefore we have the following possibilities up to reversal:

(i) 1324: $(1,1)$ and $(4,4)$ entries are 0.

(ii) 2413: $(2,2)$ and $(3,3)$ entries are 0.

(iii) 3124: $(3,3)$ and $(4,4)$ entries are 0.

(iv) 1342: $(1,1)$ and $(2,2)$ entries are 0.

Each pair of 0 diagonal entries are adjacent in the cycle. □

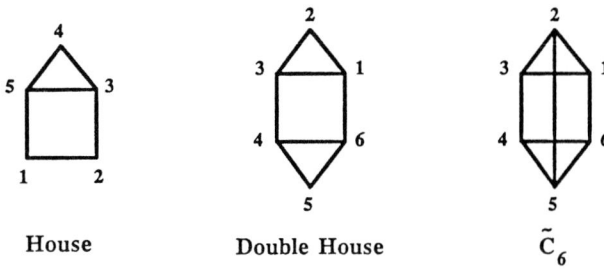

Figure 13

LEMMA 3: *If G contains a house (as shown in Figure 13) as an induced subgraph and G is a MDC-graph, then the $(1,1)$ and $(2,2)$ matrix entries are 0, and the $(4,4)$-entry is 1.*

Proof. Analogous to the proofs of lemmas 1 and 2. The only possible orders of the vertices are 13452 or 25431.

Lemma 3 can be used to give two forbidden subgraphs in any characterization of MDC graphs.

THEOREM 2: *If G is an MDC-graph, then G does not contain either a double house or $\tilde{C}_6$ (as shown in Figure 13) as induced subgraphs.*

Proof. By Lemma 3, vertices 4 and 6 are at the ends of the induced order on $\{1,2,3,4,6\}$, but vertices 1 and 3 are at the ends of the induced order on $\{1,3,4,6,5\}$, a contradiction. □

A variety of similar lemmas and theorems can be proved, but to date, no complete forbidden subgraph characterization exists.

The rules for marking provide for the inclusion of vertices (elements) with their neighborhoods (subsets) by doubling some vertices. However, situations such as described in Lemma 2 may require the deletion of one of the doubled vertices. In the **REDUCE** part of the algorithm we incorporate a set of deletion rules: Delete x if x is barred and

(i) x is part of a partial node (marked or unmarked) with a partial sibling.

(ii) x is an unmarked vertex of a Q node and x is between two marked or partially marked vertices.

(iii) x is a marked barred node between two unmarked nodes in a Q node.

Figure 14 illustrates these deletion rules.

Example of (i):

Examples of (ii):

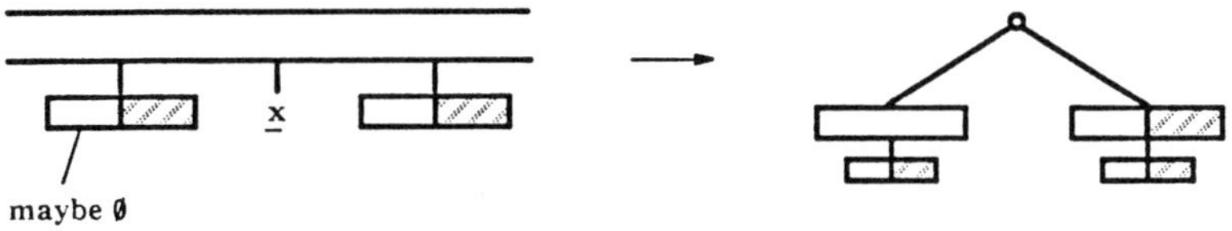

Example of (iii):

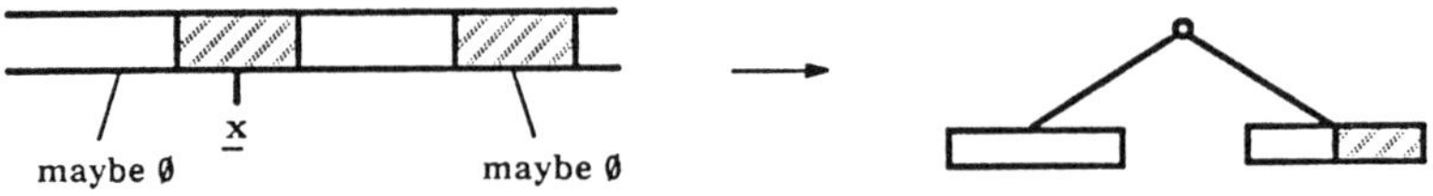

Figure 14

Figure 15 shows the application of the templates of Figure 6 and the marking and deletion rules.

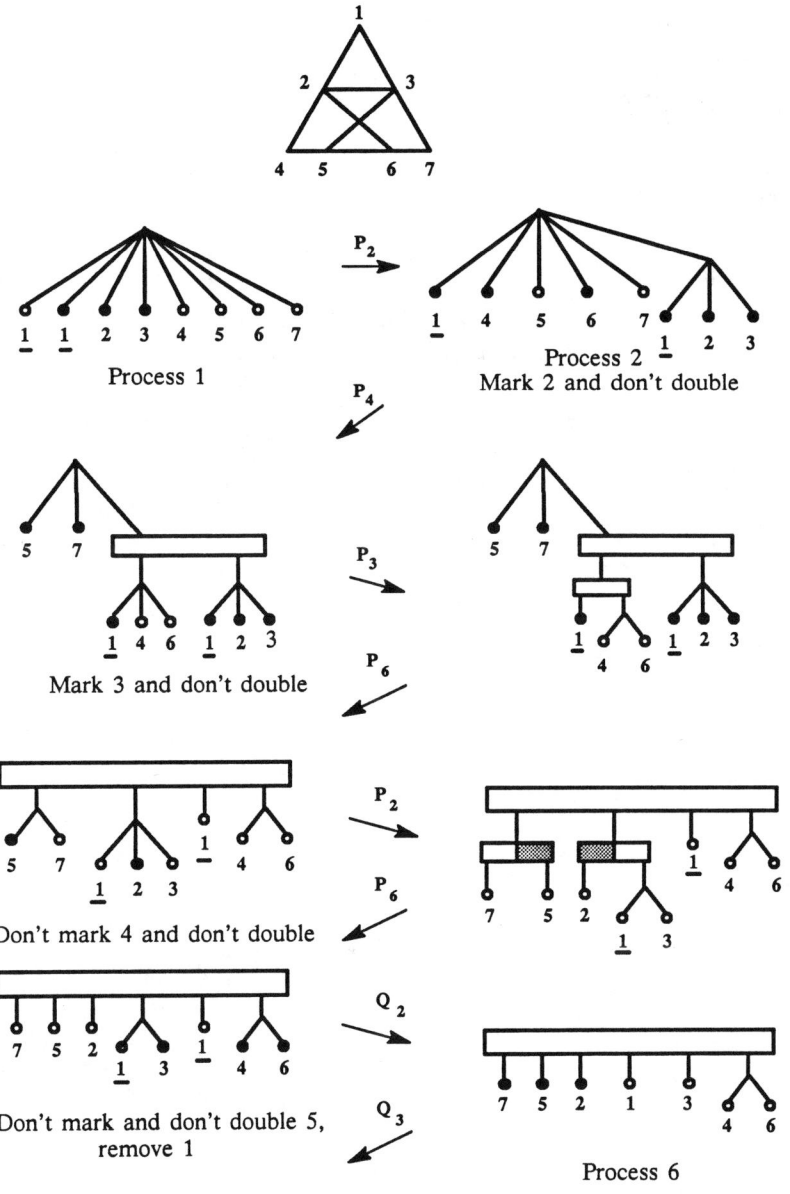

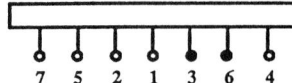

Same copy when process 7

Order: 7 5 2 1 3 6 4

1 has to be with its neighborhood
4 can't be with its neighborhood.
5 can't be with its neighborhood.
6 can't be with its neighborhood.
7 can't be with its neighborhood.
2 and 3 either way.

Figure 15

REFERENCES

[1] J.J. BARTHOLDI, III, J.B. ORLIN AND H.D. RATLIFF, *Circular ones and cyclic staffing*, in Tech. Report No. 21, Dept. of Oper. Res., Stanford Univ., 1977.

[2] K.S. BOOTH AND S.G. LUEKER, *Testing for Consecutive ones property, interval graphs, and planarity using PQ-tree algorithms*, J. Comput. Syst. Sci., 13 (1976), pp. 335-379.

[3] V. CHVÀTAL, *Perfectly ordered graphs*, in Topics on Perfect Graphs, C. Berge and V. Chvàtal Eds, North Holland, Amsterdam, 1984, pp. 63-65.

[4] V. CHVÀTAL, C.T.HOÀNG, N.V.R. MAHADEV AND D. DE WERRA, *Four Classes of Perfectly Orderable Graphs*, J. of Graph Theory, 11, No. 4 (1987), pp. 481-495.

[5] C.COOMBS AND J. SMITH, *On the detection of structures in attitude and developmental processes*, Psych. Rev., 80 (1973), pp. 337-351.

[6] M.B. COZZENS AND F.S. ROBERTS, *T-colorings of graphs and the channel assignment problem*, Congressus Numeratum, 35 (1982).

[7] M.B. COZZENS AND R. LEIBOWITZ, *Threshold dimension of graphs*, SIAM J. of Alg. and Disc. Meth., 5, No. 4 (1984).

[8] M.B. COZZENS AND N.V.R. MAHADEV, *Recognizing mixed diagonal consecutive ones graphs*, to be submitted to, SIAM J. of Discrete Math.

[9] S. FÖLDES AND P.L. HAMMER, *The Dilworth number*, Ann Discrete Math., 2 (1978), pp. 211-219.

[10] D.R. FULKERSON AND O.A. GROSS, *Incidence matrices and interval graphs*, Pacific J. Math., 15 (1965), pp. 835-855.

[11] S.P. GHOSH, *File organization: the consecutive retrieval property*, Comm. Assoc. Comput. Mach., 15 (1972), pp. 802-808.

[12] ———, *On the theory of consecutive storage of relevant records*, J. Inform. Sci., 6 (1973), pp. 1-9.

[13] ———, *File organization: consecutive storage of relevant records on a drum-type storage*, Inform. Control, 25 (1974), pp. 145-165.

[14] ———, *Consecutive storage of relevant records with redundancy*, Comm. Assoc. Comput. Mach., 18 (1975), pp. 464-471.

[15] M.C. GOLUMBIC, *Algorithmic Graph Theory and Perfect Graphs*, Academic Press, New York, 1980.

[16] U. GUPTA, *Bounds on storage for consecutive retrieval*, J. Assoc. Comput. Mach., 26 (1979), pp. 28-36.

[17] P.L. HAMMER AND N.V.R. MAHADEV, *Bithreshold graphs*, SIAM J. Discrete Math., 6 (1985), pp. 497-506.

[18] D.G. KENDALL, *Incidence matrices, interval graphs, and seriation in archeology*, Pac. J. Math., 28 (1969), pp. 565-570.

[19] W. LIPSKI, JR. AND T. NAKANO, *A note on the consecutive 1's property (infinite case)*, Comment. Math. Univ. St. Paul, 25 (1976/1977), pp. 149-152.

[20] R.D. LUCE, *Semiorders and a theory of utility discrimination*, Econometrica, 24 (1956), pp. 178-191.

[21] J.R. LUNDGREN AND J.S. MAYBEE, *Food webs with interval competition graphs*, in Graphs and Applications: Proceedings of the First Colorado Symposium on Graph Theory, Wiley, New York, 1984.

[22] D.W. MATULA, *A min-max theorem with application to graph coloring*, SIAM Rev., 10 (1968), pp. 481-482.

[23] T. NAKANO, *A characterization of intervals; the consecutive (one's or retrieval) property*, Comment. Math. Univ. St. Paul, 22 (1973a), pp. 49-59.

[24] ———, *A remark on the consecutivity of incidence matrices*, Comment. Math. Univ. St. Paul, 22 (1973b), pp. 61-62.

[25] R. OPSUT AND F. S. ROBERTS, *I-colorings, I-phasings, and I-interval assignments for graphs, and their applications*, Networks (1983).

[26] *On the fleet maintenance, mobile radio frequency, task assignments, and traffic phasing problems*, in Proceedings of the Fourth International Conference on Theory and Applications of Graphs, Wiley, New York, 1980.

[27] A.N. PATRINOS AND S.L. HAKIMI, *File organization with consecutive retrieval and related properties*, in Large Scale Dynamical Systems, R. Sacks, ed., Point Lobos, North Hollywood, California, 1976.

[28] M. PREISSMANN, D. DE WERRA AND N.V.R. MAHADEV, *A Note on Superbrittle Graphs*, Discrete Mathematics, 61 (1986), pp. 259-267.

[29] F.S. ROBERTS, *Representations of indifference relations*, Ph.D. thesis, Stanford Univ., 1968.

[30] ——————, *Indifference graphs*, in *Proof Techniques in Graph Theory*, F. Harary, ed., Academic Press, New York, 1969, pp. 139-146.

[31] ——————, *Graph Theory and its Applications to Problems of Society*, NSF-CBMS Monograph, #29, SIAM Publications, Philadelphia, 1978.

[32] ——————, *Indifference and seriation*, in *Advances in Graph Theory*, F.Harary (ed.), Proc. NY Acad. of Sciences, 1979, pp. -999.

[33] ——————, *Applications of the theory of meaningfulness to order and matching experiments*, in *Trends in Mathematical Psychology*, E.DeGreef and J.van Buggenhaut (eds.), North Holland, Amsterdam, 1984, pp. 283-292.

[34] D. SKREIN, *A relationship between triangulated graphs, comparability graphs, proper interval graphs, proper circular-arc graphs, and nested interval graphs*, Journal of Graph Theory, 6 (1982), pp. 309-316.

[35] K.E. STOFFERS, *Scheduling of traffic lights - a new approach*, Transport Res., 2 (1968), pp. 199-234.

[36] A.C. TUCKER, *Characterizing the consecutive 1's property,*, in *Proc. 2nd Chapel Hill Conf. on Combinatorial Mathematics and its Applications*, Univ. North Carolina, Chapel Hill, 1970a, pp. 472-477.

[37] ——————, *A structure theorem for the consecutive 1's property*, J. Combin. Theory, 12 (1972), pp. 153-195.

[38] A. WAKSMAN AND M.W. GREEN, *On the consecutive retrieval property in file organization*, IEEE Trans. Comput, C-23 (1974), pp. 173-174.

PROBABILISTIC KNOWLEDGE SPACES: A REVIEW*

JEAN-CLAUDE FALMAGNE[†]

Abstract. This paper outlines the essential ideas of a theory for the efficient assessment of knowledge. The key concept is that of a *knowledge space*, that is, a basic set Q of questions or problems in a given domain of information, equipped with a distinguished family $\mathcal{K}$ of subsets. The family $\mathcal{K}$ is assumed to be closed under union, and its elements are *knowledge states*. The equivalence between this concept and other useful ones is spelled out. A practical implementation of computerized, robust knowledge assessment devices requires a probabilistic framework. These developments are also briefly summarized. All the results have been presented at length elsewhere, and are either in press or published.

There is a vast category of empirical situations in which the state of a complex system has to be assessed rapidly and efficiently. The system is then tested by some expert device, which checks sequentially for the presence of some revealing features. Here are three examples.

Failure Analysis. The system is a complex piece of machinery displaying a malfunction, the nature of which has to be ascertained by a sequence of verifications.

Medical Diagnosis. The system is a patient complaining of some ailment. The physician will examine the patient and check which symptoms are present. As above, a sequence of careful verifications will take place.

Knowledge Assessment. A teacher is examining a new student, to determine, for instance, which courses should be taken at this stage of the student career. The teacher will ask one question, then another, selected as a function of the response to the first one. Gradually, the student's knowledge state will emerge.

Even though our results are potentially applicable to these and many other cases, the focus of our work is on the third Example, which has guided our choice of terminology. In the last five years, our group has developed the mathematical theory and the related computer algorithms, for a computer assessment device capable of mimicking a teacher examining a student–or even, possibly, of outperforming the human expert. This work will be briefly reviewed here.

For concreteness, we shall illustrate our concepts with the following set Q of five problems in high school mathematics:

---

*This paper was written while the author was a fellow at the Center for Advanced Studies in the Behavioral Sciences. The financial support of the National Science Foundation BNS-870086 and the Alfred P. Sloan Foundation is gratefully acknowledged. This work has been supported by NSF grant IST-817057 ([2], [3]), NSF grant IST-8418860 ([4], [5], [6], [7], [8], [9] and [13]) and ARI grant MDA903-87-K-0002 ([7], [8], [11] and the present paper).

†University of California at Irvine, Irvine Research Unit in Mathematical Behavioral Sciences, School of Social Science, Social Science Tower, 6th floor, Irvine, CA 92717

1. 378 × 605=?
3. $\frac{1}{2} \times \frac{5}{6} = ?$

2. 58.7 × 0.94 = ?
4. What is 30% of 34?

5. Gwendolyn is $\frac{3}{4}$ as old as Rebecca.
Rebecca is $\frac{2}{5}$ as old as Edwin.
Edwin is 20 years old.
How old is Gwendolyn?

Consider the task of efficiently assessing an individual's mastery of the problems in this miniature example, which is obviously meant to represent a much broader and realistic situation. Which should be the first problem to ask? If the response to that first problem is, for instance, incorrect, which should be the next one? What have we exactly learned from the responses to the first two problems? Answering such questions requires that the set of problems be structured in some way. A candidate for such a structure is a binary relation P on Q, with the following interpretation:

$$qPt \quad \text{if and only if} \quad \begin{cases} \text{the mastery of Problem } q \\ \text{can be surmised from} \\ \text{the mastery of Problem } t \ . \end{cases}$$

It is reasonable to suppose that P is a quasi order (reflexive, transitive). We shall refer to P as a *surmise relation*. A plausible surmise relation is represented in Fig. 1 by its Hasse diagram. (In this particular case, the quasi order is a partial order.)

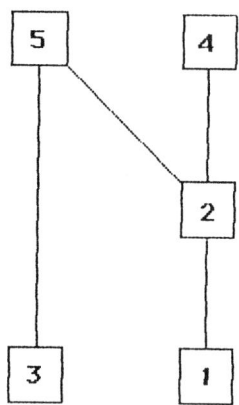

Figure 1. Hasse diagram of a surmise relation P.

Thus, the knowledge states associated with P are contained in the family

(1) $\{\emptyset, \{1\}, \{3\}, \{1,3\}, \{1,2\}, \{1,2,3\}, \{1,2,4\}, \{1,2,3,4\}, \{1,2,3,5\}, Q\}$

The obvious drawback of this formalization is that it implicitly assumes that any problem q has a unique set of antecedents (or prerequisites), namely, the set of all the problems preceding q in the quasi order. This is certainly too strong in general. Notice in this connection that the family $\mathcal{K}$ is closed under union and intersection. This illustrates a general, well known result ([1]).

THEOREM 1. (Birkhoff, 1937) *For any set Q, the formula*

$$q \preceq t \iff (t \in K \Rightarrow q \in K, \forall K \in \mathcal{F})$$

establishes a one-to-one correspondence between the set of all quasi orders $\preceq$ on Q, and the collection of all families $\mathcal{F}$ of subsets of Q which are closed under intersection and union.

(See [3] and [12], for a statement of this result in terms of a Galois connection.) This change of viewpoint in the formalization – from surmise relations to families of states – suggests how the model should be weakened. The closure under intersection is not a reasonable assumption, and should be dropped.

Definition. A *knowledge space* is a pair $(Q, \mathcal{K})$, in which Q is nonempty set of questions or problems, and $\mathcal{K}$ is a collection of subsets of Q containing $\emptyset$ and Q, and closed under union. The elements of $\mathcal{K}$ are called *(knowledge) states*. Occasionally, $\mathcal{K}$ will be called a *knowledge space on Q.*

A maximal chain of subsets of Q, all of which are states, is called a *gradation*. If any state of $\mathcal{K}$ is contained in at least one gradation, the knowledge space $(Q, \mathcal{K})$ is said to be *well graded*.

Any gradation may be regarded as a possible "learning path". An illustration of these concepts involving the five problems is given in Fig. 2. Notice that there are fifteen states, and sixteen gradations. We shall see later that this particular knowledge space is consistent with some extensive data involving close to five hundred students in a New York City high school. Notice that it is not closed under intersection since $\{4\}$ is not a state, while $\{1,4\}$ and $\{3,4\}$ are states.[1]

---

[1] A detailed analysis in terms of a stochastic model described in [8] leads to a substantial simplification of this representation, in which only eleven states and four gradations suffice to explain the data.

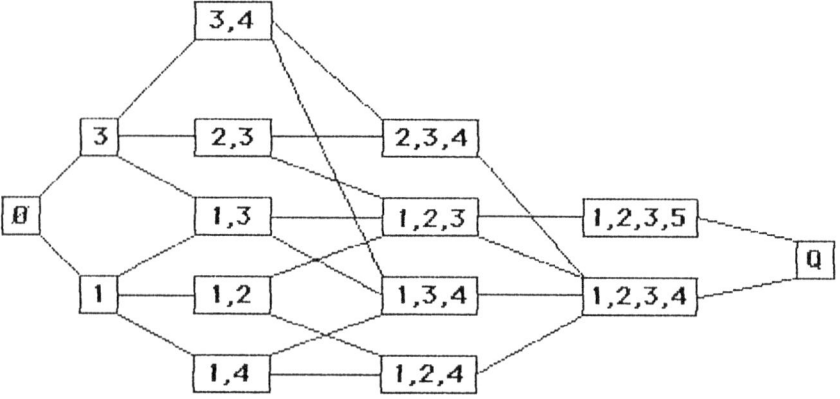

Figure 2. A knowledge space for the five problems. The states are represented by rectangles. Any gradation is represented by a sequence of five segments connecting ∅ to Q, with the inclusion going from left to right.

A number of issues arise. For instance, we objected to a model based on a quasi order because any problem had a unique set of prerequisites. Supposedly, this is no longer true for a knowledge space. How can this idea be made precise? In other words, is there a generalization of Birkhoff's Theorem along these lines? A response to these questions can be found in the next theorem, in terms of a function σ which, to each problem $q \in Q$ associates a nonempty collection of subsets of Q. Thus, the elements of $\sigma(q)$ are the possible sets of prerequisites for q. The function σ will be referred to as a *surmise function on Q* and will be assumed to satisfy a number of conditions. (For technical reasons, we suppose that $q \in C$ for any $C \in \sigma(q)$.)

Another issue concerns the practical construction of a knowledge space. In our view, much can be gathered by systematically consulting experts (that is, very experienced teachers), at least in the first phase of such a construction. However, we cannot simply ask an expert to provide a list of the possible knowledge states: whatever information has been accumulated by years of teaching is almost certainly not coded in the form of knowledge states in the sense of this paper. There is, however, one type of question that a seasoned teacher should be able to answer reliably, namely: *"Suppose that a student is not capable of solving Problems $q_1, q_2, \ldots q_n$. Are you practically certain, then, that this student would also fail Problem q_{n+1}? We assume that careless errors are excluded."* The response to all such questions define a binary relation $\mathcal{P}$ on 2^Q which, assuming that some natural conditions are satisfied, will be called an *entail relation*. Thus,

$$A\mathcal{P}B \quad \text{if and only if} \quad \begin{cases} \text{failing all the problems in } A \\ \text{entails failing all the problems in } B \end{cases}$$

It turns out that $\mathcal{P}$ provides exactly enough information to reconstruct the knowledge space completely. The next theorem summarizes our discussion.

THEOREM 2. *Let Q be a finite set. The following three concepts are equivalent:*

1. *A knowledge space $\mathcal{K}$ on Q.*
2. *A surmise function, that is, a function $\sigma : Q \to 2^{(2^Q)}$ satisfying the following three conditions*

   (i) $C \in \sigma(q) \Rightarrow q \in C$;

   (ii) $C, C' \in \sigma(q) \land C \subseteq C' \Rightarrow C = C'$;

   (iii) $q' \in C \in \sigma(q) \Rightarrow \exists C \in \sigma(q'), \ C' \subseteq C$.

3. *an entail relation, that is, a relation $\mathcal{P}$ on 2^Q, which is transitive, extends inclusion, and satisfies*

$$A\mathcal{P}B \Rightarrow A\mathcal{P}(A \cup B).$$

In other words, there exists three one–to–one correspondences linking, pairwise, the following three sets: the collection of all knowledge spaces; the collection of all surmise functions; the collection of all entail relations.

This theorem is a corollary to more general results, in terms of Galois connections, established by Doignon and Falmagne ([3]) and Koppen and Doignon ([11]). A computer algorithm for constructing the knowledge space by systematically querying the expert, based on the equivalence between 1 and 3 in Theorem 2, is currently being tested.

So far, the terms of our discussion have been deterministic. There are good reasons, however, to adopt a probabilistic framework. For instance, once a tentative knowledge space has been obtained, how can it be tested experimentally? A knowledge space may be considered as a model capable of making predictions concerning the actual responses of subjects in the course of an examination. But such data are typically regarded as being noisy (e.g. subjects make careless errors, or sometimes guess correct responses). Moreover, the standard goodness–of–fit techniques for evaluating models are statistical ones. The correct approach at this juncture is to construct a probabilistic version of a knowledge space, that would be suitable for statistical testing. The fundamental equation of such a model would be:

(2) $$\text{Prob}[R] = \sum_{K \in \mathcal{K}} \text{Prob}[R|K]\,\text{Prob}[K],$$

in which R represents any pattern of responses – e.g., a vector of zeros and ones representing the correct and incorrect responses to the questions. Thus, the axioms of the model have to specify the conditional probabilities Prob[$R|K$] of the patterns of responses, given the states, and the probabilities Prob[K] of the states.

In ([8]), a model is presented in which the probabilities of the states are obtained via a stochastic process describing the progression of the students, over time, along the gradations. The evolution of that progression depends upon the difficulty of the problems and the learning rate of a student. The difficulty of the problems are measured by parameters to be estimated from the data. For a given population of students, the learning rate is a random variable, which is assumed to have a gamma distribution (see [10]). The time required to master a problem is also distributed gamma, with two parameters depending on the item difficulty and of the learning rate of the student. The probability of choosing a particular gradation is a parameter. This model permits to obtain an explicit expression, in terms of the *incomplete beta function ratio* (c.f. [10]), of the probability that a (randomly chosen) students is in any state K of $\mathcal{K}$ at time τ. It gives a good account, from a statistical viewpoint, of the high school students data mentioned above. This application of the model led to a substantial simplification of the knowledge space originally hypothesized (c.f. Footnote 1).

Once a knowledge space has been constructed and verified, using, for instance the stochastic model just outlined the key problem remains of searching such a space efficiently to assess the state of a student.

In [6], a class of Markov processes was defined and analyzed, grounded on the following principle. On each trial, a likelihood function is specified on the set $\mathcal{K}$ of all knowledge states, representing the uncertainty of the assessor on that trial. A problem is then proposed to the student under examination, the choice of which is governed by a *questioning rule* operating on the current likelihood function. This rule is designed to provide problems which are, in some specific sense, maximally informative. The student's response to the problem is then recorded. The trial ends with a recomputation of the likelihood function, based on an *updating rule* operating on the current likelihood function and on the events of that trial, namely, the problem proposed and the response observed. Several cases of questioning rules and of updating rules are investigated. It is shown that, under fairly general conditions, the true knowledge state K. of a student can be uncovered, in the sense that its likelihood converges a.s. to one.

Another algorithm was developed [7], in the framework of discrete Markov chains, which is similar in spirit but computationally much less demanding. The basic idea of this algorithm is that, on any given trial, some of the knowledge states are "marked" as being plausible. Initially, all the knowledge states are marked. The succession of pairs of events (problem asked, response observed) gradually reduces the collection of marked knowledge states. In one version of this algorithm, the Markov chain converges to a situation in which a single knowledge state is marked, which keeps oscillating from one trial to the next in a small region of the knowledge space. The state space of this Markov chain is thus $2^{\mathcal{K}}$, a much smaller set than the simplex of all probability distributions on $\mathcal{K}$, the state space of the class of algorithms described in [6].

Computer simulations have shown these algorithms to yield robust predictions of the student's states ([13]). A large scale, realistic application of these techniques is in progress.

REFERENCES

[1] G. BIRKHOFF, *Rings of sets*, Duke Mathematical Journal 3 (1937), 443–454.

[2] E. DEGREEF, E. J.-P. DOIGNON, A. DUCAMP, AND J.-CL. FALMAGNE, *Languages for the assessment of knowledge*, Journal of Mathematical Psychology 30, (1986), 243–256.

[3] J.-P. DOIGNON AND J.-CL. FALMAGNE, *Spaces for the assessment of knowledge*, International Journal for Man–Machine Studies 23, (1985), 175–196.

[4] J.-P. DOIGNON AND J.-CL. FALMAGNE, *Parametrization of knowledge structures*, Applied Mathematics, (in press).

[5] J.-P. DOIGNON AND J.-CL. FALMAGNE, *Knowledge assessment: A set theoretical framework*, In B. Ganter, R. Wille and K.E. Wolff (Eds.) *Beiträge sur Begriffsanalyse, Vorträge der Arbeitstagung Begriffsanalyse, Darmstadt*, 1986, B.I. Wissenschaftsverlag, Mannheim, 1987, 129–140.

[6] J.-CL. FALMAGNE AND J.-P. DOIGNON, *A class of stochastic procedures for the assessment of knowledge*, British Journal of Mathematical and Statistical Psychology, (in press).

[7] J.-CL. FALMAGNE AND J.-P. DOIGNON, *A markovian procedure for the assessment of knowledge*, Journal of Mathematical Psychology, (in press).

[8] J.-CL. FALMAGNE, *A latent trait theory via a stochastic learning theory for a knowledge space*, Psychometrika, (in press).

[9] J.-CL. FALMAGNE, M. KOPPEN, M. VILLANO, L. JOHANNESEN AND J.P. DOIGNON, *Introduction to knowledge spaces: How to build, test and search them*, Submitted.

[10] N.L. JOHNSON AND S. KOTZ, *Distributions in Statistics. Continuous Univariate Distributions*, Houghton Mifflin, New York, 1970.

[11] M. KOPPEN AND J.-P. DOIGNON, *How to build a knowledge space by querying an expert*, Submitted.

[12] B. MONJARDET, *Tresses, fuseaux, préordres et topologies*, Mathématiques et Sciences humaines 30 (1979) 11–12.

[13] M. VILLANO, J.-CL. FALMAGNE, L. JOHANNESEN AND J.-P. DOIGNON, J.-P., *Stochastic procedures for assessing an individual's state of knowledge*, Proceedings of the International Conference on Computer–assisted Learning in Post–Secondary Education, Calgary, 1987 (369–371).

UNIQUENESS IN FINITE MEASUREMENT

PETER C. FISHBURN† AND FRED S. ROBERTS‡

Abstract. This article surveys recent investigations of real sequences $(d_1, \ldots, d_n)$ which arise in the theory of measurement from considerations of uniqueness for numerical representations of qualitative relations on finite sets. The sequences we discuss arise from measurement problems which include measurement of subjective probability, extensive measurement, difference measurement, and additive conjoint measurement. The measurement problems lead to sequences with fascinating combinatorial and number–theoretic properties.

The unifying mathematical framework under which we analyze uniqueness of measurement in these diverse areas involves the analysis of the sequences $(d_1, \ldots, d_n)$ as the solutions of finite systems of linear equations. Different applications are translated into different restrictions on the types of linear equations that are admissible for each area.

Two primary concerns of measurement theory are involved in the work being surveyed: (1) axioms for the qualitative relation that are necessary and sufficient, or at least sufficient, for unique representability; (2) the structure of sets of unique solutions. The latter concern leads to combinatorial and number–theoretic problems involving characterizations of unique solutions, counts of numbers of unique solutions, and extreme–value questions. Definitive results and presently open problems are described for the areas covered by the basic theory.

1. INTRODUCTION

In this paper we consider a series of problems in combinatorics and number theory which arise in the representational theory of measurement (Fishburn [1970, 1988], Krantz, et al. [1971], Narens [1985], Pfanzagl [1968], Roberts [1979]).

In measurement theory, we are frequently interested in mappings u which assign a real number to each element of a given qualitative structure $\mathcal{A}$ in such a way that certain features of the qualitative structure are preserved. We pick certain properties that we would like such a mapping to have and call these a *model*. Sometimes the model is called a *representation* and sometimes it is the mapping which is called a *representation*. The theory is concerned with three kinds of questions: *The Modeling Question*: What properties define an appropriate model?; *The Existence Question*: Given a model, what properties must the qualitative structure have in order for there to be a mapping which satisfies it?; *The Uniqueness Question*: What are the relationships among all possible mappings that satisfy the model? In this paper we shall be primarily interested in the uniqueness question.

The theory of measurement has undergone immense development during the past 20 years, as evidenced by the books cited above. The majority of this work deals with the three questions we have posed for infinite qualitative relational structures, which under assumed continuity conditions tend to have "nice" answers to the existence and uniqueness questions for a variety of models. There is a growing literature on the existence question for finite structures, thanks in part to the

---

†AT&T Bell Laboratories, Murray Hill, New Jersey 07974.
‡Department of Mathematics, Rutgers University, New Brunswick, New Jersey 08903. Dr. Roberts would like to thank the National Science Foundation for its support under grant number IST-86-04530 to Rutgers University.

breakthrough made by Kraft, Pratt, and Seidenberg [1959] in axiomatizing finite-set subjective probability with the help of solution theory for finite systems of linear inequalities. However, apart from a few contributions (Luce [1967], Roberts [1985], Roberts and Rosenbaum [1988]), relatively little attention has been paid to the uniqueness question for finite structures. We shall concentrate on this question.

Suppose the qualitative structure A has a finite set of elements, say $\{a_1, \ldots, a_n\}$. If $u_i = u(a_i)$, we shall study a sequence of numbers $(d_1, \ldots, d_n)$ where d_i is defined in a simple way from the sequence $(u_1, \ldots, u_m)$. In measurement theory, we are concerned with conditions that the $(u_1, \ldots, u_m)$ must satisfy when u is a mapping which satisfies the model. As we shall see, these conditions often translate into two types of conditions on $(d_1, \ldots, d_n)$, inequalities of the form

(1) $$\Sigma \epsilon_j d_j > 0, \quad \epsilon_j \in \{0, 1, -1\},$$

and equalities of the form

(2) $$\Sigma \epsilon_j d_j = 0, \quad \epsilon_j \in \{0, 1, -1\}.$$

Moreover, the desired uniqueness properties of the mapping u will frequently translate into the relationship that if $(d'_1, \ldots, d'_n)$ also satisfies the same sets of inequalities and equalities, then there is a number $\lambda > 0$ for which

$$d'_j = \lambda d_j, \quad j = 1, 2, \ldots, n.$$

In other words, the numbers d_j are unique up to a set of proportionally (similarity) transformations. In this case, we shall call the sequence $(d_1, \ldots, d_n)$ *unique*.

In this paper, we shall be concerned with a number of fundamental problems concerning the properties of the unique solutions $d = (d_1, \ldots, d_n)$ corresponding to a variety of basic measurement problems. The **axiomatization problem** will be concerned with giving axioms which guarantee that d exists and/or is unique. These axioms will end up leading to classes of sequences with fascinating combinatorial and number–theoretic properties. For each class of unique sequences (i.e., for each of a number of measurement problems), we shall be interested in the **counting problem**: Count the number of unique sequences d of given length n which are in the class. We shall want to obtain either exact counts or recurrences, and we shall also be interested in asymptotic counts. A third class of problems which will concern us consists of a variety of **extremization problems** concerning each class of unique sequences. In particular, for fixed n, we shall want to know the largest possible value of a term d_j in a unique sequence $(d_1, \ldots, d_n)$ in the class; the largest possible sum of values $\sum_{i=1}^{n} d_i$ in such a sequence $(d_1, \ldots, d_n)$ in the class; and the largest possible number of one–term extensions of such a sequence to another unique sequence in the class. Still other problems of interest, called **inclusion problems**, will be concerned with the appearance of specific terms in such sequences. For instance, given terms $k_1, \ldots, k_p$, we shall ask for the length of the shortest unique sequence $(d_1, \ldots, d_n)$ in the class which includes the terms $k_1, \ldots, k_p$ and for the length of the shortest such sequence which includes $k_1, \ldots, k_p$ consecutively in a subsequence.

In the next section, we introduce different measurement problems, including subjective probability measurement, extensive measurement, difference measurement, and additive conjoint measurement, and show how each one can be formulated in terms of sequences $(d_1, \ldots, d_n)$. Section 3 formally defines the type of uniqueness we investigate and states a basic theorem for uniqueness in terms of linearly independent equations. Section 4 illustrates the fundamental concerns of uniqueness theory described earlier for a simple example of extensive measurement. Similar concerns are then addressed for more general formulations of extensive measurement and subjective probability (Section 5), difference measurement (Section 6), and additive conjoint measurement (Section 7). Open problems that merit further study are identified along the way. These are collected in the final section with concluding remarks.

Most of the results reported later are established in papers that have not yet appeared in print. We refer to these by authors' initials, such as [FR1], and group them as "primary references" before the list of other references.

2. The Fundamental Measurement Problems of Interest

This section introduces a number of basic existence and uniqueness problems of measurement theory. We translate the existence problems into the existence of sequences d satisfying inequalities and equalities like (1) and (2) and translate the uniqueness problems into the uniqueness of these sequences d up to multiplication by a positive constant λ.

2.1 Subjective Probability. In measurement of subjective probability, we have a finite Boolean algebra $\mathcal{A} = \{A, B, \ldots\}$ and a comparative probability relation $>$ on $\mathcal{A}$, where $A > B$ is interpreted to mean that A is judged "more probable" than B. We seek a probability measure P on $\mathcal{A}$ so that for all A, B in $\mathcal{A}$,

$$(3) \qquad A > B \iff P(A) > P(B).$$

One of the main reasons that the measurement theory literature has a large gap when it comes to the study of finite uniqueness is that the functions used in finite representations tend to have somewhat loose and unruly uniqueness properties. In studying the uniqueness of a measure P satisfying (3), for example, we shall generally be able to say little more than that P is unique up to the satisfaction of a system of inequalities/equalities to be defined. On the other hand, there are many interesting realizations of finite qualitative relational structures in which the mapping of interest has "nice" uniqueness properties, and it is precisely these realizations that we focus on in this review. A small example in the case of subjective probability measurement illustrates our theme.

Suppose $\mathcal{A}$ consists of all subsets of $\{1, 2, 3, 4\}$. Let $\sim$ denote the *symmetric complement* of $>$, i.e.,

$$A \sim B \quad \text{if not} \quad (A > B) \quad \text{and not} \quad (B > A).$$

Note that (3) requires that

$$A \sim B \iff P(A) = P(B).$$

Suppose for $>$ on $\mathcal{A}$ that

(4)
$$\begin{aligned}\{3\} &\sim \{1,2\} \\ \{4\} &\sim \{1,3\} \\ \{2,3\} &\sim \{1,4\}.\end{aligned}$$

Then the mapping P is unique in the absolute sense since

(5) $\quad (P(\{1\}), P(\{2\}), P(\{3\}), P(\{4\})) = (1/10,\ 2/10,\ 3/10,\ 4/10)$

is the only probability solution to the system

(6)
$$\begin{aligned}P(\{3\}) &= P(\{1\}) + P(\{2\}), \\ P(\{4\}) &= P(\{1\}) + P(\{3\}), \\ P(\{2\}) + P(\{3\}) &= P(\{1\}) + P(\{4\}).\end{aligned}$$

To show how the subjective probability measurement problem translates into the form we have described, let $a_i = \{i\}$. Then if P satisfies (3), let $u_i = u(a_i) = P(a_i)$ and let $d_i = u_i$. Note that (3) holds if and only if whenever $A > B$, we have (1), and whenever $A \sim B$, we have (2), where $\epsilon_j = 1$ if $j \in A \setminus B$, -1 if $j \in B \setminus A$, and 0 otherwise. Note that in addition to (1) and (2), the properties of a probability measure imply that $d_j \geq 0$ and $d_j > 0$ for some j. Moreover, they imply that $\sum_j d_j = 1$. However, the latter requirement can be imposed by normalization. Conversely, given any solution d to (1) and (2) with $d_j \geq 0$ and $d_j > 0$ for some j, if the solution is normalized so $\sum_j d_j = 1$, and if P is defined from d by

$$P(A) = \sum_{j \in A} d_j \quad \text{for each} \quad A \in \mathcal{A},$$

then P is a probability measure which satisfies (3).

How unique is P? As we have seen above, P can be unique. On the other hand, it need not be. Take for instance $\mathcal{A}$ to be all subsets of $\{1,2\}$ and define $>$ on $\mathcal{A}$ by letting $\{1,2\} > \{1\} > \{2\} > \emptyset$. Then for any $1 > \alpha > 1/2$, $P(\{1\}) = \alpha$ and $P(\{2\}) = 1 - \alpha$ defines a measure of subjective probability satisfying (3).

In many circumstances, we wish P to be unique. Some sufficient conditions for P to be unique are given in Roberts [1979] for infinite Boolean algebras and in Van Lier [VL] and [FR1] for finite ones. If P satisfying (3) is unique, then nonnegative d with at least one positive component and satisfying (1) and (2) is unique up to multiplication by a positive constant, i.e., it is unique in the sense we have defined uniqueness of a sequence d; conversely, if such d are unique in our sense, then P is unique (in the absolute sense). Unique sequences d arising from subjective probability measurement were introduced in [FO] and also studied in [FMR].

2.2 Extensive Measurement. The classic measurement problem called extensive measurement focuses on an attribute such as mass or length of objects in a set A. We consider a binary relation $>$ on A, representing a comparison with respect to the attribute, for instance heavier than or longer than, and a binary operation o on A, representing combination. We then seek a real–valued mapping u on A so that for all $a, b \in A$,

(7) $$a > b \iff u(a) > u(b)$$

and

(8) $$u(a \circ b) = u(a) + u(b).$$

We can reformulate this type of measurement problem in the finite situation as follows. Suppose $X = \{1, \ldots, n\}$ and we aggregate objects as subsets of X and make comparisons between aggregates, perhaps disjoint, to reflect the extent to which they possess the attribute. Let $\mathcal{B}$ denote the set of all unordered pairs $\{A, B\}$ of aggregates with $A \neq B$ so that only such unordered pairs admit comparison. Also, let $A > B$ for $\{A, B\} \in \mathcal{B}$ signify that A has more of the attribute. Then $A \sim B$, where $\sim$ is the symmetric complement of $>$, means that A and B have the same amount of the attribute.

A mapping u from X into the positive reals is said to satisfy *positive additive extensive measurement* if

(9) $$A > B \iff \sum_{j \in A} u(j) > \sum_{j \in B} u(j), \quad \text{for all} \quad \{A, B\} \in \mathcal{B}.$$

Here we simply take $d_j = u(j)$ for all j, and note that $d_j > 0$ for all j.

Suppose for simplicity that all comparisons between disjoint aggregates are feasible, so that $\mathcal{B} = \{\{A, B\} : A, B \subseteq \{1, \ldots, n\} \text{ and } A \cap B = \emptyset\}$. We then have a situation similar to that for subjective probability except that $d_j > 0$ for all j. In particular, (9) is readily translated into (1) and (2) exactly as for subjective probability.

Additional restrictions on $\mathcal{B}$ that limit the types of vectors ϵ used in (1) and (2) may be plausible in some situations. A very restrictive form for $\mathcal{B}$ is adopted in Section 4 to illustrate our basic concerns.

In extensive measurement defined by (7) and (8), one can prove under sufficiently strong conditions that if u exists, then it is unique up to multiplication by a positive constant. (See Roberts and Luce [1968].) Thus, it is a reasonable goal to seek conditions under which u satisfying (9), and hence d satisfying the corresponding (1) and (2), is unique up to multiplication by a positive constant.

2.3 Difference Measurement. A third type of measurement problem, which arises for instance in the measurement of temperature differences and of utility differences, is called difference measurement. Various aspects of finite difference measurement, including numerous references to the literature of difference measurement, are described in [FMR]. The finite problem is also studied in [FOR]. For simplicity here, we only describe the form of difference measurement called absolute difference measurement, referring to [FMR] or the books by Fishburn [1970], Krantz, et al. [1971] or Roberts [1979] for a description of the closely related algebraic difference measurement. For simplicity, we can describe finite absolute difference measurement as follows. Assume that a set $X = \{a_1, a_2, \ldots, a_{n+1}\}$ is linearly ordered by a binary "less than" relation $<_1$ as

$$a_1 <_1 a_2 <_1 \cdots <_1 a_{n+1}$$

and that $u : X \to \mathbf{R}$ preserves $<_1$ by

$$u(a_1) < u(a_2) < \cdots < u(a_{n+1}).$$

(The assumption that we have a linear ordering, rather than a strict weak ordering which would allow ties, is easy to do away with.) Given $<_1$, we consider comparisons between ordered pairs in the set $\mathcal{C} = \{(a_i, a_j) : 1 \leq i < j \leq n+1\}$ and let $<_2$ with symmetric complement $\sim_2$ be a binary relation on $\mathcal{C}$. For $(a,b), (c,d) \in \mathcal{C}, (a, b) <_2 (c, d)$ signifies that the "absolute difference" between a and b is less than the "absolute difference" between c and d. Similarly, $(a, b) \sim_2 (c, d)$ denotes equal "absolute differences".

The model of absolute difference measurement calls for a function $u : X \to \mathbf{R}$ that preserves $<_1$ and also preserves differences in the sense that, for all (a, b) and (c, d) in $\mathcal{C}$,

(10) $$(a, b) <_2 (c, d) \iff |u(a) - u(b)| < |u(c) - u(d)|.$$

Under certain assumptions, including a solvability condition (see Krantz, et al. [1971]), one can prove that a u satisfying (10) is unique up to a positive affine transformation $u \to \alpha u + \beta$ with $\alpha > 0$. It seems natural to consider conditions for finite structures under which u satisfies this type of uniqueness.

Let $d_j = u(a_{j+1}) - u(a_j), j = 1, \ldots, n$. Preservation of $<_1$ is then tantamount to $d_j > 0$ for all j, and, with $i < j$ and $k < l$, preservation of $<_2$ is tantamount to

(11) $$(a_i, a_j) <_2 (a_k, a_l) \iff \sum_{c=i}^{j-1} d_c < \sum_{c=k}^{l-1} d_c.$$

If the summation indices in the inequality in the right side of (11) overlap, then identical d_c can be cancelled to produce disjoint sets of indices. It follows that $(a_i, a_j) <_2 (a_k, a_l)$ corresponds to (1) with $\epsilon_c = 1$ for $c = i, \ldots, j-1, \epsilon_c = -1$ for $c = k, \ldots, l-1$, and $\epsilon_c = 0$ otherwise, where $1 \leq i < j \leq k < l \leq n+1$. Similarly, $(a_i, a_j) \sim_2 (a_k, a_l)$ corresponds to equation (2) with ϵ defined in the same way. Moreover, u is unique to a positive affine transformation if and only if d is unique up to multiplication by a positive constant, i.e., is unique in the sense we have defined.

In addition to the basic difference representation for $\mathcal{C}$, we consider a similar representation in which difference comparisons must have a common end point. This restriction limits $<_2$ and $\sim_2$ to comparisons between (a_i, a_j) and (a_j, a_k) with $1 \leq i < j < k \leq n+1$. Since only three a_i are involved in each comparison, we refer to the restricted difference model as the *ternary model*. In the ternary model, (1) and (2) all have $\epsilon_c = 1$ for $c = i, \ldots, j-1, \epsilon_c = -1$ for $c = j, \ldots, k-1$, and $\epsilon_c = 0$ otherwise.

2.4 Additive Conjoint Measurement. Additive conjoint measurement arises when we compare multidimensional alternatives, and the attribute with respect to which they are being compared, for instance utility, is additive over dimensions. We consider a binary relation $<$, "is less preferred than", on a product set $X = X_1 \times X_2 \times \cdots \times X_m$ with $m \geq 2$. In *additive conjoint measurement*, we seek a mapping $u : X \to \mathbf{R}$ so that $a < b \iff u(a) < u(b)$ holds and so that u is additive, i.e., we seek functions $u_i : X_i \to \mathbf{R}$ so that for all $(a_1, \ldots, a_m), (b_1, \ldots, b_m) \in X$,

$$(12) \qquad (a_1, \ldots, a_m) < (b_1, \ldots, b_m) \iff \sum_{i=1}^{m} u_i(a_i) < \sum_{i=1}^{m} u_i(b_i).$$

Various sets of axioms that are sufficient for the existence of such mappings are described in Fishburn [1970], Krantz, et al. [1971] and Roberts [1979]. Scott [1964] gives necessary and sufficient conditions for such mappings in the finite case. [FR2] begins the study of the finite uniqueness problem for additive conjoint measurement. Many of the theorems giving sufficient conditions for additive conjoint measurement carry with them the result that the u_i are *unique to similar positive affine transformations*, which means that if $(u_1, \ldots, u_m)$ satisfies (12), then so does $(v_1, \ldots, v_m)$ if and only if there are real numbers $\alpha > 0$ and $\beta_1, \ldots, \beta_m$ such that, for all $i \in \{1, \ldots, m\}$,

$$v_i(a_i) = \alpha u_i(a_i) + \beta_i \quad \text{for all} \quad a_i \in X_i.$$

However, if this does not obtain and $(u_1, \ldots, u_m)$ and $(v_1, \ldots, v_m)$ satisfy (12), it must still be true that

$$u_i(a_i) < u_i(b_i) \iff v_i(a_i) < v_i(b_i)$$

for all i and all $a_i, b_i \in X_i$.

Suppose all X_i are finite sets, with $X_i = \{a_{i1}, a_{i2}, \ldots, a_{i,n_i+1}\}$. Suppose (12) holds and assume (with a slight loss of generality) that u_i linearly orders X_i as

$$u_i(a_{i1}) < u_i(a_{i2}) < \cdots < u_i(a_{i,n_i+1}).$$

Let

$$(13) \qquad d_{ij} = u_i(a_{i,j+1}) - u_i(a_{ij})$$

for $j = 1, \ldots, n_i$ and $i = 1, \ldots, m$, and let

(14) $\qquad d = (d_{11}, \ldots, d_{1n_1}; \; d_{21}, \ldots, d_{2n_2}; \ldots; \; d_{m1}, \ldots, d_{mn_m}).$

Note that $d_{ij} > 0$ for all i, j. We refer to subsequence $(d_{i1}, \ldots, d_{in_i})$ as *block i*, and to avoid trivial factors assume that $n_i \geq 1$ for every i.

With $\sim$ the symmetric complement of $<$, $(a_{1j_1}, \ldots, a_{mj_m}) < (a_{1k_1}, \ldots, a_{mk_m})$ yields

$$\sum_{i=1}^{m}[u_i(a_{ij_i}) - u_i(a_{ik_i})] < 0$$

and $(a_{1j_1}, \ldots, a_{mj_m}) \sim (a_{1k_1}, \ldots, a_{mk_m})$ yields

$$\sum_{i=1}^{m}[u_i(a_{ij_i}) - u_i(a_{ik_i})] = 0.$$

In turn, these inequalities and equalities translate into inequalities and equalities

$$\sum_{i=1}^{m}\sum_{j=1}^{n_i} \epsilon_{ij} d_{ij} > 0$$

and

$$\sum_{i=1}^{m}\sum_{j=1}^{n_i} \epsilon_{ij} d_{ij} = 0,$$

where all ϵ_{ij} for block i are either in $\{0, 1\}$ or in $\{0, -1\}$, all nonzero ϵ_{ij} for a block are contiguous, and some coefficients are $+1$ while others are -1. Moreover, the u_i are unique up to similar positive affine transformations if and only if the positive d defined by (13) and (14) is unique up to multiplication by a positive constant, i.e., is unique in the sense we have defined.

2.5 Non–Transitive Applications. To show that our considerations can apply to any asymmetric relation $>$ that is not necessarily transitive and may even have cycles, we mention a specialization of an alternative to expected utility theory based on Fishburn [1982].

In the expected utility problem, we consider the set P_0 of all finite–support probability distributions $p, q, r, \ldots$ on a set X of decision outcomes and a binary relation $>$, "is preferred to", on the set P_0. We seek a mapping $u : X \to \mathbb{R}$ so that for all $p, q \in P_0$,

$$p > q \iff \sum_{X} p(a)u(a) > \sum_{X} q(a)u(a).$$

The existence question for this model is answered by three simple axioms (Jensen [1967], Fishburn [1970]). The uniqueness question is answered by proving that u is unique up to a positive affine transformation $u \to \alpha u + \beta, \alpha > 0$.

In our alternative formulation of expected utility theory, $>$ applies to the family $\mathcal{A}^*$ of all *nonempty* subsets of $\{1, 2, \ldots, n\}$. We can think of $A \in \mathcal{A}^*$ as a lottery on the elements of A that assigns probability $1/|A|$ to each element, i.e., the "even–chance gamble" on A; but we will not incorporate these probabilities in the modeling question. The model calls for a real–valued function ϕ on $\mathcal{A}^* \times \mathcal{A}^*$ such that, for all $A, B \in \mathcal{A}^*$,

$$A > B \iff \phi(A, B) > 0,$$

with ϕ *skew–symmetric* and *biadditive*. The former means that $\phi(B, A) = -\phi(A, B)$. The latter means that if A and B are nonempty and disjoint, then $\phi(A \cup B, C) = \phi(A, C) + \phi(B, C)$ for the first argument, and similarly for the second argument. It follows that with ϕ_{ij} defined as $\phi(\{i\}, \{j\})$,

$$\phi(A, B) = \sum_{i \in A} \sum_{j \in B} \phi_{ij},$$

with $\phi_{ij} = -\phi_{ji}$ by skew–symmetry, and hence $\phi_{ii} = 0$. Thus the representation can be written as

(15) $$A > B \iff \sum_{i \in A} \sum_{j \in B} \phi_{ij} > 0$$

with $\phi_{ij} = -\phi_{ji}$ for all $i, j \in \{1, \ldots, n\}$. Note also that

(16) $$A \sim B \iff \sum_{i \in A} \sum_{j \in B} \phi_{ij} = 0.$$

Because of biadditivity, we hope for a stronger uniqueness result than in the expected utility model; namely, we hope that ϕ_{ij} is unique up to multiplication by a positive constant as in the related model based on P_0 described in Fishburn [1982]. Since there are no restrictions on the signs of the ϕ_{ij} other than those imposed by skew–symmetry, the model allows preference cycles. For example, $\{1\} > \{2\} > \{3\} > \{1\}$ corresponds to $\phi_{12} > 0, \phi_{23} > 0$, and $\phi_{31} > 0$.

Because of skew–symmetry, the representation has $\binom{n}{2}$ independent quantities. We take these as

$$d_{ij} = \phi_{ij} \quad \text{for} \quad 1 \leq i < j \leq n,$$

with the understanding that each instance of ϕ_{ji} for $i < j$ is replaced by $-\phi_{ij}$, or $-d_{ij}$. Then the right sides of (15) and (16) have the form of (1) and (2), with $\epsilon_{ij} \in \{1, 0, -1\}$. Moreover, the ϕ_{ij} are unique up to multiplication by a positive constant if and only if the vector $d = (d_{12}, d_{13}, \ldots, d_{1n}, d_{23}, \ldots, d_{2n}, \ldots, d_{n-1,n})$ is unique in the sense we have defined.

3. Finite Uniqueness by Linearly Independent Equations

Throughout the rest of this paper, we shall focus on real vectors of the form

$$d = (d_1, \ldots, d_n)$$

which arise from measurement mappings in ways like those indicated in the previous section. In other words, d is the solution to a set S_1 of inequalities of the form (1) and a set S_2 of equalities of the form (2). Either by definition (as with probabilities) or by construction (as with positive differences) the d_j will be nonnegative or strictly positive except for the case of Section 2.5. In all cases, $d_j \neq 0$ will be presumed for some j.

We say that a vector d which is the solution to sets S_1, S_2 of inequalities and equalities is *unique* or is a *unique solution* if for every $d' = (d'_1, \ldots, d'_n)$ that satisfies S_1, S_2, there is a number $\lambda > 0$ for which

$$d'_j = \lambda d_j \quad \text{for} \quad j = 1, 2, \ldots, n.$$

In other words, d is unique up to a set of proportionality (similarity) transformations. If d is unique, we shall usually express it in *smallest–integer format*, i.e., as a vector of integers so that no termwise–smaller vector of integers is a solution. Since the coefficients ϵ_{ij} are $1, 0,$ or -1, we can always express a unique solution as a vector with all rational components and hence we can find one in smallest–integer format.

Given appropriate hypotheses, the following theorem ([FO], [FMR]) tells us what is needed for d to be unique. We recall that m linear equations are *linearly independent* if none of them is implied by the other $m - 1$ under the operations of multiplication by real numbers and addition.

THEOREM 1. *Suppose $n \geq 1$ and $d = (d_1, \ldots, d_n)$ is the solution to a set S_1 of inequalities of the form*

$$\sum_{j=1}^{n} \epsilon_{ij} d_j > 0$$

and a set S_2 of equations of the form

$$\sum_{j=1}^{n} \epsilon_{ij} d_j = 0,$$

with all ϵ_{ij} coefficients in $\{1, 0, -1\}$, and suppose that the inequalities in S_1 imply that some d_j has a fixed nonzero sign. Then d is unique (up to multiplication by a positive constant) if and only if S_2 contains a subset of $n - 1$ linearly independent equations.

The theorem effectively avoids the existence question by presuming a d solution so as to focus on uniqueness. We shall say more about existence later. Note that the inequalities do not enter into uniqueness, except to the extent that we suppose that they require some d_j to be nonzero.

Suppose S_1 requires $d_1 > 0$. If $n = 1$, then the conclusion of Theorem 1 is immediate: S_2 is empty and $d = (d_1)$ is unique. If $n = 2$ with $d_1 > 0$, the only possibilities for S_2 are $d_2 = 0, d_1 + d_2 = 0$, and $d_1 - d_2 = 0$, and each of these ensures uniqueness. When $n \geq 3$, the $n - 1$ linearly independent equations for uniqueness from S_2 may force $d_j = 0$ for some values of j, and for each other d_j for $j > 1$ they imply that there is a unique nonzero λ_j such that $d_j = \lambda_j d_1$.

To illustrate the uses of this theorem in the various measurement representations, let us consider first subjective probability measurement. Recall the set of $\sim$ relations (4). Here, the corresponding equations of (6) involving P become

$$d_3 = d_1 + d_2$$
$$d_4 = d_1 + d_3$$
$$d_2 + d_3 = d_1 + d_4.$$

These three equations are linearly independent and hence, by Theorem 1, have a unique solution. In smallest–integer format, the solution is $(1, 2, 3, 4)$. By normalizing, we translate this into the unique solution for P given by (5). In subjective probability measurement, all $\epsilon \in \{1, 0, -1\}^n$ with $\epsilon_j \neq 0$ for some j are candidates for equations in S_2. However, not all sets of $n - 1$ linearly independent equations of the form (2) identify a unique d for subjective probability since an admissible solution must have $d_j \geq 0$ for all j. On the other hand, each nonnegative d solution to a set of $n - 1$ linearly independent equations with $d_j > 0$ for some j does give an instance of unique subjective probability. We normalize d so that $\Sigma d_j = 1$, define $P(A)$ from d by taking $P(A) = \sum_{j \in A} d_j$, and define $>$ on $\mathcal{A}$ by $A > B$ if $P(A) > P(B)$.
Then $(\mathcal{A}, >)$ has a unique subjective probability measure. Conversely, if $(\mathcal{A}, >)$ has a unique subjective probability measure, then Theorem 1 tells us that there must be $A_i \sim B_i$ for $i = 1, \ldots, n - 1$ whose corresponding equations for (2) are linearly independent and have a nonnegative d solution with $d_j > 0$ for some j.

The case of extensive measurement is similar to that for subjective probability measurement except that $d_j > 0$ for all j. Thus $\epsilon \in \{1, 0, -1\}^n$ is a candidate for S_2 so long as $\epsilon_j = 1$ and $\epsilon_k = -1$ for some j and k. Now a set of $n - 1$ linearly independent equations corresponds to a unique instance for positive additive extensive measurement provided that it has a strictly positive d solution.

In the case of difference measurement, recall that $d_j = u(a_{j+1}) - u(a_j) > 0$ for all j. The candidates for S_2 can all be written as $\Sigma \epsilon_c d_c = 0$ with

$$\epsilon_c = 1 \quad \text{for} \quad c = i, \ldots, j - 1$$
$$\epsilon_c = -1 \quad \text{for} \quad c = k, \ldots, l - 1$$
$$\epsilon_c = 0 \quad \text{otherwise,}$$

for some $1 \leq i < j \leq k < l \leq n + 1$. A set of $n - 1$ linearly independent equations of this type that have a positive d solution gives an instance of the u representation in which u is unique up to a positive affine transformation. For instance, consider the equations

$$d_1 = d_2 + d_3$$
$$d_2 = d_3$$
$$d_1 + d_2 = d_4.$$

These three linearly independent equations have, in smallest-integer format, the unique solution (2,1,1,3). The corresponding $\sim_2$ relations from Section 2.3 are the following:

$$(a_1, a_2) \sim (a_2, a_4)$$
$$(a_2, a_3) \sim (a_3, a_4)$$
$$(a_1, a_3) \sim (a_4, a_5),$$

and the corresponding mapping u unique up to a positive affine transformation can be expressed by taking $u(a_1) = 0$ and $u(a_{j+1}) = d_j + u(a_j)$. Thus, we get

$$(u(a_1), u(a_2), u(a_3), u(a_4), u(a_5)) = (0, 2, 3, 4, 7).$$

In additive conjoint measurement, the candidates for S_2 involve ϵ_{ij} so that all ϵ_{ij} in block i are either in $\{0,1\}$ or in $\{0,-1\}$, all nonzero ϵ_{ij} for a block are contiguous, and some coefficients are $+1$ while others are -1. It then follows that $d = (d_{11}, \ldots, d_{1n_1}; d_{21}, \ldots, d_{2n_2}; \ldots; d_{m1}, \ldots, d_{mn_m})$ is unique if and only if there are $\Sigma n_i - 1$ linearly independent equations in S_2. To illustrate, note that (1,2,2,1;1,3) is a unique solution, since it is a solution to the five linearly independent equations

$$d_{11} = d_{21}$$
$$d_{14} = d_{21}$$
$$d_{11} + d_{12} = d_{22}$$
$$d_{13} + d_{14} = d_{22}$$
$$d_{12} + d_{13} = d_{21} + d_{22}.$$

The corresponding $\sim$ relations on $X_1 \times X_2$ and the corresponding mapping u unique up to similar positive affine transformations are left to the reader.

Turning finally to the skew-symmetric, biadditive version of expected utility, we recall that equations for S_2 come directly from $\sim$ comparisons. Thus, we see that ϕ and hence $d = (d_{12}, d_{13}, \ldots, d_{1n}, d_{23}, \ldots, d_{2n}, \ldots, d_{n-1,n})$ is unique if and only if there is a subset of $\binom{n}{2} - 1$ linearly independent equations of the form

$$\sum_{1 \leq i < j \leq n} \epsilon_{ij} d_{ij} = 0,$$

with $\epsilon_{ij} \in \{1, 0, -1\}$. Given $\{2\} > \{1\}$, an example of a unique d for $n = 4$ is

$$(d_{12}, d_{13}, d_{14}, d_{23}, d_{24}, d_{34}) = (-1, -1, 2, 1, -1, 1).$$

Five $\sim$ comparisons that imply this d are

$$\{3\} \sim \{1, 2\}$$
$$\{3\} \sim \{2, 4\}$$
$$\{4\} \sim \{2, 3\}$$
$$\{2\} \sim \{1, 4\}$$
$$\{1\} \sim \{2, 3, 4\}.$$

4. Very Restricted Extensive Measurement

To illustrate aspects of finite measurement, we consider a somewhat artificially restricted case of positive additive extensive measurement in which comparisons are limited to the pairs in

$$\mathfrak{B}_0 = \{\{A, B\} : A, B \subseteq \{1, \ldots, n\}, A \cap B = \emptyset, |A| \leq 2, |B| \leq 2, |A \cup B| \leq 3\}.$$

Thus comparisons between nonempty aggregates can be made only between singletons and between a singleton $\{i\}$ and doubleton $\{j, k\}$ that does not contain i.

4.1 Axioms. With $>$ and its symmetric complement $\sim$ confined to the pairs in $\mathfrak{B}_0$, we investigate the four kinds of fundamental problems mentioned in Section 1, beginning with the **axiomatization problem**. We first state axioms for $>$ and $\sim$ that are necessary and sufficient for positive $u(j)$ that satisfy (9) or, equivalently, for the existence of positive d_j such that, for all $\{A, B\} \in \mathfrak{B}_0$,

(17) $$A > B \iff \sum_{j \in A} d_j > \sum_{j \in B} d_j.$$

Only two axioms are needed. The first is a necessary positivity assumption:

B1. $\{j\} > \emptyset$ *for each j in* $\{1, \ldots, n\}$.

The second is an additivity condition. Let $(A_1, \ldots, A_m) =_0 (B_1, \ldots, B_m)$ mean that $\{A_i, B_i\} \in \mathfrak{B}_0$ for $i = 1, \ldots, m$ and, for each $j \in \{1, \ldots, n\}$,

$$\left| \{i : 1 \leq i \leq m \text{ and } j \in A_i\} \right| = \left| \{i : 1 \leq i \leq m \text{ and } j \in B_i\} \right|.$$

In other words $(A_1, \ldots, A_m) =_0 (B_1, \ldots, B_m)$ if each $\{A_i, B_i\}$ is a comparison pair and each object appears in the same number of A_i as in B_i. The additivity condition is

B2. *For all $m \geq 2$ and all $A_i, B_i \in \{C : C \subseteq \{1, \ldots, n\}$ and $|C| \leq 2\}$, if $(A_1, \ldots, A_m) =_0 (B_1, \ldots, B_m)$ and either $A_i > B_i$ or $A_i \sim B_i$ for every $i < m$, then not $(A_m > B_m)$.*

If B2 fails (with ... and $A_m > B_m$) then its balance of objects between the A_i and B_i by $=_0$ produces a $0 > 0$ contradiction from the desired representation. Hence B2 is necessary for the representation.

The sufficiency of B1 and B2 for the model (17) is established as follows. Given $>$ and $\sim$ that satisfy B2, we use (17) to generate a set of linear inequalities for all $>$ statements ($\Sigma \epsilon_{ij} d_j > 0$ from $A > B$) and a set of linear equations for all $\sim$ statements ($\Sigma \epsilon_{ij} d_j = 0$ from $A \sim B$). A standard theorem for the existence of a solution to a finite system of linear inequalities/equalities (see, for example,

Fishburn [1970], Theorem 4.2) then shows that there is a d solution for our sets if and only if B2 holds. Finally, B1 is used to conclude that $d_j > 0$ for all j.

Given the existence of a positive u solution for (9), equivalently a positive d solution for (17), Theorem 1 and the restrictions of $\mathfrak{B}_0$ say that d is unique if and only if there are $n-1$ linearly independent equations obtained from $\sim$ of the form

$$d_i = d_j \ (i \neq j) \quad \text{and} \quad d_i = d_j + d_k \ (i \neq j \neq k \neq i).$$

The existence of such a set can be proposed as a third axiom to complete our necessary and sufficient conditions for the existence of a mapping u that satisfies (9) and is unique up to proportionality transformations.

However, a simpler uniqueness condition can be stated once we understand the nature of the set of all unique d solutions. Assume $n \geq 2$ and for convenience fix $\min d_j$ at 1. Then it is easily verified that d is a unique solution for (17) if and only if there is a permutation $x_1, \ldots, x_n$ of $d_1, \ldots, d_n$ such that

$$x_1 \leq x_2 \leq \cdots \leq x_n,$$
$$x_1 = \cdots = x_t = 1 \quad \text{for some} \quad 2 \leq t \leq n,$$

and, for every i such that $t < i \leq n$,

$$x_i = x_j + x_k \quad \text{for some distinct} \quad j \quad \text{and} \quad k \quad \text{less than } i.$$

This leads to the uniqueness axiom

B3. *For all* $i, i' \in \{1, \ldots, n\}$, *if* $\{i\} > \{i'\}$ *then* $\{i\} \sim \{j, k\}$ *for some distinct* $j, k \in \{1, \ldots, n\} \setminus \{i\}$,

and to the following representation–uniqueness theorem.

THEOREM 2. *The relation $>$ on $\mathfrak{B}_0$ satisfies B1, B2 and B3 if and only if there is a unique (up to multiplication by a positive constant) positive d that satisfies (17) for all $\{A, B\} \in \mathfrak{B}_0$.*

Let E_n be the set of all n-term nondecreasing sequences $x_1, x_2, \ldots, x_n$ of positive integers that satisfy

$$x_1 = x_2 = 1,$$
$$x_i = x_j + x_k \quad \text{for some} \quad j \neq k \quad \text{whenever} \quad x_i > 1.$$

We refer to sequences in $E_2 \cup E_3 \cup \ldots$ as *elementary sequences*. As just noted, the set E_n of n-term elementary sequences represents the set of unique solutions for the $\mathfrak{B}_0$ model for n in the sense that, up to proportionality, d is a unique solution if and only if its components form a permutation of a sequence in E_n. Similar characterizations by sequences of positive integers in smallest–integer format will also be used in ensuing sections for solution sets encountered there.

4.2 Numbers of Solutions. The E_n characterization gives us access to the **counting problem** since it provides a convenient base for counting the number of different unique solutions. Table 1 gives both $|E_n|$ for $n \geq 2$ and the number of different solutions obtained from permutations of the components of each member of E_n. The table shows that even in a very restricted context there can be a large

TABLE 1

Counts for restricted extensive case (E_n)
and all sub–Fibonacci sequences (F_n)

n	$|F_n|$	$|E_n|$	$|E_n$ with permutations $|$
2	1	1	1
3	2	2	4
4	4	4	23
5	10	10	256
6	31	31	4647
7	127	120	128,262
8	711	578	5,128,503

number of unique realizations of the representation for modest values of n. We do not have a workable recurrence for $|E_n|$ to extend the table's values to larger n, nor do we have an order–of–magnitude approximation of $|E_n|$ for large n, but such things could be considered in this and other contexts.

4.3 Extreme Values. An example of an **extremization problem** for elementary sequences is to determine the largest value of x_n for all sequences in E_n. The answer is obvious:

$$\max\{x_n : (x_1,\ldots,x_n) \in E_n\} = f_n,$$

where f_n is the nth Fibonacci number, i.e., the nth term in the Fibonacci sequence $f_1, f_2, f_3, \ldots$ defined by $f_1 = f_2 = 1$ and $f_j = f_{j-1} + f_{j-2}$ for all $j \geq 3$. There is a unique member of E_n that yields this maximum, namely $(f_1, f_2, \ldots, f_n)$, and this member of E_n also clearly maximizes the sum of the x_i. By the well–known Fibonacci identity

$$f_1 + f_2 + \cdots + f_n = f_{n+2} - 1, \quad n \geq 1,$$

the maximum sum is $f_{n+2} - 1$.

We refer to a nondecreasing integer sequence $x_1, x_2, \ldots, x_n$ as a *sub–Fibonacci sequence* if $x_1 = x_2 = 1$ and $x_j \leq x_{j-1} + x_{j-2}$ for $j = 3, \ldots, n$ and let F_n denote the set of all n–term sub–Fibonacci sequences. It is easy to see that $E_n \subseteq F_n$. Table 1 notes that $n = 7$ is the smallest n at which $|E_n| < |F_n|$. The smallest sub–Fibonacci sequence not in E_7 is 1122447. After $n = 6$, $|E_n|/|F_n|$ appears to decrease rapidly and probably approaches 0, but we leave this open.

An extremization problem for E_n that involves number theory is to determine

$$e_n^* = \max\left\{\left|\{y : (x_1, \ldots, x_n, y) \in E_{n+1}\}\right| : (x_1, \ldots, x_n) \in E_n\right\},$$

the most one–term extensions of a sequence from E_n that are in E_{n+1}. Calculations show that e_2^* through e_7^* are 2,2,3,4,5 and 7, respectively, and the maximum of 7 extensions in E_8 arises from each of the E_7 sequences 1123567, 1123578 and 1123589, and from no others. Examples show that $e_8^* \geq 9$, $e_9^* \geq 12$, $e_{10}^* \geq 15$, $e_{11}^* \geq 18$ and $e_{12}^* \geq 22$. We also have an idea of the large-n behavior of e_n^*. In the following theorem, and later, $o(1)$ denotes a function of n that approaches 0 as $n \to \infty$.

THEOREM 3. $e_n^* \geq n^2(1 + o(1))/4$.

Proof. Consider the E_n sequence

$$f_1, f_2, \ldots, f_{m-1}, f_m, f_m + 1, f_m + 2, \ldots, f_m + n - m.$$

This is in E_n because $f_m + 1 = f_m + f_1$ and $f_m + k + 1 = (f_m + k) + f_1$. For large n let $m = \lfloor n/2 \rfloor$, and let $k(n)$ denote the smallest value of k for which $f_k + (f_m + n - m) < f_{k+1} + f_m$, i.e., for which

$$f_{k-1} > n - m = \lceil n/2 \rceil.$$

Then the intervals $[f_{k(n)} + f_m, f_{k(n)} + f_m + n - m]$, $[f_{k(n)+1} + f_m, f_{k(n)+1} + f_m + n - m]$, $\ldots$, $[f_{m-1} + f_m, f_{m-1} + f_m + n - m]$ are disjoint, and every integer in these intervals gives a one–term extension to E_{n+1} of our original sequence. Therefore

$$e_n^* \geq (n - m + 1)(m - k(n)) \geq \frac{n}{2}\left(\frac{n}{2} - 1 - k(n)\right).$$

Since $k(n)/n \to 0$ as $n \to \infty$, $e_n^* \geq (n/2)^2(1 + o(1))$. □

The reader can readily prove also that $e_n^* \leq n^2/2$, for $n \geq 2$, and we conjecture that $e_n^* \leq n^2/4$ for all $n \geq 3$.

4.4 Inclusion of Subsequences. We turn finally to an **inclusion problem** involving number theory: Determine the smallest n such that a given nondecreasing sequence $k_1, k_2, \ldots, k_m$ of positive integers is a subsequence of some elementary sequence in E_n. It is clear that such an n always exists. Let

$$e(k_1, \ldots, k_m) = \min\{n : \text{there exists } (x_1, \ldots, x_n) \in E_n \text{ such that}$$
$$(k_1, \ldots, k_m) \text{ is a subsequence of } (x_1, \ldots, x_n)\}.$$

We comment only on $e(k)$ for a single positive integer $k \geq 2$.

Clearly, a shortest elementary sequence that contains k must end in k. If $k = f_n$ for the Fibonacci number f_n, then it follows from earlier remarks that $e(k) = n$.

Similarly, if $k = f_i + f_n$ for $i < n$, then $e(k) = n+1$. It is not generally true however that $e(k) = n+1$ when $f_n < k < f_{n+1}$. In particular, with $13 = f_7 < 20 < f_8 = 21$, calculation shows that no 8-term elementary sequence ends in 20, and we find that $e(20) = 9$. Since $e(21) = e(f_8) = 8$, the sequence $e(2), e(3), \ldots$ is *not* monotonic nondecreasing.

We conclude this section by proving an upper bound on $e(k)$ that limits the discrepancy between $e(k)$ and n when k is near to f_n.

THEOREM 4. *If $f_n < k < f_{n+1}$ then $n + 1 \leq e(k) \leq n - 1 + \lfloor n/2 \rfloor$.*

Proof. Clearly, $x_n \leq f_n$, so if $k > f_n$ then $e(k) \geq n + 1$. To establish the upper bound, we use Zeckendorf's theorem (see e.g., Brown [1964] or Hoggatt [1969]) which asserts that every positive integer k equals a sum of distinct numbers in $f_2, f_3, f_4, \ldots$ using no two consecutive f_i. Given $f_n < k < f_{n+1}$, it follows that

$$k = \sum_{i \in I} f_i \quad \text{with} \quad |I| \leq \lfloor n/2 \rfloor.$$

Suppose $I = \{a_1 < a_2 < \cdots < a_{|I|}\}$. Form an elementary sequence composed of $f_1, f_2, \ldots, f_n$ and $(f_{a_1} + f_{a_2}), [(f_{a_1} + f_{a_2}) + f_{a_3}], \ldots, [(f_{a_1} + \cdots + f_{a_{|I|-1}}) + f_{a_{|I|}}] = k$, with the later $|I|-1$ numbers inserted in their natural order into the first n Fibonacci numbers. By construction, this sequence is in E_t with $t \leq n + \lfloor n/2 \rfloor - 1$, and it ends in k. □

The upper bound in Theorem 4 is not the best possible for integers in (f_n, f_{n+1}) when $n \geq 8$ since, for example, $8 - 1 + \lfloor 8/2 \rfloor = 11$ but $\max e(k)$ for $21 < k < 34$ occurs at $k \in \{32, 33\}$, where $e(k) = 10$. We leave it as a challenge to determine the smallest k, if any, at which $f_n < k < f_{n+1}$ and $e(k) \geq n + 3$.

5. SUBJECTIVE PROBABILITY AND EXTENSIVE MEASUREMENT

Throughout this section we consider the opposite extreme from the very restricted context of the preceding section by admitting qualitative comparisons between all A and B in the set $\mathcal{A}$ of all subsets of $\{1, \ldots, n\}$. Our results here apply both to subjective probability and to unrestricted additive positive extensive measurement (where no real loss in generality obtains if $A \cap B = \emptyset$ is presumed in comparisons). However, because these results were presented in [FO] and [FR1] in terms of subjective probability, we adhere to this context in our interpretations.

5.1 Axioms. We begin with remarks on the **axiomatization problem** of finding axioms for the existence and uniqueness of a probability measure P on $\mathcal{A}$ that satisfies the representation (3), i.e.,

$$A > B \iff P(A) > P(B), \quad \text{for all} \quad A, B \in \mathcal{A},$$

to further illustrate the axiomatic approach to unique measurement with finite sets. While similar remarks could be made in our ensuing discussions of difference measurement and additive conjoint measurement, we shall not discuss axiomatics there in order to focus on aspects of unique solutions. A complete account of axioms for existence in those contexts is given in Fishburn [1970].

Some years ago de Finetti [1931] set forth four axioms as transparent necessary conditions for the existence of a probability measure P on $\mathcal{A}$ that satisfies the representation of the preceding paragraph. With $\sim$ the symmetric complement of $>$, and $A \gtrsim B$ if $A > B$ or $A \sim B$, de Finetti's axioms can be written as follows. For all $A, B, C \in \mathcal{A}$:

P1. $>$ on $\mathcal{A}$ is asymmetric and negatively transitive;

P2. $\{1, \ldots, n\} > \emptyset$;

P3. $A \gtrsim \emptyset$;

P4. $(A \cup B) \cap C = \emptyset \Rightarrow [A \gtrsim B \iff A \cup C \gtrsim B \cup C]$.

The question of whether these axioms were also sufficient for the existence of a representing P was first answered by Kraft, Pratt and Seidenberg [1959], who showed that they were insufficient when $n \geq 5$. The problem is that the independence or additivity axiom P4 does not go far enough. To enrich it, but retain necessity, let $(A_1, \ldots, A_m) =_0 (B_1, \ldots, B_m)$ have the meaning given after B1 in Section 4 with the amendment that $A_i, B_i \in \mathcal{A}$ for all i. The strong independence axiom proposed by Kraft, Pratt and Seidenberg is tantamount to

P4*. For all $m \geq 2$, if $(A_1, \ldots, A_m) =_0 (B_1, \ldots, B_m)$ and $A_j \gtrsim B_j$ for each $j < m$, then $B_m \gtrsim A_m$.

Their main theorem says that $(\mathcal{A}, >)$ has a representing probability measure P if, and only if, P1, P2, P3 and P4* hold for $>$ on $\mathcal{A}$.

Within this setting, we consider three further conditions, each of which implies that the representing P is unique. The differences among these conditions lie both in their intuitive directness and in the variety of unique P that they characterize.

UP1. *For all* $i, j \in \{1, \ldots, n\}$, *if* $\{i\} > \{j\}$, *then* $\{i\} \sim \{j\} \cup A$ *for some* $A \in \mathcal{A}$.

UP2. *For all* $i, j \in \{1, \ldots, n\}$, *if* $\{i\} > \{j\} > \emptyset$, *then* $\{i\} \sim A$ *for some* $A \in \mathcal{A}$ *for which* $\{i\} > \{k\}$ *for all* $k \in A$.

UP3. *There are* $A_1, \ldots, A_{n-1}, B_1, \ldots, B_{n-1} \in \mathcal{A}$ *such that* $A_j \cap B_j = \emptyset$ *and* $A_j \sim B_j$ *for* $j = 1, \ldots, n-1$, *and such that the equations*

$$\sum_{\{i:\, i \in A_j\}} r_i = \sum_{\{i:\, i \in B_j\}} r_i$$

are linearly independent.

The last of these is obviously motivated by Theorem 1, and although its intuitive appeal is minimal it does serve to characterize all P that are unique for some $(\mathcal{A}, >)$.

The other conditions are much more direct but characterize sets of unique P that are much more restricted than the general set for UP3.

UP1 is tantamount (given P1–P4) to a uniqueness condition proposed by Van Lier [VL], UP2 is from [FR1], and UP3 was first stated in Fishburn [1986]. The following composite theorem summarizes results from Kraft, Pratt and Seidenberg [1959], [VL] and [FR1].

THEOREM 5. *There exists a unique probability measure P on $\mathcal{A}$ that satisfies $A > B \iff P(A) > P(B)$, for all $A, B \in \mathcal{A}$, if the axioms in any one of T1, T2 and T3 hold:*

$$T1 = \{P1\text{-}P4, UP1\}$$
$$T2 = \{P1\text{-}P4, UP2\}$$
$$T3 = \{P1, P2, P3, P4^*, UP3\}.$$

Moreover, the axioms in T3 are all necessary for unique existence, while the axioms in $\{P1\text{-}P4, UP3\}$ do not imply the existence of a representing P when $n \geq 5$. Finally, T1 $\Rightarrow$ T2, and T2 $\not\Rightarrow$ T1 when $n \geq 5$; T2 $\Rightarrow$ T3, and T3 $\not\Rightarrow$ T2 whenever $n \geq 4$.

The first part of the theorem shows that de Finetti's P4 can be used in place of the strong P4* when UP1 or UP2 is used. The next part shows that P4* in T3 cannot be replaced by the weaker P4 when $n \geq 5$. And the concluding part shows that some of the unique P allowed by UP3 do not satisfy UP2 when $n \geq 4$, and that some of the unique P allowed by UP2 do not satisfy UP1 when $n \geq 5$. Of particular relevance for ensuing discussion is the fact that, when $\{i\} > \{j\} > \emptyset$, UP1 requires $\{i\}$ to be equally as probable as some event that contains j, while UP2 only requires $\{i\}$ to be equally as probable as an event that contains only points that are less probable than i.

5.2 Characterization by Integer Sequences. To describe the types of unique P measures that arise under UP1, UP2 and UP3, we revert to representation by integer sequences d. For simplicity, we consider only unique solutions in which all $p_i = P(\{i\})$ are positive and write the corresponding d in nondecreasing smallest–integer format. For example, (1,2,3,4,6), or 12346, represents a unique P for $n = 5$ for which some permutation of the p_i is 1/16, 2/16, 3/16, 4/16, 6/16, and 22233 represents a unique P for $n = 5$ for which some permutation of the p_i is 2/12, 2/12, 2/12, 3/12, 3/12.

Assume $n \geq 2$ henceforth, and let V_n, R_n and P_n denote the sets of n-term nondecreasing positive integer sequences (in smallest–integer format) that correspond to unique probability measures P with $p_i > 0$ for all i that arise under T1 (UP1), T2 (UP2) and T3 (UP3), respectively. Thus, for instance, $x_1, \ldots, x_n$ is in V_n if and only if it is a nondecreasing positive smallest–integer format sequence and there is some $>$ on $\mathcal{A}$ that satisfies P1–P4 and UP1 and whose uniquely representing P is such that $P(\{1\}), \ldots, P(\{n\})$ gives a permutation of $x_1/\sum x_i, \ldots, x_n/\sum x_i$. The following corollary is a direct consequence of the latter part of Theorem 5.

COROLLARY 1. $V_n \subseteq R_n \subseteq P_n$ for all n; $V_n \subset R_n$ if $n \geq 5$; $R_n \subset P_n$ if $n \geq 4$.

We examine R_n, V_n and P_n in that order in the next few subsections.

5.3 Regular Sequences. We refer to a nondecreasing sequence $x_1, \ldots, x_n$ of positive integers as a *regular sequence* if

$$x_1 = x_2 = 1$$

and

$$x_j \leq \sum_{i=1}^{j-1} x_i \quad \text{for} \quad j = 3, \ldots, n.$$

It can be shown [FO] that if $x_1, x_2, \ldots, x_n$ is a regular sequence, then every positive integer from 1 to $\sum x_i$ is a partial sum of the x_i. The following result says that the unique solutions that arise under UP2 are precisely characterized by regular sequences.

THEOREM 6 ([FO]). R_n *is the set of all n-term regular sequences.*

Thus $R_2 = \{11\}, R_3 = \{111, 112\}$, and $R_4 = \{1111, 1112, 1113, 1122, 1123, 1124\}$.

Turning to extremization concerns, we note that the sequence in R_n with the largest term and sum is

$$1, 1, 2, 4, \ldots, 2^{n-3}, 2^{n-2},$$

with $\sum x_i = 2^{n-1}$. If this sequence is extended by one term, x_{n+1} can be any integer from 2^{n-2} to 2^{n-1}.

With respect to counting, the characterization of R_n in Theorem 6 leads by direct calculation to values for $|R_2|, |R_3|, \ldots$ of 1, 2, 6, 27, 192, 2280, 47097, ... The following theorem gives an order-of-magnitude value of $|R_n|$ for all n.

THEOREM 7 ([FO]). $|R_n| = 2^{n^2(1+o(1))/2}$.

It is observed in [FMR] that the number of n-term sequences that are permutations of sequences in R_n is also $2^{n^2(1+o(1))/2}$.

5.4 Van Lier Sequences. We define a regular sequence $1, 1, x_3, \ldots, x_n$ as a *Van Lier sequence* if, whenever $j < k \leq n$, there is a set $A \subseteq \{1, \ldots, n\}$ such that

$$j \notin A \quad \text{and} \quad x_k - x_j = \sum_{i \in A} x_i.$$

Examples are 1124 and 112358 ($= f_1, f_2, \ldots, f_6$). The first of these is Van Lier because $4 - 2 = 1 + 1, 4 - 1 = 1 + 2$ and $2 - 1 = 1$. The second is Van Lier because [FRM] every initial subsequence of the Fibonacci sequence is a Van Lier sequence. The smallest regular sequence that is not Van Lier is 11245 (because 5–2 is not a sum of terms from (1,1,4,5)), and this is the only sequence in R_5 that is not Van Lier.

As might be anticipated, our interest in Van Lier sequences comes from their connection to UP1.

THEOREM 8 ([FR1]). *V_n is the set of all n-term Van Lier sequences.*

The first few $|V_n|$ for $n = 2, 3, \ldots$ are 1, 2, 6, 26, 164, 1529, ... We do not presently have an order–of–magnitude value of $|V_n|$ similar to that for $|R_n|$ in Theorem 7. However, we do have the following conjecture.

Conjecture 1. $|V_n|/|R_n| \to 0$ *as* $n \to \infty$.

While this conjecture remains unsettled, [FRM] shows that for large n, $|V_n|/|R_n| \leq \lambda$ for some $\lambda < 1$ (e.g., $\lambda = 0.75$).

Our next result makes connections among V_n, R_n and F_n (sub–Fibonacci n-term sequences). We say that a nondecreasing sequence $x_1, \ldots, x_n$ of positive integers has a *gap* at x_j for $3 \leq j \leq n - 1$ if $(x_1 + x_2 + \cdots + x_{j-1}) + 1 < x_{j+1}$, that is, if there is an integer t such that

$$\sum_{i=1}^{j-1} x_i < t < x_{j+1}.$$

If the sequence does not have a gap at x_j for $j = 3, \ldots, n - 1$, then it is *without gaps*.

THEOREM 9 ([FRM]). *Every regular sequence without gaps is a Van Lier sequence; every sub–Fibonacci sequence is Van Lier.*

The second part of the theorem ($F_n \subseteq V_n$) follows from the first part and the elementary observation that every sub–Fibonacci sequence is regular without gaps. The first instance of a Van Lier sequence that is not sub–Fibonacci is 1124. With the latter part of Theorem 9 at hand, we now have $E_n \subseteq F_n \subseteq V_n \subseteq R_n$, where E_n is the set of elementary sequences from Section 4.

Our next theorem involves one–term extensions. Given $x_1, \ldots, x_n$ in V_n, we refer to $x_1, \ldots, x_n, y$ as a *regular extension* if $x_n \leq y$ (an integer) $\leq \sum x_i$, and call it a *Van Lier extension* if, in addition, it is in V_{n+1}.

THEOREM 10 ([FRM]). *Suppose $(x_1, \ldots, x_n) \in V_n$. Then every one–term regular extension of $x_1, \ldots, x_n$ is a Van Lier extension if and only if $x_1, \ldots, x_n$ is without gaps. If $x_1, \ldots, x_n$ has a gap at x_j and*

$$y = x_j + t + \sum_{i=j+2}^{n} x_i \quad \text{with} \quad \sum_{i=1}^{j-1} x_i < t < x_{j+1},$$

then $x_1, \ldots, x_n, y$ is not a Van Lier extensions. Moreover, if $x_1, \ldots, x_n$ has gaps at x_j and x_k with $j \neq k$, then the sets of y integers defined in this way for the two gaps are disjoint.

It is also true that a regular extension y that differs from those defined in Theorem 10 may give a regular extension that is not Van Lier. For example, (1, 1, 2, 3, 6, 9, 13, 23) is in V_8. It has two gaps, and four of its $36 (= 1 + 1 + \cdots + 13 + 1)$

regular extensions are not Van Lier extensions. The y values for the latter are 27, 37, 50 and 53, but only 50 and 53 are values defined as in Theorem 10.

Similar to the definition of e_n^* in Section 4, let

$$v_n^* = \max\left\{|\{y : (x_1,\ldots,x_n,y) \in V_{n+1}\}| : (x_1,\ldots,x_n) \in V_n\right\},$$

the maximum number of one–term Van Lier extensions of a Van Lier sequence in V_n. It is easily checked that $v_n^* = f_{n+1}$ for the first few n. We know more.

THEOREM 11 ([FRM]). *Every one–term regular extension of* $f_1, f_2, \ldots, f_{n-1}, x_n$ *with* $f_{n-1} \leq x_n \leq f_n$ *is in* V_{n+1}. *Therefore*

$$v_n^* \geq f_1 + \cdots + f_{n-1} + 1 = f_{n+1}.$$

Investigation of v_n^* for small n leads to the conjectures that the bound in Theorem 11 is exact, and that the number of one–term Van Lier extensions of $(x_1,\ldots,x_n) \in V_n$ is precisely f_{n+1} if and only if $x_i = f_i$ for all $i < n$ and $f_{n-1} \leq x_n \leq f_n$. However, it turns out that the exactness of the bound holds only for $n \leq 7$. At $n = 8$ the Van Lier sequence 1, 1, 2, 4, 6, 10, 13, 16 has $35 = f_9 + 1$ one–term Van Lier extensions. The latter conjecture holds for $n \leq 6$, but fails at $n = 7$ since the Van Lier sequence 1, 1, 2, 4, 6, 10, 13 has $21 = f_8$ one–term Van Lier extensions but violates the properties indicated.

5.5 Unique Probability Sequences. Conjecture 1 is suggested by the next theorem.

THEOREM 12 ([FO]). $|R_n|/|P_n| \to 0$ *as* $n \to \infty$.

This shows that there are many more nondecreasing integer sequences in the solution set P_n for UP3 than in the regular set R_n for large n. The proof of Theorem 12 follows from the observation that 1234 is in $P_4 \setminus R_4$ (its three equations are $p_3 = p_1 + p_2$, $p_4 = p_1 + p_3$ and $p_1 + p_4 = p_2 + p_3$) and then that, as m gets large, the number of regular m-term extensions of 1234 overwhelms the number of regular m-term extensions for each of the six sequences in R_4.

We refer to sequences $P_n \setminus R_n$ as *irregular sequences*. There are two 4–term irregular sequences, 1223 and 1234, and 75 irregular sequences for $n = 5$. Table 2 summarizes counts for small n of the sequence types considered thus far. The blank spaces in the table have not been determined at present.

A sample of irregular sequences in $P_5 \setminus R_5$ includes

$$(1,1,3,3,5), \quad (2,2,2,3,3), \quad (4,5,6,7,8), \quad (1,3,6,8,10).$$

We invite readers to specify four linearly independent equations (in p_i or d_i or x_i) that correspond to each solution. The middle two sequences show that x_1 must exceed 1 in the smallest–integer format for some cases.

TABLE 2

Counts for restricted extensive (E_n), sub–Fibonacci (F_n),
Van Lier (V_n), Regular (R_n) and Irregular ($P_n \setminus R_n$)
Sequences with n terms

n	$|E_n|$	$|F_n|$	$|V_n|$	$|R_n|$	$|P_n \setminus R_n|$	$|P_n|$
2	1	1	1	1	0	1
3	2	2	2	2	0	2
4	4	4	6	6	2	8
5	10	10	26	27	75	102
6	31	31	164	192		
7	120	127	1529	2280		
8	578	711	21,439	47,097		

The last of the preceding solutions, (1, 3, 6, 8, 10), shows that the smallest $p_i > 0$ in an irregular unique solution, which is $p_1 = x_1/\sum x_i = 1/28$ for this case, may be substantially smaller than the smallest $p_i > 0$ for R_n, which is $p_1 = 1/2^{n-1}$ for the regular sequence $1, 1, 2, 4, \ldots, 2^{n-2}$. More specifically, we have

THEOREM 13 ([FO]). $\min\{x_1/\sum x_i : (x_1, \ldots, x_n) \in P_n\}/2^{n-1} \to 0$.

The minimum value of $x_1/\sum x_i$ for P_5 is 1/28, for sequence (1,3,6,8,10). The smallest $x_1/\sum x_i$ for P_6 and P_7 presently known are 1/64 and 1/192, respectively. These arise from $(1, 4, 7, 10, 20, 22) \in P_6$ and $(1, 5, 14, 18, 36, 44, 74) \in P_7$.

Another extremization problem for irregular sequences examined in [FO] is to determine the largest ratio x_2/x_1 for sequences in P_n. Since $x_2/x_1 = 1$ for all regular sequences, it might be supposed that we can do somewhat better than this with irregulars.

THEOREM 14 ([FO]). $\max_{P_n}(x_2/x_1) \geq 2^{n-4} + 2^{n-6} + 2^{(n-4)/2}$ for $n \geq 5$.

Sequence $(1,4,5,6,8) \in P_5$ verifies the theorem at $n = 5$.

A third extremization problem, which is considered in [FMR], is to determine the maximum value of x_n/x_1 for sequences in P_n. We know from [FO] that $\max x_4/x_1 =$

4, $\max x_5/x_1 = 10$, $\max x_6/x_1 \geq 28$ and $\max x_7/x_1 \geq 74$ for sequences in P_4, P_5, P_6 and P_7, respectively. In general, the max ratio grows more rapidly than the max ratio for regular sequences, which is $2^{n-2}/1$.

THEOREM 15 ([FMR]). $\max_{P_n}(x_n/x_1) \geq (2.12)^{n-2}$ for all $n \geq 5$.

Finally, we remark that not much is known about $|P_n|$ at present besides the upper bound
$$|P_n| \leq 3^{n^2(1+o(1))}.$$
See Section 5 of [FO] for a proof and further discussion.

6. DIFFERENCE MEASUREMENT

Recall from Section 2 that for absolute difference measurement with base set $X = \{a_1, a_2, \ldots, a_{n+1}\}$ we assume that $u(a_1) < \cdots < u(a_{n+1})$ for a real-valued function u on X and define $d = (d_1, \ldots, d_n)$ by $d_j = u(a_{j+1}) - u(a_j)$ for $j = 1, \ldots, n$. With $d_j > 0$ for all j, we write unique d solutions in the smallest-integer format for $d_1, d_2, \ldots, d_n$. Unlike the preceding section, we do *not* rearrange the d_j to be nondecreasing because permutations of unique d's in the difference context are not necessarily unique solutions.

6.1 Two Types of Sequences. We consider two types of unique d solutions according to restrictions on linear equations in the d_j that are admissible for S_2 in Theorem 1. The more restrictive type, called *type A*, applies to the ternary model described in Section 2.3. The other, *type B*, applies to general nonoverlapping difference comparisons. Thus, d is a type A sequence if and only if it is a solution to $n-1$ linearly independent equations of the form

$$d_i + d_{i+1} + \cdots + d_{j-1} = d_j + d_{j+1} + \cdots + d_{k-1} \quad (1 \leq i < j < k \leq n+1);$$

and d is a type B sequence if and only if it is a solution to $n-1$ linearly independent equations of the form

$$d_i + \cdots + d_{j-1} = d_k + \cdots + d_{l-1} \quad (1 \leq i < j \leq k < l \leq n+1).$$

Every type A sequence is a type B sequence, but not conversely. For example, as noted in Section 3, 2113 is a type B sequence. But it is not type A because the third independent equation $d_1 + d_2 = d_4$, used to obtain 2113, cannot be written in the type A form. On the other hand, some permutations of 2113, including 1123 and 3211, are type A. Moreover, 1213 is of neither type since it can only be obtained if $d_2 = d_1 + d_3$ is used. However, $d_2 = d_1 + d_3$ is not a type B equation. The generalization of type B that allows any such equation between sums of d_j is covered by the preceding section.

Let A_n and B_n be the sets of n-term type A sequences and type B sequences, respectively, in smallest-integer format. Enumeration [FMR] for the first few n gives

$(|A_2|, |B_2|) = (1, 1),$ $\quad (|A_3|, |B_3|) = (3, 3),$ $\quad (|A_4|, |B_4|) = (16, 20),$
$(|A_5|, |B_5|) = (148, 266),$ $\quad (|A_6|, |B_6|) \geq (1937, 5295).$

Order-of-magnitude upper bounds are specified in the next theorem.

THEOREM 16 ([FMR]). $|A_n| \leq n^{2n(1+o(1))}$ and $|B_n| \leq n^{3n(1+o(1))}$.

We do not have interesting lower bounds for these sets. The upper bound on $|B_n|$ in conjunction with Theorem 7 shows that $|B_n|/|R_n| \to 0$ as n gets large, where R_n is the set of n-term regular sequences as defined in the preceding section.

6.2 Extremization Problems. Before we look at suitably defined regular sequences for types A and B, we consider general extremization problems for A_n and B_n that correspond to the problems for P_n that gave rise to Theorems 13–15.

Recall first of all that $\min_{R_n}(x_1/\sum x_i) = 1/2^{n-1}$ and, by Theorem 13, that $\min_{P_n}(x_1/\sum x_i)/2^{n-1} \to 0$. Since $1124\ldots 2^{n-2}$ is both type A and type B in the present context, we know that $\min_{A_n \cup B_n}(d_i/\sum d_j) \leq 1/2^{n-1}$. However, we have been unable to do better than this under the equations allowed for types A and B and conjecture that $1/2^{n-1}$ is best-possible.

Conjecture 2A. $\min_{A_n}(\min d_i/\sum d_j) = 1/2^{n-1}$.

Conjecture 2B. $\min_{B_n}(\min d_i/\sum d_j) = 1/2^{n-1}$.

We have verified these through $n = 5$ and partly, by computer search, for $n = 6$. Their resolution could substantially enhance our understanding of A_n and B_n.

For any sequence $d = (d_1,\ldots,d_n)$, we define the *penultimate minimum* d_i, or *pin* d_i for short, as the value of d'_2 when $d'_1 d'_2 \ldots d'_n$ is a nondecreasing permutation of $d_1 d_2 \ldots d_n$. Consider the n-term sequence for $n = 2m$ and $m \geq 2$ given by

$$2^{m-2}c, 2^{m-3}c, \ldots, 4c, 2c, c, c, 1, c+1, 2c+2, 4c+4, \ldots, 2^{m-2}c + 2^{m-2}.$$

Apart from indeterminate c, this is a type A sequence built inside–out in each direction away from $c, c, 1$. To the left, $2c = c + c$, $4c = 2c + c + c, \ldots$; to the right, $c + 1 = (c) + (1), 2c + 2 = (c) + (1) + (c+1), \ldots$. A final type A equation that equates the sum of all terms preceding 1 to the sum of those from 1 on yields $c = 2^{m-1}$ and provides a proof of the following relative of Theorem 14.

THEOREM 17 ([FMR]). $\max_{A_n} (\text{pin } d_i/\min d_i) \geq 2^{\lfloor n/2-1 \rfloor}$;

$$\max_{B_n} (\text{pin } d_i/\min d_i) \geq 2^{\lfloor n/2-1 \rfloor}.$$

Our third extremization problem is to maximize the ratio $d_j/\min d_i$ for each $j \leq \lceil n/2 \rceil$, i.e., to get as large a number as possible in position j (or $n-j$) relative to the smallest component of d. Unlike Conjectures 2A and 2B and Theorem 17, our A and B results differ.

THEOREM 18 ([FMR]). For all $n \geq 6$:

$$\max_{A_n}(d_1/\min d_i) \geq 2^{n-2}, \quad \max_{B_n}(d_1/\min d_i) \geq 2^{n-2};$$

$$\max_{A_n}(d_j/\min d_i) \geq 2^{n-3} \quad \text{for} \quad j = 2,\ldots,\lceil n/2 \rceil;$$

$$\max_{B_n}(d_j/\min d_i) \geq 2^{n-3} + 2^{n-j-2} - 2 \quad \text{for} \quad j = 2,\ldots,\lceil n/2 \rceil.$$

The max ratios for A and B are the same at $n = 5$, where they are 8,4 and 4 for positions 1, 2 and 3, respectively. The type B sequence (2,10,1,2,3,6) shows that $\max_{B_6}(d_2/\min d_i) \geq 10$, but it appears impossible to have $\max_{A_6}(d_2/\min d_i)$ greater than 8.

6.3 Regular Type A and Type B Sequences. It is useful to introduce regular solutions for difference measurement just as for subjective probability measurement. Because unique solutions in difference measurement are sensitive to the order of the d_j, we modify our previous definition to reflect this feature. A unique d of type A or B is *regular* if it can be constructed inside–out from a pair of adjacent 1's by adding one term at a time whose value is a sum of one or more *contiguous* terms already in place, which in the case of type A must be immediately adjacent to the new term. Otherwise d is *irregular*. For example, 61124 for $n = 5$ is a type B regular:

11	$(d_2 = d_3)$
112	$(d_4 = d_2 + d_3)$
1124	$(d_5 = d_2 + d_3 + d_4)$
61124	$(d_1 = d_4 + d_5)$.

This sequence is not a type A regular since 6 is not the sum of contiguous terms *immediately to its right*. However, it is a type A irregular since 6 is obtained from the type A equation $d_1 + d_2 = d_3 + d_4 + d_5$. Unlike type B regulars, all type A regulars must be nondecreasing in each direction away from (1,1). [FOR] refers to type A regulars as two–sided generalized Fibonacci sequences because their construction generalizes the construction of the Fibonacci sequence through the property $f_{n+1} = f_n + f_{n-1}$.

Let α_n and β_n be the number of n-term *regular* sequences of type A and type B, respectively. The following composite theorem [FMR, FOR] notes various facts about these numbers.

THEOREM 19. $\alpha_1 = \alpha_2 = 1$ and, for all $n \geq 2$,

$$\alpha_{n+1} = 2n\,\alpha_n - (n-1)^2 \alpha_{n-1}.$$

Moreover, $\alpha_n = n^{n(1+o(1))}$ and $\beta_n = n^{2n(1+o(1))}$ for all n, and $\alpha_n/|A_n| \to 0$ as $n \to \infty$.

We also suspect that $\beta_n/|B_n| \to 0$, but lack proof.

Conjecture 3. $\beta_n/|B_n| \to 0$ as $n \to \infty$.

It is noted in [FOR] that the nonlinear recurrence for type A regulars implies that

$$\left(\sqrt{n} + \frac{1}{2}\right)^2 - 1/\sqrt{n} < \frac{\alpha_{n+1}}{\alpha_n} < \left(\sqrt{n} + \frac{1}{2}\right)^2 \quad \text{for} \quad n \geq 2.$$

We also have the following asymptotic relationship for α_n.

THEOREM 20 ([FOR]). $\alpha_n \sim \dfrac{K}{2}\sqrt{\dfrac{e}{\pi}}\dfrac{e^{2\sqrt{n}}(n-1)!}{n^{1/4}}$, where

$$K = e^{-1} - \int_0^1 \left[\left(\exp\left(\frac{1}{1-y}\right)\right)/(1-y)\right]\,dy = 0.148495\ldots.$$

6.4 Specific Terms in Regular Sequences. In [FOR] and [FMR] we investigate the appearance of specified positive integers in regular type A and type B sequences. For any sequence $k_1,\ldots,k_m$ of positive integers, let $a(k_1,\ldots,k_m)$ [respectively $b(k_1,\ldots,k_m)$] be the smallest n such that some *permutation* of $k_1,\ldots,k_m$ is a subsequence of a regular type A [respectively type B] sequence with n terms. If no regular sequence of the given type contains a permutation of $k_1,\ldots,k_m$, then $a(k_1,\ldots,k_m)$ or $b(k_1,\ldots,k_m)$ is undefined.

The following composite theorem pertains solely to type A regulars.

THEOREM 21 ([FOR]). $a(k_1,\ldots,k_m)$ *is always defined for* $m \leq 4$ *but can be undefined for* $m \geq 5$. *For smaller values of* m,

$$a(k) = \lceil \log_2 k \rceil + 2 \qquad \text{for all } k \geq 2,$$
$$a(k, k+1, k+2, k+3) \geq \lfloor k/3 \rfloor + 6 \qquad \text{for all } k \geq 4,$$

and if $k_1 \leq k_2 \leq k_3 \leq k_4$, *then* $a(k_1,k_2,k_3,k_4) - a(k_2,k_3,k_4)$ *can be arbitrarily large.*

Proof comments. Given $k_1 \leq k_2 \leq k_3 \leq k_4$, the sequence

$$k_2, k_1, 1, \ldots, 1, k_3, k_4 \quad \text{with} \quad k_4 \quad \text{central} \quad 1\text{'s}$$

is a type A regular, so $a(k_1,\ldots,k_4)$ is always defined. This form of sequence leads readily to the conclusion that no type A regular contains a permutation of 4, 5, 6, 7, 8. A similar construction, but with some central terms larger than 1, is used to establish $a(k, k+1, k+2, k+3) \geq \lfloor k/3 \rfloor + 6$ for all $k \geq 4$. Next, observe that the sequence
$$2^t + 2, 2^t + 1, 1, 1, 1, 1, 2, 4, 8, \ldots, 2^t$$
is a type A regular which contains $k+1, k+2$ and $k+3$ when $k+1 = 2^t$. This fact and the preceding result imply that $a(k, k+1, k+2, k+3) - a(k+1, k+2, k+3)$ can be arbitrarily large. The other conclusion of Theorem 21, that $a(k) = \lceil \log_2 k \rceil + 2$ for $k \geq 2$, uses the maximum type A regular sequence $(1, 1, 2, 4, \ldots, 2^{n-2})$ for $a(k) \geq \lceil \log_2 k \rceil + 2$, and then uses a type A regular permutation of this maximum sequence, with k appended at one end if it is not a power of 2, to show that $a(k) \leq \lceil \log_2 k \rceil + 2$.

To bring type B regulars into the picture, we note first that $a(4,5,6) > b(4,5,6)$. Since 541126 is a type B regular, $b(4,5,6) \leq 6$. However, the same sequence is not a type A regular, and it is easily seen that no six–term regular type A sequence can contain all of 4, 5 and 6. Straightforward extensions of this example show that for every $m \geq 3$ there is a $k_1, \ldots, k_m$ at which $a(k_1, \ldots, k_m) > b(k_1, \ldots, k_m)$. The question of whether this is true also when $m = 2$ remains open.

Conjecture 4. $a(k_1, k_2) = b(k_1, k_2)$ *for all positive integers k_1 and k_2.*

Equality at $m = 1$, $a(k) = b(k)$, follows from Theorem 21 and the $b(k)$ part of our next theorem.

THEOREM 22 ([FMR]). $b(k_1, \ldots, k_m)$ *is always defined. Moreover,*

$$b(k) = \lceil \log_2 k \rceil + 2 \quad \text{for all } k \geq 2,$$

$$b(k, k+1, k+2, k+3) \leq \lceil \log_2 k \rceil + 7 \quad \text{for all } k \geq 2,$$

and $a(k_1, k_2, k_3, k_4) - b(k_1, k_2, k_3, k_4)$ can be arbitrarily large.

The question of whether $a(k_1, k_2, k_3) - b(k_1, k_2, k_3)$ can be arbitrarily large remains open. We have also been unable to settle the following conjectures, which we recommend as attractive candidates for further study.

Conjecture 5A. *If $k_1 \leq k_2$, then $a(k_1, k_2) \leq a(k_2) + 1$.*

Conjecture 5B. *If $k_1 \leq k_2$, then $b(k_1, k_2) \leq b(k_2) + 1$.*

Since $b(k_1, k_2) \leq a(k_1, k_2)$ [every type A regular is a type B regular] and $a(k_2) = b(k_2)$, if Conjecture 5A is true then 5B is also true.

7. ADDITIVE CONJOINT MEASUREMENT

Since the uniqueness results in [FR2] apply primarily to two–factor additive conjoint measurement, where $X = X_1 \times X_2$, we focus on this case here. In terms of Section 2.4, or (13) and (14), each sequence d consists of two blocks, one for successive positive differences of u_1 values for X_1, and the other for successive positive differences of u_2 values for X_2. For notational convenience, let m and n denote the number of successive differences in blocks 1 and 2, respectively, and use d_i for block 1 and e_i for block 2 so that

$$d = (d_1, \ldots, d_m; e_1, \ldots, e_n).$$

With no loss of generality, it is assumed throughout this section that $m \geq n \geq 1$.

7.1 Two Types of Sequences. We consider two types of unique d solutions according to restrictions on the $m + n - 1$ linearly independent equations of the form

(18) $$\sum_{i \in I} d_i = \sum_{j \in J} e_j$$

that are needed for uniqueness via S_2 in Theorem 1. The more restrictive type, which adheres to the usual binary comparisons of conjoint measurement, requires each I and J in (18) to be nonempty *intervals* of integers from their respective index sets $\{1,\ldots,m\}$ and $\{1,\ldots,n\}$. We let $C(m,n)$ denote the set of all positive unique d in smallest–integer format obtained from equations like (18) thus restricted.

The more general solution type allows I and J in (18) to be any nonempty subsets of their index sets. This is similar to the general case for subjective probability except that all comparisons are *between* the two blocks rather than within a single block. It arises from the following problem: Suppose that A_1 is a finite set of red pebbles and A_2 is a finite set of blue pebbles, and we wish to assess the relative weights of all pebbles in $A_1 \cup A_2$. We are given a two–pan scale that allows us to determine which of two disjoint sets of pebbles weighs more (or whether they weigh the same), with the restriction that pebbles of the same color are not allowed in both pans. We let $G(m,n)$ denote the set of all positive unique d in smallest–integer format obtained from equations like (18) with no restrictions on I and J apart from nonemptiness. Clearly, $C(m,n) \subseteq G(m,n)$.

It is obvious that $C(1,1) = G(1,1) = \{(1;1)\}$ and that $C(2,1) = G(2,1) = \{(1,1;1)\}$. On the other hand, $G(3,1)$ contains $(1, 1, 1; 2)$ from the equations

$$d_1 + d_2 = e_1$$
$$d_1 + d_3 = e_1$$
$$d_2 + d_3 = e_1,$$

while $(1, 1, 1; 2)$ is not in $C(3,1)$ because $d_1 + d_3 = e_1$ is not admissible for the more restrictive type. Other nontrivial examples that readers can verify are:

$$(2,1,1,2;2,3) \quad \text{is in} \quad C(4,2);$$
$$(1,2,2,3,4,5;9) \quad \text{is in} \quad G(6,1).$$

7.2 Sequences with Interval Restrictions. While $G(m,1)$ contains an enormous number of unique d when m is large, its interval–restricted companion $C(m,1)$ is quite slim.

THEOREM 23 ([FR2]). $C(m,1) = \{(1,1,\ldots,1;1)\}$.

Thus, there is exactly one way to get m linearly independent equations like (18) with a positive solution when $J = \{1\}$ and each I is an interval of integers in $\{1,\ldots,m\}$.

$C(m,2)$ is more interesting. We have $|C(2,2)| = 5$, $|C(3,2)| = 19$, $|C(4,2)| = 79$, and $|C(5,2)| \geq 335$, but do not have an approximation of $|C(m,2)|$ for large m. The following theorem gives more information about $C(m,2)$.

THEOREM 24 ([FR2]). If $(d_1,\ldots,d_m; e_1, e_2) \in C(m,2)$, then e_1 and e_2 are relatively prime. When e_1 and e_2 are relatively prime positive integers, $(1,1,\ldots,1; e_1, e_2)$ for every $m \geq e_1 + e_2 - 1$ is in $C(m,2)$.

It follows that when $e_1 = e_2$, all solutions in $C(m,2)$ for this case have form $(d_1, \ldots, d_m; 1, 1)$ with $d_i \in \{1, 2\}$ for all i and $d_i = 1$ for some i, and each such vector is a solution. The number of such solutions is clearly $2^m - 1$.

The latter part of Theorem 24 says that, given relatively prime e_1 and e_2, there is a unique $d = (d_1, \ldots, d_m; e_1, e_2)$ in $C(m, 2)$ for which m is as small as $e_1 + e_2 - 1$. We conjecture that this is best–possible.

Conjecture 6. If e_1 and e_2 are relatively prime and $m < e_1 + e_2 - 1$, then no $(d_1, \ldots, d_m; e_1, e_2)$ is in $C(m, 2)$.

We have verified the conjecture for $m \leq 5$, but its status for larger m remains open.

Very little is presently known about $C(m, n)$ for $n \geq 3$.

7.3 The Special Case of G(m,1). Very little is also known about $G(m, n)$ for $n \geq 2$. However, we have two results for $G(m, 1)$ in addition to some initial counts, namely $|G(2,1)| = 1$, $|G(3,1)| = 2$, $|G(4,1)| = 11$, $|G(5,1)| = 169$, and $|G(6,1)| = 6639$.

First, let K_m denote the largest number of distinct d_i values in any $(d_1, \ldots, d_m; e_1) \in G(m, 1)$.

THEOREM 25 ([FR2]). $(K_2, K_3, K_4, K_5, K_6) = (1, 1, 2, 3, 5)$, and $K_m = m$ for all $m \geq 7$.

A unique d with three d_i values for $m = 5$ is $(1, 1, 2, 2, 3; 5)$, and a unique d with five d_i values for $m = 6$ is $(1, 2, 2, 3, 4, 5; 9)$. At $m = 7$, we find that $(1, 2, 3, 4, 5, 6, 7; 10)$ is in $G(7, 1)$. More generally, $(1, 2, 3, \ldots, m-1, m; m+3)$ is in $G(m, 1)$ for all $m \geq 7$.

Our second result for $G(m, 1)$ involves the maximum possible value $e_1^{(m)}$ of e_1 for any d in $G(m, 1)$. For example, we know that $e_1^{(5)} = 5$ and $e_1^{(6)} = 9$; these values are realized by the first two sequences in the preceding paragraph. Let $(c_1, c_2, c_3, \ldots)$ be the denumerable sequence defined by

$$c_i = 1 \quad \text{for} \quad 1 \leq i \leq 5;$$
$$c_{2i} = c_{2i+1} - c_{i-1} \quad \text{for all} \quad i \geq 3;$$
$$c_{2i+1} = c_{2i-1} + c_{2i-2} \quad \text{for all} \quad i \geq 3.$$

The first part of this sequence is

$$1, 1, 1, 1, 1, 2, 2, 3, 4, 5, 8, 9, 16, 17, 31, 33, 62, 64, 123, 126.$$

One of its interesting properties is the identity

$$c_{2i} = c_i + c_{i+1} + \cdots + c_{2i-3} \quad \text{for} \quad i \geq 3.$$

In addition, it can also be shown that there is a constant c_0, with $0.9479939 < c_0 < 0.9479940$, such that

$$c_{2i+1} \sim c_0 \, 2^{i-3}.$$

This asymptotic and the following theorem give an idea of how large e_1 can be for $G(m, 1)$.

THEOREM 26 ([FR2]). $e_1^{(m)} \geq c_{2m+1}$.

We know that the lower bound c_{2m+1} is the exact value of $e_1^{(m)}$ for $m \leq 6$, and suspect that this is true for larger m.

7.4 Biregular Sequences. In correspondence to notions of regular sequences used in Section 5.3 and 6.3, we say that a two-block sequence $(d_1, \ldots, d_m; e_1, \ldots, e_n)$ is *biregular* if it can be built up one term at a time in the following way, beginning with one 1 in each block:

$$\ldots \leftarrow 1 \rightarrow \ldots \quad \Big| \quad \ldots \leftarrow 1 \rightarrow \ldots \; .$$

Given the starting 1's, add one term at a time (to either block) that is *adjacent* to the terms already specified for that block and whose value equals a nonempty sum of terms already specified for the other block. If the latter sum must consist of contiguous terms, then we say that the biregular sequence is *interval-restricted*. An example of the construction of an interval-restricted biregular sequence with $(m, n) = (5, 4)$ is

```
                    1 | 1
                  1 1 | 1
                  1 1 | 1 2
                3 1 1 | 1 2
              2 3 1 1 | 1 2
              2 3 1 1 | 6 1 2
            2 3 1 1 7 | 6 1 2
            2 3 1 1 7 | 6 1 2 14.
```

If the final 14 is changed to 10, we get a biregular sequence that is not interval-restricted.

Let $C_b(m, n)$ be the set of all interval-restricted biregular sequences with m terms in block 1 and n terms in block 2, and let $G_b(m, n)$ be the corresponding set without the interval restriction. By the construction it is clear that $C_b(m, n) \subseteq C(m, n)$ and $G_b(m, n) \subseteq G(m, n)$.

Also let $\gamma(m, n) = |C_b(m, n)|$ and $\delta(m, n) = |G_b(m, n)|$. It is easily seen that $\gamma(m, 1) = \delta(m, 1) = 1$ for all m, and that $\gamma(m, 2) = \delta(m, 2)$ for all m. In addition, the values of $\gamma(m, 2)$ for m from 2 through 8 are 2, 19, 69, 243, 841, 2859 and 9573. The next theorem gives order-of-magnitude values of γ and δ for all $m \geq n$.

THEOREM 27 ([FR2]). *For each fixed $n \geq 1$,*

$$\gamma(m, n) = \binom{n+1}{2}^{m(1+o(1))} \qquad m = n, n+1, \ldots;$$

$$\delta(m, n) = (2^n - 1)^{m(1+o(1))} \qquad m = n, n+1, \ldots \; .$$

7.5 Extreme Values for Biregular Sequences. We conclude our summary of technical results by considering the maximum value of the largest term in all biregular sequences for (m, n). Since the construction of a biregular sequence that has the largest possible term at (m, n) will always take the sum of all terms in the other block when it adds a new term, this extremization problem applies simultaneously to $C_b(m,n)$ and $G_b(m,n)$.

Formally, let

$$\mu(m,n) = \max_{d \in C_b(m,n)} [\max\{d_1, \ldots, d_m, e_1, \ldots, e_n\}].$$

Determination of μ comes down to the specification of an optimal strategy for adding new terms. When $m = n$, the optimal strategy is to alternate between blocks:

$$1 \mid 1, \ 11 \mid 1, \ 11 \mid 12, \ 311 \mid 12, \ 311 \mid 125, \ 8311 \mid 125, \ldots.$$

The appearance of Fibonacci numbers here is no accident.

THEOREM 28 ([FR2]). *For each $n \geq 1$, $\mu(n,n) = f_{2n-1}$ and $\mu(n+1,n) = f_{2n}$.*

Our next result gives exact values of μ for $n \leq 4$.

THEOREM 29 ([FR2]). *For $n \in \{1, 2, 3, 4\}$ and all $m \geq n$,*

$$\mu(m, 1) = 1$$
$$\mu(m, 2) = m$$
$$\mu(m, 3) = \begin{cases} m(m+4)/4 & \text{for even } m \\ [m(m+4) - 1]/4 & \text{for odd } m \end{cases}$$

$$\mu(m, 4) = \begin{cases} t^3 + 4t^2 + 3t & \text{if } m = 3t \\ t^3 + 5t^2 + 6t + 1 & \text{if } m = 3t + 1 \\ t^3 + 6t^2 + 10t + 4 & \text{if } m = 3t + 2. \end{cases}$$

General bounds on μ are given in our final theorem.

THEOREM 30 ([FR2]). *For all $m \geq n \geq 4$, if $m = (n-1)t + b$ with $0 \leq b \leq n-2$, then*

$$(t+2)^b (t+1)^{n-1-b} \leq \mu(m,n) \leq \left[t + \frac{b + 2(n-2)}{n-1} \right]^{n-1}.$$

It follows that, for fixed $n \geq 2$,

$$\mu(m,n) \sim [m/(n-1)]^{n-1},$$

as might be guessed from Theorem 29.

8. Discussion

The research and its results that we have reviewed in this paper were motivated by a striking lack of knowledge about uniqueness properties of quantitative models for finite qualitative structures in the representational theory of measurement. Our first task was to understand how the most orderly types of uniqueness that are frequently encountered for infinite structures, namely uniqueness up to positive affine transformations and up to proportionality transformations, arise for finite structures. As explained in Sections 2 and 3, we now know that these types of uniqueness for finite structures are tantamount to the existence of a suitably rich system of linearly independent equations that correspond to indifference or equality comparisons within the qualitative structure. Theorem 1 gives a terse summary of this finding, and it is this theorem that drives the rest of our work.

Apart from axiomatization problems, illustrated in Sections 4 and 5, we have focused on structures of sets of unique solutions to systems of linearly independent equations delineated by Theorem 1. Our understanding of these structures has been enhanced by the use of finite integer sequences in smallest–integer format to characterize unique solutions. Such characterizations suggest interesting questions to ask, and they facilitate analyses of counting, extremization and inclusion problems.

As we amply demonstrate in the paper, solution sets that arise from various measurement contexts via restrictions on equations allowed for S_2 in Theorem 1 are a surprisingly fertile source of intriguing and often difficult problems involving linear algebra, combinatorics, and number theory. Despite the numerous results presented, we have been able to raise many more questions than we can answer and suspect that we have barely begun.

The six numbered conjectures in the paper suggest basic open problems that are important, challenging, and might be settled within a reasonable time. We gather these here with reference to their section and problem type.

1. (Section 5, counting). $|V_n|/|R_n| \to 0$ as $n \to \infty$.

2. (Section 6, extremization). For type A and type B sequences, the minimum of $(\min d_i / \sum d_j)$ over all n-term sequences is $\geq 1/2^{n-1}$.

3. (Section 6, counting). $\beta_n/|B_n| \to 0$ as $n \to \infty$.

4. (Section 6, inclusion). $a(k_1, k_2) = b(k_1, k_2)$ for all positive integers k_1 and k_2.

5. (Section 6, inclusion). $k_1 \leq k_2 \Rightarrow a(k_1, k_2) \leq a(k_2) + 1$ and $b(k_1, k_2) \leq b(k_2) + 1$.

6. (Section 7, extremization/algebra). For relatively prime e_1 and $e_2, (d_1, \ldots, d_m; e_1, e_2) \in C(m, 2) \Rightarrow m \geq e_1 + e_2 - 1$.

Many other open problems have been noted in passing, and many more are indirectly suggested by the lack of definitiveness in some of our theorems. We conclude with a secondary list.

Section 4. Does $|E_n|/|F_n| \to 0$? Is $e_n^* \leq n^2/4$ for all $n \geq 3$? Determine $e(k)$. Give order-of-magnitude values for $|E_n|$ and $|F_n|$.

Section 5. Give order-of-magnitude values for $|V_n|$ and $|P_n|$. Does $v_n^* - f_{n+1} \to \infty$?

Section 6. Find good lower bounds for $|A_n|$ and $|B_n|$.

Section 7. Give an asymptotic approximation for $|C(m,2)|$. Is the c_{2m+1} bound for $e_1^{(m)}$ exact?

Primary References

[FMR] FISHBURN, P.C., H.M. MARCUS-ROBERTS & F.S. ROBERTS, Unique finite difference measurement, SIAM Journal on Discrete Mathematics 1 (1988), 334-354.

[FO] FISHBURN, P.C. & A.M. ODLYZKO, Unique subjective probability on finite sets, Preprint, AT&T Bell Laboratories, Murray Hill, NJ, 1986 (Journal of the Ramanujan Mathematical Society, to appear).

[FOR] FISHBURN, P.C., A.M. ODLYZKO & F.S. ROBERTS, Two-sided generalized Fibonacci sequences, Preprint, AT&T Bell Laboratories, Murray Hill, NJ, 1987 (Fibonacci Quarterly, to appear).

[FR1] FISHBURN, P.C., & F.S. ROBERTS, Axioms for unique subjective probability on finite sets, Preprint, AT&T Bell Laboratories, Murray Hill, NJ 1987 (Journal of Mathematical Psychology, 33 (1989), in press).

[FR2] FISHBURN, P.C. & F.S. ROBERTS, Unique finite conjoint measurement, Mathematical Social Sciences 16 (1988), 107-143.

[FRM] FISHBURN, P.C., F.S. ROBERTS & H.M. MARCUS-ROBERTS, Van Lier sequences, Preprint, AT&T Bell Laboratories, Murray Hill, NJ, 1987 (Discrete Applied Mathematics, to appear).

[VL] VAN LIER, L., A simple sufficient condition for the representability of a finite qualitative probability by a probability measure, Discussion paper 8708, Centre d'Economie Mathematique et d'Econometrie, Université Libre de Bruxelles, Brussels, 1987 (Journal of Mathematical Psychology, to appear).

REFERENCES

BROWN, J.L., JR. (1964), Zeckendorf's theorem and some applications, The Fibonacci Quarterly 2, 163-168.

DE FINETTI, B. (1931), Sul significato soggettivo della probabilità, Fundamenta Mathematicae 17, 298-329.

FISHBURN, P.C. (1970), Utility Theory for Decision Making, New York: Wiley.

FISHBURN, P.C. (1982), Nontransitive measurable utility, Journal of Mathematical Psychology 26, 31-67.

FISHBURN, P.C. (1986), The axioms of subjective probability, Statistical Science 1, 335-345.

FISHBURN, P.C. (1988), Nonlinear Preference and Utility Theory, Baltimore, MD: Johns Hopkins University Press.

HOGGATT, V.E., JR. (1969), Fibonacci and Lucas Numbers, Boston: Houghton Mifflin.

JENSEN, N.E. (1967), An introduction to Bernoullian utility theory. I. Utility functions, Swedish Journal of Economics 69, 163-183.

KRAFT, C.H., J.W. PRATT & A. SEIDENBERG (1959), Intuitive probability on finite sets, Annals of Mathematical Statistics 30, 408-419.

KRANTZ, D.H., R.D. LUCE, P. SUPPES & A. TVERSKY (1971), *Foundations of Measurement*, Vol. 1, New York: Academic Press.

LUCE, R.D. (1967), *Sufficient conditions for the existence of a finitely additive probability measure*, Annals of Mathematical Statistics **38**, 780–786.

NARENS, L. (1985), *Abstract Measurement Theory*, Cambridge, MA: MIT Press.

PFANZAGL, J. (1968), *Theory of Measurement*, New York: Wiley.

ROBERTS, F.S. (1979), *Measurement Theory with Applications to Decisionmaking, Utility and the Social Sciences*, Reading, MA: Addison–Wesley.

ROBERTS, F.S. (1985), *Issues in the theory of uniqueness in measurement*, Graphs and Orders (I. Rival, ed.), 415–444. Amsterdam: Reidel.

ROBERTS, F.S. & R.D. LUCE (1968), *Axiomatic thermodynamics and extensive measurement*, Synthese **18**, 311–326.

ROBERTS, F.S. & Z. ROSENBAUM (1988), *Tight and loose value automorphisms*, Discrete Applied Mathematics **22**, to appear.

SCOTT, D. (1964), *Measurement models and linear inequalities*, Journal of Mathematical Psychology **1**, 233–247.

Conceptual Scaling

Bernhard Ganter and Rudolf Wille *

1 Introduction

Scaling and measurement are usually based on numerical methods; this is, for instance, indicated by Stevens' definition: "measurement is the assignment of numerals to objects or events according to rules" (see [15], p. 22), or by Torgerson's statement: "measurement of a property involves the assignment of numbers to systems to represent that property" (see [16], p. 14). In contrast to this, conceptual scaling uses first of all set-theoretical methods to explore conceptual patterns in empirical data. Conceptual scaling has been developed in the frame of formal concept analysis, a theory based on a mathematization of conceptual hierarchies (see [18]). In this paper we give an introductory survey on conceptual scaling which concentrates on scales of ordinal type. First we recall basic notions and results of formal concept analysis and demonstrate them by an example (section 2). Then ideas of conceptual measurement are discussed in focussing on the question of measurability by standardized scales of ordinal type (section 3 and 4). These ideas are applied to the scaling of data contexts to derive conceptual hierarchies for the data (section 5). Finally, these scalings are used to introduce and to study a general notion of dependency between attributes which covers special notions like functional and linear dependency (section 6). Mathematically we presuppose some basic knowledge of order and lattice theory which can be found in [1] and [6].

2 Concept Lattices

Formal concept analysis is based on a set-theoretic model for conceptual hierarchies. This model mathematizes the philosophical understanding of a concept as a unit of thoughts consisting of two parts: the extension and the intension (comprehension); the extension covers all objects (or entities) belonging to the concept while the intension comprises all attributes (or properties) valid for all those objects (cf. [17]). In the set-theoretic model we fix a set G, the elements of which are called *objects*, and a set M, the elements of which are called *attributes*; furthermore we assume a binary relation I between G and M where $(g,m) \in I$ (resp. gIm) is read: the object g has the attribute m. The triple (G, M, I) which is called a *(formal) context* is the basic structure of formal concept analysis. For a context (G, M, I), the most frequently used operators are defined as follows:

$$A' := \{m \in M | gIm \text{ for all } g \in A\} \text{ for } A \subseteq G,$$

$$B' := \{g \in G | gIm \text{ for all } m \in B\} \text{ for } B \subseteq M.$$

---
*Technische Hochschule Darmstadt, West Germany

	Large Blade	Medium Blade	Small Blade	Manicure Blade	Screwdriver/Caplifter	Corkscrew	Can Opener/Screwdriver	Phillips Screwdriver	Fine Screwdriver	Scissors	Key Ring	Toothpick	Tweezers	Sew Blade/File	Wood Saw	Inch-Metric Rule/Fish Scaler	Reamer	Magnifier
CLASSIC			×	×						×	×	×	×					
SPARTAN	×	×			×	×	×				×						×	
NEW TINKER	×	×			×		×	×			×	×					×	
CAMPER	×	×			×	×	×				×	×	×		×		×	
CLIMBER	×	×			×	×	×			×	×	×	×				×	
EXPLORER	×	×			×	×	×	×		×	×	×	×				×	×
OUTDOORSMAN	×	×			×	×	×			×	×	×	×	×	×		×	
CHAMPION	×	×			×	×	×	×	×	×	×	×	×	×	×	×	×	×

Figure 1: Swiss Army Officers' Knives

Now, a *(formal) concept* of a context (G, M, I) is defined to be a pair (A, B) with $A \subseteq G$, $B \subseteq M$, $A' = B$, and $B' = A$; A and B are called the *extent* and the *intent* of the concept (A, B), respectively. The hierarchy of concepts is given by the relation *"subconcept-superconcept"* which has to be defined for a context by

$$(A_1, B_1) \leq (A_2, B_2) :\Leftrightarrow A_1 \subseteq A_2 \; (\Leftrightarrow B_1 \supseteq B_2).$$

The set of all concepts of (G, M, I) with this order relation is a complete lattice called the *concept lattice* of (G, M, I) and denoted by $\mathfrak{B}(G, M, I)$. Thus, to any set S of concepts of (G, M, I) there exists always a greatest subconcept, the *infimum* of S in $\mathfrak{B}(G, M, I)$, and a smallest superconcept, the *supremum* of S in $\mathfrak{B}(G, M, I)$. More precise information is given by the following theorem:

Basic Theorem on Concept Lattices (cf. [18]) *Let (G, M, I) be a context. Then $\mathfrak{B}(G, M, I)$ is a complete lattice in which infimum and supremum can be described as follows:*

$$\bigwedge_{j \in J}(A_j, B_j) = (\bigcap_{j \in J} A_j, (\bigcup_{j \in J} B_j)''), \quad \bigvee_{j \in J}(A_j, B_j) = ((\bigcup_{j \in J} A_j)'', \bigcap_{j \in J} B_j).$$

In general, a complete lattice L is isomorphic to $\mathfrak{B}(G, M, I)$ if and only if there are mappings $\gamma : G \longrightarrow L$ and $\mu : M \longrightarrow L$ such that γG is supremum-dense in L (i.e. $L = \{\bigvee X | X \subseteq \gamma G\}$), μM is infimum-dense in L (i.e. $L = \{\bigwedge X | X \subseteq \mu M\}$), and $gIm \Leftrightarrow \gamma g \leq \mu m$ for all $g \in G$ and $m \in M$. In particular, $L \cong \mathfrak{B}(L, L, \leq)$.

We demonstrate the basic notions and results of formal concept analysis by an example. The table in fig.1 (from [2], p. 132) describes a context: its objects are

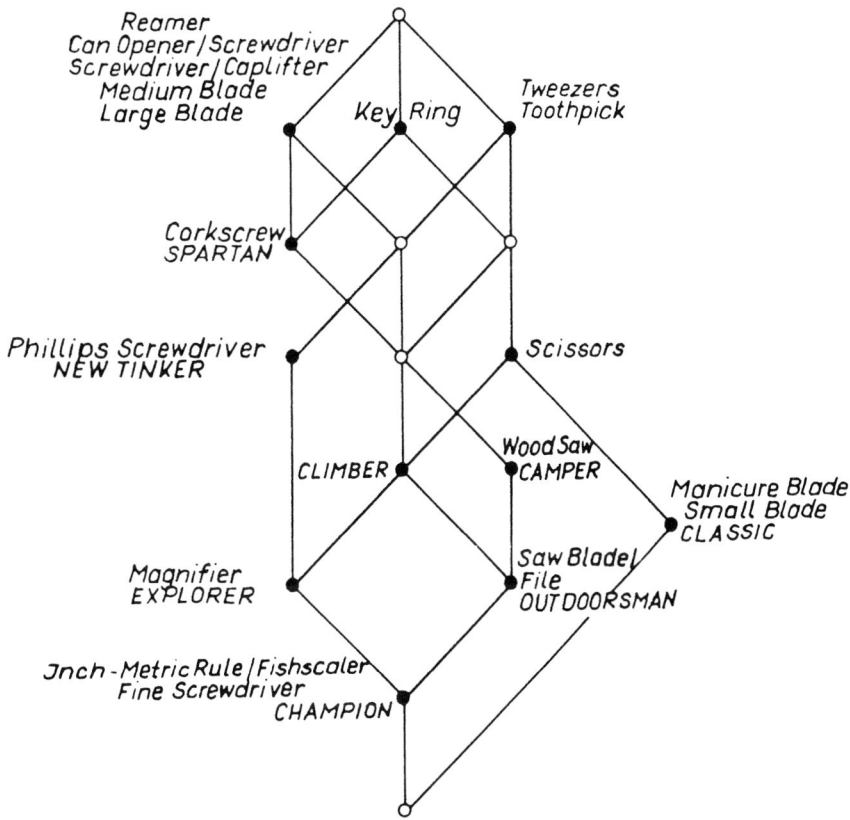

Figure 2: The concept lattice of the context in figure 1

types of Swiss Army Officers' Knives, its attributes are the possible components of the knives, and its relation is given by the crosses in the table. A concept of this context is for instance the pair consisting of the extent {Classic, Climber, Explorer, Outdoorsman, Champion} and the intent {Scissors, Key Ring, Toothpick, Tweezers}; a name of this concept may be "Knife with Scissors". All concepts of the context can be seen in fig.2 which shows the concept lattice of the context by a line diagram; a little circle labelled by a name of an object g represents $\gamma g := (\{g\}'', \{g\}')$, i.e. the smallest concept with g in its extent, and a little circle labelled by a name of an attribute m represents $\mu m := (\{m\}', \{m\}'')$, i.e. the greatest concept with m in its intent. The labelling allows to determine the extent and intent for any concept (A, B) since $A = \{g \in G | \gamma g \leq (A, B)\}$ and $B = \{m \in M | \mu m \geq (A, B)\}$ by the Basic Theorem. The equivalence $gIm \Leftrightarrow \gamma g \leq \mu m$ yields that the context can also be read from the diagram.

By the Basic Theorem, the lattice structure of a finite concept lattice $L := \mathfrak{B}(G, M, I)$ is already determined by a *reduced context* $(G_r, M_r, I \cap (G_r \times M_r))$ with $G_r \subseteq G$ and $M_r \subseteq M$ so that the mappings γ and μ are bijections from G_r onto the set $J(L)$ of all join-irreducible elements of L and from M_r onto the set $M(L)$

Figure 3: The dichotomic context

of all meet-irreducible elements of L, respectively. In our example such a reduced context is given by $G_r := \{$Classic, Spartan, New Tinker, Camper, Explorer, Outdoorsman, Champion$\}$ and $M_r := \{$Large Blade, Small Blade, Phillips Screwdriver, Scissors, Key Ring, Toothpick, Wood Saw$\}$. Notice that quite different contexts may have isomorphic concept lattices.

Not all information of the table in fig.1 is used to form concepts; for instance, the concept "Knife with no Scissors" does not occur in the concept lattice of fig.2. If one understands the table in fig.1 as a formal context, then the attributes are only used positively and combined by conjunction. If one wants to form concepts also with the negation of attributes, one has to extend the table by the negated attributes as it is done in fig.3. Of course, the dichotomic context in fig.3 has more concepts than the context in fig.1; this can be seen in fig.4. The derivation of the dichotomic context and its concept lattice is a simple example of conceptual scaling of a many valued context, which will be extensively discussed in section 5. Methods to determine the concept lattice of a given context and to draw adequate line diagrams can be found in [7, 19, 22]; there are also computer programs based on these methods (see [3, 8, 14]).

3 Scales and Scale Measures

Scaling is the development of formal patterns and their use for analyzing empirical data. In conceptual scaling these formal patterns consist of formal contexts and their concept lattices which have a clear structure and which reflect some meaning. Such a context is called a *scale* and, in general, denoted by $S := (G_S, M_S, I_S)$; the elements of G_S and M_S are called *scale values* and *scale attributes*, respectively. In this section we restrict to the case where the empirical data are given in the form of a context

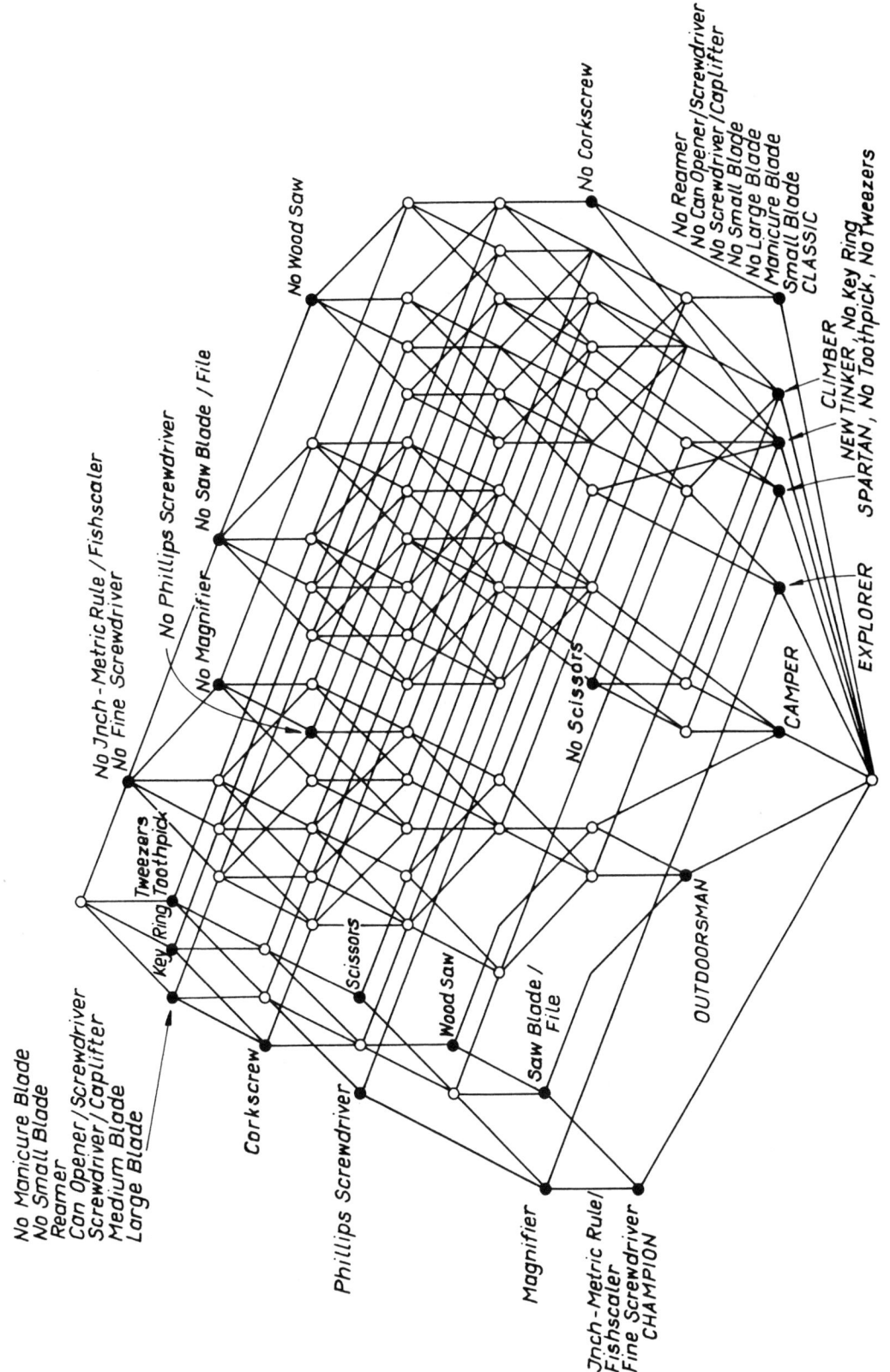

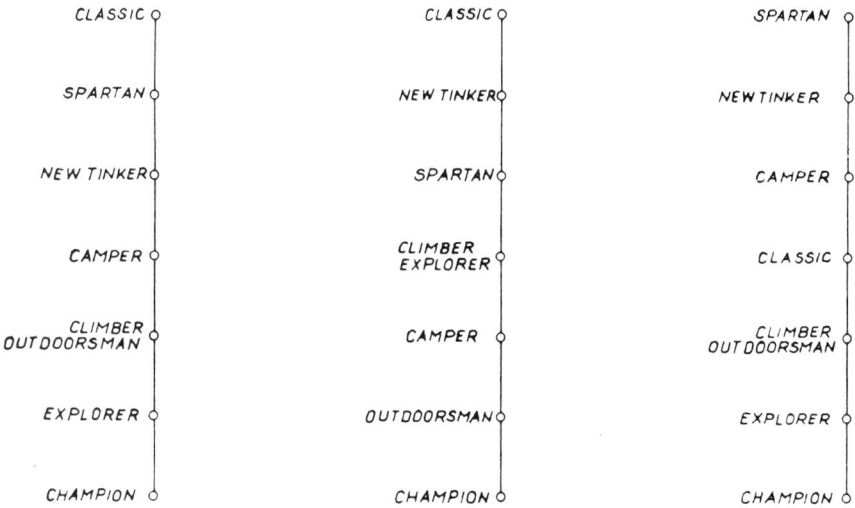

Figure 5: Three surjective measures onto the ordinal scale O_7

$K := (G, M, I)$. Empirical contexts are connected with scales by scale measures: a *scale measure* or, more precisely, an *S-measure* of K is defined to be a mapping σ from G into G_S such that, for every extent A of S, the preimage $\sigma^{-1}A$ is an extent of K. An S-measure σ of K is called *full* if every extent of K is the preimage of an extent of S under σ (cf. [18, 19, 9]). In measurement theory one usually assumes relational systems of a fixed type as empirical data so that scale measures are defined as homomorphisms between relational systems (cf. [13]); conceptual measurement, however, is based on a type-free notion of a scale measure which reflects the idea of a continuous map between topological spaces.

Conceptual scaling is in the first place of ordinal nature although other types of measurements may also be treated within conceptual scaling. In this paper we shall restrict our explanations to conceptual measurement with different scales of ordinal type. Let us begin with the discussion of some scale measures of the empirical data from section 2. Rankings of the objects are given by scale measures into the *one-dimensional ordinal scale*

$$O_n := (\{1, 2, \ldots, n\}, \{1, 2, \ldots, n\}, \leq);$$

the concept lattice of O_n is an n-element chain. Since the preimages of the extents of a one-dimensional ordinal scale have to form a chain under set-inclusion, the scale measures into one-dimensional ordinal scales correspond to chains in the concept lattice of the empirical context. Thus, one can easily read from fig.2 that there are 22 surjective O_7-measures of the context of fig.1, three of them are shown in fig.5. There does not exist a full measure into a one-dimensional scale because the empirical context is ordinally of higher dimension, i.e. its concept lattice is not a chain. To obtain full scale measures, we may use the *k-dimenssional grid scale*

$$G_{n_1, n_2, \ldots, n_k} := (\mathbf{n}_1 \times \mathbf{n}_2 \times \cdots \times \mathbf{n}_k, \mathbf{n}_1 \times \{1\} \cup \mathbf{n}_2 \times \{2\} \cup \cdots \cup \mathbf{n}_k \times \{k\}, \nabla)$$

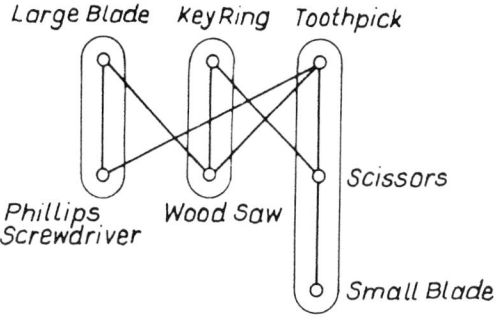

Figure 6: The ordered set of meet–irreducible concepts of the context in figure 1.

where **n** is the set $\{1, 2, \ldots, n\}$ with the natural order and

$$(v_1, v_2, \ldots, v_k) \nabla (m, j) :\Leftrightarrow v_j \leq m \ ;$$

the concept lattice of $G_{n_1,n_2,\ldots,n_k}$ is isomorphic to the direct product of the chains $\mathbf{n}_1, \mathbf{n}_2, \ldots, \mathbf{n}_k$. To find full scale measures into grid scales, a general method is to decompose the ordered set of all meet-irreducible concepts of the empirical context into chains; then a full scale measure into a grid scale is given by assigning to an object g, for each chain, the number of concepts in the chain not containing g in their extent plus one. Fig. 6 shows the ordered set of all meet-irreducible concepts of our example with a partition into chains; the corresponding scale measure is visualized in fig.7 by a (join-) embedding of the concept lattice of fig.2 into a direct product of chains of natural numbers. The values of the grid scale which do not correspond to concepts of the empirical context indicate non-trivial implications between the attributes of the empirical context (cf. [4]). The value $(4, 2, 1)$, for instance, represents the attribute set {Phillips Screwdriver, Key Ring}; the greatest value below $(4, 2, 1)$ corresponding to a concept is the value $(2, 2, 1)$ which represents the attribute set {Phillips Screwdriver, Scissors, Key Ring}. This yields that {Phillips Srewdriver, Key Ring}$\Longrightarrow$ {Scissors}, i.e. every knife which has a Phillips screwdriver and a key ring has also scissors. The implications with a one-element premise, as {Small Blade}$\Longrightarrow${Key Ring} which is indicated by the scale value $(1, 3, 3)$, can already be read from the ordered set $M(L)$ in fig.6. If one restricts to the sublattice determined by the image of the scale measure as it is visualized in fig.8, only implications with more than one attribute in their premise are indicated by lattice elements not corresponding to a concept of the empirical context.

Classification is another basic meaning of data analysis besides ranking. Conceptual scaling yields classifications of objects by forming preimages of scale extents under scale measures. Such classifications consist of classes which are describable

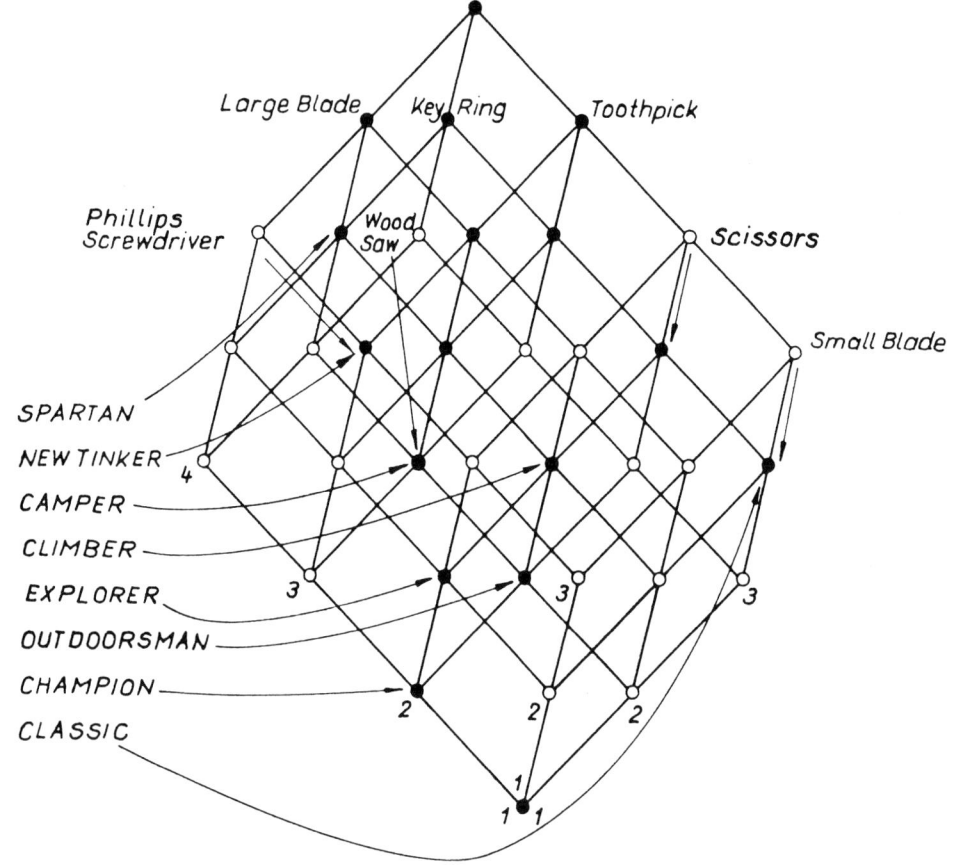

Figure 7: A full measure into the grid scale $G_{3,3,4}$

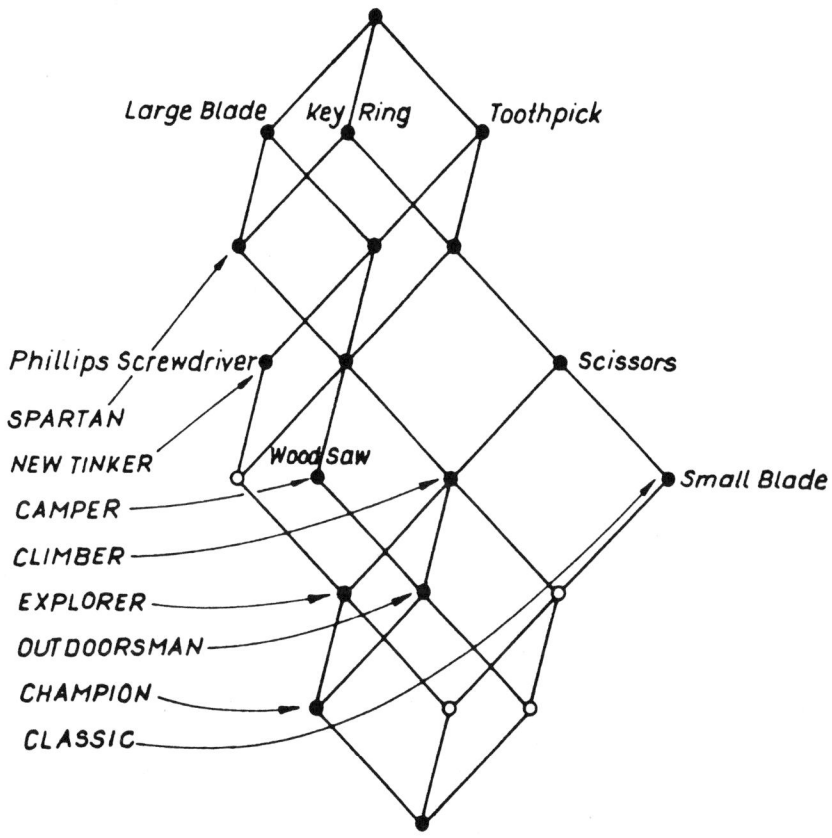

Figure 8: The sublattice of the concept lattice of $G_{3,3,4}$ generated by the image of the scale measure of fig.7

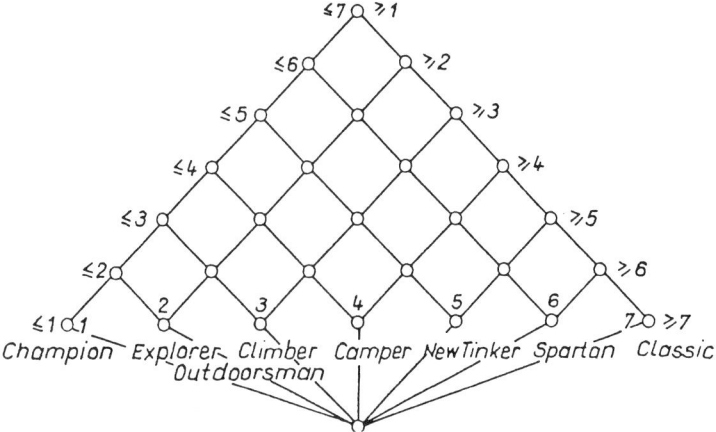

Figure 9: A surjective measure onto the interordinal scale I_7

by the attributes of the empirical context. The most elementary scale measures for classification are the *nominal scales*

$$N_n := (\{1, 2, \ldots, n\}, \{1, 2, \ldots, n\}, =);$$

the concept lattice of N_n consists of an n-element antichain with a common upper and lower bound. The scale measures into nominal scales correspond to partitions of the empirical objects into extents. Thus, one can read from fig.2 that the context of fig.1 admits only surjective scale measures into N_1 and N_2; in this way we obtain only one proper partition consisting of the set {Classic} and its complement. Many more partitions into extents has the dichotomic context of fig.3. In such a case it is more interesting to determine a hierarchical system of classification than only one single partition. This can be performed by measures into *one-dimensional interordinal scales*

$$I_n := (\{1, 2, \ldots, n\}, \{1, 2, \ldots, n\} \times \{1, 2\}, \Diamond)$$

where $i \Diamond (j, 1) :\Leftrightarrow i \leq j$ and $i \Diamond (j, 2) :\Leftrightarrow i \geq j$; the concept lattice of I_7 is shown in fig.9 together with a scale measure of the dichotomic context (the symbol $\leq j$ stands for the attribute $(j, 1)$ and dually $\geq j$ for $(j, 2)$.) This measure which is the dichotomic version of the first example in fig.5 emphasizes a left-right-structure with its inherent hierarchical classification of the empirical objects. A full scale measure of the dichotomic context into an interordinal scale of higher dimension can be constructed analogue to the measure described by fig.7.

Now, we introduce scales of ordinal type in general. Let us first define some context constructions which are used to form scales. For contexts $K := (G, M, I)$ and $K_j := (G_j, M_j, I_j)$ $(j = 1, 2, \ldots, k)$, we use the abbreviations $\dot{G}_j := G_j \times \{j\}$, $\dot{M}_j := M_j \times \{j\}$ and $\dot{I}_j := \{((g, j), (m, j)) | (g, m) \in I_j\}$ to define

$$K^c := (G, M, (G \times M) \setminus I),$$

the *complementary context* of K,

$$K^d := (M, G, I^{-1}),$$

the *dual context* of K,

$$K_1 \dot\cup \cdots \dot\cup K_k := (\dot{G}_1 \cup \cdots \cup \dot{G}_k, \dot{M}_1 \cup \cdots \cup \dot{M}_k, \dot{I}_1 \cup \cdots \cup \dot{I}_k),$$

the *disjoint union* of $K_1, \ldots, K_k$,

$$K_1 + \cdots + K_k := (\dot{G}_1 \cup \cdots \cup \dot{G}_k,$$
$$\dot{M}_1 \cup \cdots \cup \dot{M}_k,$$
$$\dot{I}_1 \cup \cdots \cup \dot{I}_k \cup \bigcup_{i \neq j} \dot{G}_i \times \dot{M}_j),$$

the *direct sum* of $K_1, \ldots, K_k$,

$$K_1 \boxtimes \cdots \boxtimes K_k := (G_1 \times \cdots \times G_k, \dot{M}_1 \cup \cdots \cup \dot{M}_k, \nabla),$$

the *semiproduct* of $K_1, \ldots, K_k$,

where $(g_1, \ldots, g_k) \nabla (m, j) \Leftrightarrow g_j I_j m$,

$$K_1 \times \cdots \times K_k := (G_1 \times \cdots \times G_k, M_1 \times \cdots \times M_k, \nabla),$$

the *direct product* of $K_1, \ldots, K_k$,

where $(g_1, \ldots, g_k) \nabla (m_1, \ldots, m_k)$

$:\Leftrightarrow g_j I_j m_j$ for some j,

$$K^{\underline{0}} := K + (\{\underline{0}\}, \emptyset, \emptyset),$$
$$K^{\underline{1}} := K \,\dot{\cup}\, (\{\underline{1}\}, \emptyset, \emptyset),$$
$$K^{\overline{0}} := K \,\dot{\cup}\, (\emptyset, \{\overline{0}\}, \emptyset),$$
$$K^{\overline{1}} := K + (\emptyset, \{\overline{1}\}, \emptyset),$$

and, if $G = G_1 = G_2$,

$$K_1 \mid K_2 := (G, \dot{M}_1 \cup \dot{M}_2, \dot{I}_1 \cup \dot{I}_2),$$

the *apposition* of K_1 and K_2,

likewise, if $M = M_1 = M_2$,

$$\frac{K_1}{K_2} := (\dot{G}_1 \cup \dot{G}_2, M, \dot{I}_1 \cup \dot{I}_2),$$

the *subposition* of K_1 and K_2.

For defining and analyzing scales of ordinal type we use the following constructions of ordered sets (see [1]): $P_1 + \cdots + P_k$ is the *cardinal sum* of the ordered sets $P_1, \ldots, P_k$, $P_1 \times \cdots \times P_k$ is the *direct product* of $P_1, \ldots, P_k$, and $\mathfrak{P}(\mathbf{n})$ is the *power set* of $\mathbf{n} := \{1, 2, \ldots, n\}$ ordered by set-inclusion.

The following table lists standardized scales of ordinal type which have been proved useful up to now.

Standardized Scales of Ordinal Type			
Symbol	Definition	Name	Basic Meaning
O_P	$(P, P, \leq)$	ordinal scale	hierarchy
O_n	$(\mathbf{n}, \mathbf{n}, \leq)$	one-dimensional ordinal scale	ranking
N_n	$(\mathbf{n}, \mathbf{n}, =)$	nominal scale	partition
$M_{n_1, n_2, \ldots, n_k}$	$O_{n_1+n_2+\cdots+n_k}$	multiordinal scale	partition with rankings
$M_{m,n}$	O_{n+m}	biordinal scale	two-class ranking
B_n	$(\mathfrak{P}(\mathbf{n}), \mathfrak{P}(\mathbf{n}), \subseteq)$	n-dimensional boolean scale	attribute dependency
$G_{n_1, n_2, \ldots, n_k}$	$O_{n_1} \boxtimes \cdots \boxtimes O_{n_k}$	k-dimensional grid scale	multiple ranking
O_P^{cd}	$(P, P, \not\leq)$	contrary ordinal scale	hierarchy and independence
N_n^c	$(\mathbf{n}, \mathbf{n}, \neq)$	complementary nominal scale	partition and independence
D	$(\{0,1\}, \{0,1\}, =)$	dichotomic scale	dichotomy
D_k	$\underbrace{D \boxtimes \cdots \boxtimes D}_{k\text{-times}}$	k-dimensional dichotomic scale	multiple dichotomy
I_P	$O_P \mid O_P^d$	interordinal scale	betweenness
I_n	$O_n \mid O_n^d$	one-dimensional interordinal scale	linear betweenness
C_P	$O_P^{cd} \mid O_P^c$	convex ordinal scale	convex ordering

For working with the standardized scales the following equalities can be useful:

$$O_{P_1 + \cdots + P_k} = O_{P_1} \dot{\cup} \cdots \dot{\cup} O_{P_k}$$
$$O_{P_1 + \cdots + P_k}^{cd} = O_{P_1}^{cd} + \cdots + O_{P_k}^{cd}$$
$$O_{P_1 \times \cdots \times P_k}^{cd} = O_{P_1}^{cd} \times \cdots \times O_{P_k}^{cd}$$
$$I_{P_1 + \cdots + P_k} = I_{P_1} \dot{\cup} \cdots \dot{\cup} I_{P_k}$$
$$C_{P_1 + \cdots + P_k} = C_{P_1} + \cdots + C_{P_k}$$
$$C_{P_1 \times \cdots \times P_k} = O_{P_1}^{cd} \times \cdots \times O_{P_k}^{cd} \mid O_{P_1}^c \times \cdots \times O_{P_k}^c$$

4 Measurability

Most important in measurement is the problem: by which scales can a given empirical structure be measured? In conceptual measurement we ask more specific whether an empirical context $\mathbb{K} := (G, M, I)$ admits (full) $\mathbb{S}$-measures for significant scales $\mathbb{S} := (G_\mathbb{S}, M_\mathbb{S}, I_\mathbb{S})$. The definition of an $\mathbb{S}$-measure yields directly that a mapping σ from G into $G_\mathbb{S}$ is an $\mathbb{S}$-measure if and only if $\mathbb{K}$ has the same extents as $\mathbb{K} \mid \mathbb{K}_\sigma$ where $\mathbb{K}_\sigma := (G, M_\mathbb{S}, I_\sigma)$ with $gI_\sigma m :\Leftrightarrow \sigma(g)I_\mathbb{S} m$; σ is a full $\mathbb{S}$-measure if and only if $\mathbb{K}$ and $\mathbb{K}_\sigma$ have the same extents. The context $\mathbb{K}_\sigma$ may be understood as another version of the subscale of $\mathbb{S}$ based on the image $\sigma(G)$; in general, a *subscale* $\mathbb{S}_T$ of $\mathbb{S}$ based on a subset T of $G_\mathbb{S}$ is the context $(T, M_\mathbb{S}, I_\mathbb{S} \cap (T \times M_\mathbb{S}))$. Since σ is an $\mathbb{S}$-measure of $\mathbb{K}$ if and only if σ is an $\mathbb{S}_{\sigma(G)}$-measure of $\mathbb{K}$, every scale measure can be replaced by a surjective one which has the same preimages of extents. This observation is basic for approaching the measurability problem; a first step is given by the following proposition:

Proposition 1 *Let σ be an S-measure of $\mathbf{K}$. Then*

$$(A, A') \mapsto (\sigma^{-1}(A), \sigma^{-1}(A)')$$

describes a $\wedge$-preserving map from $\mathfrak{B}(\mathbf{S})$ into $\mathfrak{B}(\mathbf{K})$; this map is injective if σ is surjective.

From this proposition we obtain a lattice-theoretical characterization of scale measures using a basic result on Galois connections (cf. [6]):

Proposition 2 *(cf. [18]) Let $\mathbf{S}$ be a scale in which $v \neq w$ implies $\{v\}' \neq \{w\}'$ for all $v, w \in G_\mathbf{S}$. Then, for an S-measure σ of $\mathbf{K} := (G, M, I)$,*

$$(A, A') \mapsto \tilde{\sigma}(A, A') := (\sigma(A)'', \sigma(A)')$$

describes a $\vee$-preserving map $\tilde{\sigma}$ from $\mathfrak{B}(\mathbf{K})$ into $\mathfrak{B}(\mathbf{S})$; in particular, $\tilde{\sigma}(\gamma g) = \gamma_\mathbf{S} \sigma(g)$ for all $g \in G$. Conversely, if φ is a $\vee$-preserving map from $\mathfrak{B}(\mathbf{K})$ into $\mathfrak{B}(\mathbf{S})$ such that for each $g \in G$ there is a $\bar{\varphi}(g) \in G_\mathbf{S}$ with $\varphi(\gamma g) = \gamma_\mathbf{S} \bar{\varphi}(g)$, then $\bar{\varphi}$ is an S-measure of $\mathbf{K}$. There is a one-to-one correspondence between the S-measures σ (resp. $\bar{\varphi}$) and the specific $\vee$-preserving maps $\tilde{\sigma}$ (resp. φ). σ is full if and only if $\tilde{\sigma}$ is injective.

For the characterization of scale measures of a given context $\mathbf{K}$ it is useful to know of which type are the subscales of the considered scales. For comparing scales we introduce a notion of equivalence: two scales $\mathbf{S}_1$ and $\mathbf{S}_2$ are called *equivalent* if there is an isomorphism φ from $\mathfrak{B}(\mathbf{S}_1)$ onto $\mathfrak{B}(\mathbf{S}_2)$ such that φ induces a bijection from $\gamma_{\mathbf{S}_1} G_{\mathbf{S}_1}$ onto $\gamma_{\mathbf{S}_2} G_{\mathbf{S}_2}$. The following table describes equivalences for the subscales of some standardized scales of ordinal type:

scale	subscales equivalent to
$\mathbf{O}_n$	one-dimensional ordinal scales
$\mathbf{N}_n$	one-dimensional nominal scales
$\mathbf{M}_{n_1,\ldots,n_k}$	multiordinal scales
$\mathbf{O}_\mathbf{p}^{cd}$	contrary ordinal scales
$\mathbf{N}_n^c$	complementary nominal scales
$\mathbf{I}_n$	one-dimensional interordinal scales
$\mathbf{C}_\mathbf{P}$	convex ordinal scales

The next table gives necessary and sufficient conditions for a context $\mathbf{K} := (G, M, I)$ to admit surjective measures onto specific scales:

scale S	condition for admitting a surjective S-measure
O_n	chain of n non-empty extents
N_n	partition of G into n extents
$M_{n_1,\ldots,n_k}$	partition of G into k chains of n_1 up to n_k extents
O_P^{cd}	isomorphic copy of P formed by extents which have as unions again extents; furthermore, for each object g there is a smallest of the fixed extents containing g and each of the fixed extents corresponds in this way to some object
N_n^c	partition of G into n extents which have as unions again extents
I_n	chain of n non-empty extents the complement of which are again extents
C_P	isomorphic copy of P formed by extents which have as unions and as complements again extents; furthermore, for each object g there is a smallest of the fixed extents containing g and each of the fixed extents corresponds in this way to some object

The conditions in the above table can be easily extended to obtain characterizations for full measures. Since subscales of ordinal scales need not to be ordinal again, the measurability into ordinal scales cannot be characterized via surjective measures. On the other hand, each context admits even full O_P-measures for suitable ordered sets P which can be determined via the ordered set of meet-irreducible concepts of the given context. This is demonstrated by the next two propositions.

Proposition 3 *Let $K := (G, M, I)$ be a finite context. For a bijection $\iota : \mathbf{n} \longrightarrow M$ a full B_n-measure σ of K is given by $\sigma(g) := \mathbf{n} \setminus \iota^{-1}\{g\}'$ for $g \in G$.*

The Boolean scales are equivalent to special grid scales; in general, $G_{n_1,\ldots,n_k}$ is equivalent to $O_{n_1 \times \cdots \times n_k}$. A full measure into a grid scale offers a scheme of ordinal dimensions for the interpretation of an empirical context. For a finite context K it is especially interesting to determine the smallest number k such that K admits a full measure into a k-dimensional grid scale; k is called the *grid dimension* of K.

Proposition 4 *(cf. [18]) Let $K := (G, M, I)$ be a finite context. For an order-preserving bijection $\iota : \mathbf{n}_1 + \cdots + \mathbf{n}_k \longrightarrow M(\mathfrak{B}(K))$ a full $G_{n_1+1,\ldots,n_k+1}$-measure σ_ι of K is given by $\sigma_\iota(g) := (v_1 + 1, \ldots, v_k + 1)$ for $g \in G$ where v_j is the greatest number in $\mathbf{n}_j$ such that g is not in the extent of $\iota(v_j)$. If τ is any full $G_{m_1,\ldots,m_l}$-measure of K then there exist always a bijection $\iota : \mathbf{n}_1 + \cdots + \mathbf{n}_k \longrightarrow M(\mathfrak{B}(K))$ with $k \leq l$ and order-preserving surjections $\nu_j : \{1, \ldots, m_j\} \longrightarrow \{1, \ldots, n_j + 1\}$ for $j = 1, \ldots, k$ such that $\sigma_\iota(g) = (\nu_1(w_1), \ldots, \nu_1(w_k))$ for $g \in G$.*

Corollary 1 *The grid dimension of K equals the width of $M(\mathfrak{B}(K))$.*

The full measures into semiproducts of (one-dimensional) scales are also interesting for other types of scales. Besides grid scales we consider here only semiproducts of nominal scales, in particular dichotomic scales. For the characterization of full measures in such scales we introduce the following notion: a context $K := (G, M, I)$ is called *atomistic* if γg is an atom of $\mathfrak{B}(K)$ for all $g \in G$.

Proposition 5 *A finite context* K *admits a full scale measure into a semiproduct of nominal scales if and only if* K *is atomistic.* K *admits a full scale measure into the k-dimensional dichotomic scale if and only if* K *is atomistic and there are at most k pairs of complementary extents to which all extents of meet-irreducible concepts of* K *belong.*

Direct products of scales are also interesting in conceptual measurement as it becomes clearer in section 6. Here we give only a characterization of full measures into some type of direct product scales which indicate a connection to order dimension; let us recall that the *order dimension* of an ordered set **P** is the smallest number of chains the direct product of which allows an order embedding of **P**.

Proposition 6 *(cf. [20]) A finite context* K *admits a full scale measure into a direct product of k contrary one-dimensional ordinal scales if and only if* K *is isomorphic to a context* $(P, P, \not\geq)$ *for some ordered set* **P** *of order dimension at most k.*

5 Scaled Contexts

A formal context, as defined in section 2, has a natural conceptual structure. Empirical data models, however, often arise in a form which does not *a priori* fall under this data type. They frequently use many-placed relations and operations. It seems natural to generalize the definition to that of a many-valued context (see [18, 10]): a *many-valued context* is a quadruple (G, M, W, I), where G, M and W are sets and I is a ternary relation between G, M and W (i.e. $I \subseteq G \times M \times W$) such that $(g, m, v) \in I$ and $(g, m, w) \in I$ always imply $v = w$; the elements of G, M and W are called *objects*, *(many-valued) attributes* and *attribute values*, respectively.

An attribute m of a many-valued context (G, M, W, I) may be considered as a partial map of G into W, which suggests to write $m(g) = w$ rather than $(g, m, w) \in I$, and to define the *domain* of m by

$$dom(m) := \{g \in G | (g, m, w) \in I \text{ for some } w \in W\}.$$

The attribute m is said to be *complete* if $dom(m) = G$, and a many-valued context is *complete* if all its attributes are. (G, M, W, I) is called an *n-valued context* if W has cardinality n. One-valued contexts correspond to the contexts defined in section 2.

Figure 10 shows an example taken from [11], it gives the rank-ordering of different Jazz styles according to personal constructs of a single test-person. We may interpret this data set as a many-valued context, with the styles ($G = \{$New Orleans Jazz, Ragtime,...,Swing$\}$) as object, the personal constructs ($M = \{$cheerful to aggressive – sad to melancholy, ... , white music – black music$\}$) as many-valued attributes, and with value set $W := \{1, 2, \ldots, 9\}$. The obvious interpretation is to read the ternary relation I from the table: the value written in row g and column m is just $m(g)$. This example yields a complete 9-valued context.

In general, there is no immediate, "automatic" way to associate a conceptual structure with a given many-valued context. The reason is that the notion is to general to reflect the structural information about the data set which is needed to do a conceptual analysis. Therefore a refined and enriched model is neccessary. This will be formalized below as a *scaled context*. We shall not abandon the notion of a many-valued context; our basic view is that empirical data are often represented in

Jazz styles	C1: cheerful to agressive — sad to melancholic	C2: wound up, wild — cool, damped	C3: emotional — intellectual	C4: regular rhythm — irregular rhythm	C5: typical of an area — typical of a breed of men	C6: suitable for film music — requires greater attention	C7: familiar, harmonic melody — unfamiliar, unharmonic melody	C8: short "songs" — longer compositions	C9: dance music — listening music	C10: long word — short word	C11: historically older — newer	C12: white music — black music
New Orleans Jazz	6	7	3	3	2	4	4	4	3	9	2	6
Ragtime	7	4	6	2	6	1	2	2	4	4	6	4
Free Jazz	1	6	8	9	8	9	9	9	9	5	9	1
Chicago Jazz	2	5	7	7	1	5	5	5	5	8	5	8
Bebop	4	2	2	5	4	7	6	6	6	2	3	9
Hard Bop	3	1	4	6	5	8	7	8	7	3	7	7
Cool Jazz	8	9	9	8	9	6	8	7	8	6	8	5
Dixieland	5	3	5	1	3	3	1	1	1	7	1	2
Swing	9	8	1	4	7	2	3	3	2	1	4	3

Figure 10: A ranking of Jazz styles by a non–expert

this form, from which in a formalized process of interpretation (called *scaling*) a scaled context can be derived. Among other advantages, this allows several interpretations of the same data set, e.g. "rougher" and "finer" analysis. The aim of the scaling process is to obtain a one-valued *derived context* with the same objects as (G, M, W, I), whose extents can be considered as the "meaningful" subsets of G with respect to an interpretation.

The first step of scaling is to uncover the structure of each attribute's value set. In the definition of a many-valued context, the values are just elements of a set; in practice, the values are often implictely structured, and sometimes it is tacitely assumed that some attribute values imply others. If e.g. a many-valued attribute has values "big" and "very big", two interpretations are possible: it may either be meant that "very big" implies "big", i.e. that every very big object ist also big. Or "big" is an abbreviation for "big, but not very big". In the second case, the concept of all big objects would not cover the very big ones. So, in the first step of scaling, our aim is to make precise which subsets T of the value-set $m(G)$ of an attribute m are "meaningful" or "concept-constituting" in the sense that the set of all objects having values in T is considered as an extent.

If we look at our example, we find that the values (of all attributes) are numbers, and it is stated that they indicate rankings of the objects. A natural assumption is that the values are taken from an ordinal scale, and that the extents of such a scale may be taken as concept-constituting entities. To be concret, if we interprete the values $1, 2, \ldots, 9$ of the attribute "emotional – intellectual" as the objects of the one-dimensional ordinal scale O_9, then we can form the concept of the "more emotional" Jazz styles (attribute value ≤ 4, say). But we cannot form the concept of the Jazz styles whose value for the emotional – intellectual attribute is one of $\{1, 2, 5, 7\}$, since $\{1, 2, 5, 7\}$ is not an extent of the ordinal scale.

Formally, the first step of scaling consists of assigning to each attribute $m \in M$ a scale $S_m := (G_m, M_m, I_m)$ with $m(G) \subseteq G_m$. (Sometimes it is useful to remove the condition that $m(G) \subseteq G_m$ and to introduce a mapping $\nu_m : m(G) \longrightarrow S_m$ instead. To keep notations short, we shall discuss only the simpler model.) The choice of these scales is a matter of interpretation, the task is to select S_m in such a way that every extent $U \subseteq G_m$ induces a meaningful set $U \cap m(G)$ of values of m and, conversely, that every meaningful set of attribute values is obtained in this way.

The second step of scaling is to decide how the different many-valued attributes can be combined to describe concepts. For qualitative data, as in our example, a simple conjuction of attributes may suffice. In a mathematical context, phrases such as "those quadrangles in which the height equals the width" must be permitted as more complex concept-constituting attribute combinations. Formally, we apply some *product operator* $\prod$ which composes the given scales to a common scale

$$S := \prod_{m \in M} S_m = \left(\underset{m \in M}{\times} G_m, N, J \right).$$

We shall not be very precise about the nature of such product operators, frequently used examples are the *semiproduct* and the *direct product*, introduced in section 3, but there is also the possibility of defining algebraic scale compositions, cf. [21]. It is assumed that the set of objects of the composed scale S is the cartesian product of the value sets of the scales S_m, and, moreover, that for each $i \in M$ the i-th projection π_i, defined by $\pi_i((g_m)_{m \in M}) := g_i$, is an S_m-measure of S. The simple case of just allowing conjunctions of the attribute values, as mentioned above, is obtained when using the semiproduct operator. We then speak of *plain scaling*.

A many-valued context $\mathbb{K} := (G, M, W, I)$ together with the scale $\mathbb{S}$ is called a *scaled context* and is denoted by $(\mathbb{K}; \prod_{m \in M} \mathbb{S}_m)$. To understand the interplay between $\mathbb{K}$ and $\mathbb{S}$, it is helpful to think of the many-valued context as given by a rectangular table, the rows of which are indexed by the objects, columns by attributes, and where the entries are the values, as in figure 10. Suppose that $\mathbb{K}$ is complete, then every row can be read as a tuple $(m(g))_{m \in M}$ of attribute values, and each such tuple is an element of $\bigtimes_{m \in M} G_m$. But these are just the objects of the commen scale $\mathbb{S}$, so that to each row of $\mathbb{K}$ there corresponds an object of the scale $\mathbb{S}$. If each row of $\mathbb{K}$ is replaced by the corresponding row of $\mathbb{S}$, we derive a (formal) context for the scaled context $(\mathbb{K}; \prod_{m \in M} \mathbb{S}_m)$. This is made precise by the following definition:

Let $\mathbb{K}$ be a complete many-valued context scaled by

$$\mathbb{S} := \prod_{m \in M} \mathbb{S}_m = (\bigtimes_{m \in M} G_m, N, J).$$

We define the *derived context* as a one-valued context $(G, N, \tilde{J})$, where the objects are the objects of $\mathbb{K}$, the attributes are those of $\mathbb{S}$, and the incidence is given by

$$g \tilde{J} n :\Leftrightarrow (m(g))_{m \in M} J n \text{ in } \mathbb{S}.$$

The concepts of the derived context are called the *concepts of the scaled context* $(\mathbb{K}; \prod_{m \in M} \mathbb{S}_m)$. By $\mathfrak{U}(\mathbb{K}; \prod_{m \in M} \mathbb{S}_m)$ we denote the set of extents of the derived context.

In case that $\mathbb{K} := (G, M, W, I)$ is not complete, the definition is a little more technical. Again, the derived context will have the objects G of $\mathbb{K}$ and the attributes N of $\mathbb{S}$. We first give an intuitive description, and then a formal definition of the incidence $\tilde{J}$ between G and N. Let g be an object and n be an attribute. In the complete case, we have identified the g-row of (G, M, W, I) with the object $(m(g))_{m \in M}$ of $\mathbb{S}$. In the non-complete case, there may be empty cells in this row. If we complete the row by filling values in the empty cells, we obtain an object of $\mathbb{S}$ which has the attribute n, or has not. This will, of course, usually depend on the values which were filled in, but it may happen that *every* completion has n, independently of the supplemented values. Only in this case we define $g \tilde{J} n$. The formal definition is:

Let $\mathbb{K}$ be a many-valued context scaled by

$$\mathbb{S} := \prod_{m \in M} \mathbb{S}_m = (\bigtimes_{m \in M} G_m, N, J).$$

We define the *derived context* as a one-valued context $(G, N, \tilde{J})$, where the objects are the objects of G, the attributes are those of $\mathbb{S}$, and the incidence is given by

$$g \tilde{J} n :\Leftrightarrow h J n \text{ for all objects } h := (h_m)_{m \in M} \text{ of } \mathbb{S} \text{ satisfying}$$
$$h_m = m(g) \text{ whenever } m(g) \text{ is defined.}$$

The concepts of the derived context are called the *concepts of the scaled context* $(\mathbb{K}; \prod_{m \in M} \mathbb{S}_m)$ and, as above, $\mathfrak{U}(\mathbb{K}; \prod_{m \in M} \mathbb{S}_m)$ denotes the set of extents of the derived context.

For defining dependencies we need to know how one may restrict a product operator Π to a subfamily of scales. For a subset R of M we define the derived context of $(\mathbb{K}; \prod_{m \in R} \mathbb{S}_m)$ to be the context $(G, N_R, \tilde{J}_R)$, where N_R is the set of all attributes $n \in N$ for which $(g_m)_{m \in M} J n$ iff $(h_m)_{m \in M} J n$ for all elements of $\bigtimes_{m \in M} G_m$ with $g_r = h_r$ for all $r \in R$. $\tilde{J}_R$ is defined as the restriction $\tilde{J}_R := \tilde{J} \cap G \times N_R$.

	C10≤6	C1≤2	C12≤8	C2≤4	C4≤8	C1≤6	C2,C3≤8	C1≤8	C4≤6	C11≤8	C10≤7	C8≤7	C3≤4	C3≤5	C5≤2	C11≤5	C12≤3	C12≤6	C12≤7
New Orleans Jazz			x			x	x	x	x		x	x	x	x	x			x	x
Ragtime	x		x		x		x	x	x	x	x							x	x
Free Jazz	x	x	x	x	x	x	x				x						x	x	x
Chicago Jazz			x	x	x	x	x	x		x		x				x	x		
Bebop	x		x		x	x	x	x	x	x	x	x	x		x				
Hard Bop	x			x	x	x	x	x	x	x		x	x						x
Cool Jazz	x		x		x				x		x	x	x					x	x
Dixieland			x		x	x	x	x	x	x	x		x			x	x	x	x
Swing	x		x		x			x		x	x	x	x	x	x			x	x

Figure 11: The reduced derived context from plain ordinal scaling

	C10≤6	C1≤2	C2≤4	C4≤8	C1≤6	C2,C3≤8	C1≤8	C4≤6	C10≤7	C3≤4	C3≤5
New Orleans Jazz					x	x	x	x	x	x	x
Ragtime	x			x		x	x	x	x		
Free Jazz	x	x	x	x	x	x			x		
Chicago Jazz			x	x	x	x	x				
Bebop	x			x	x	x	x	x	x	x	
Hard Bop	x		x	x	x	x	x	x		x	
Cool Jazz	x			x				x	x	x	
Dixieland				x	x	x	x	x	x	x	
Swing	x			x			x		x	x	x

	C12≤8	C11≤8	C8≤7	C5≤2	C11≤5	C12≤3	C12≤6	C12≤7
New Orleans Jazz	x		x	x			x	x
Ragtime	x	x					x	x
Free Jazz	x					x	x	x
Chicago Jazz	x	x	x		x	x		
Bebop	x	x	x	x				
Hard Bop		x	x					x
Cool Jazz	x		x				x	x
Dixieland	x	x			x	x	x	x
Swing	x	x	x	x			x	x

Figure 12: Subcontexts obtained by splitting the attribute set

Now let us apply the definitions to the example of the Jazz styles. We have already mentioned that the twelve many-valued attributes are all of ordinal nature, which can naturally be interpreted by an one-dimensional ordinal or interordinal scale. For ordinal scaling, we use the scale O_9 for all attributes. We decide for plain scaling, thus the composed scale will be the twelvefold semiproduct of O_9:

$$\mathbb{S} = O_9 \talloblong O_9 \talloblong O_9 \talloblong O_9 \talloblong O_9 \talloblong O_9 \talloblong O_9 \talloblong O_9 \talloblong O_9 \talloblong O_9 \talloblong O_9 \talloblong O_9.$$

The 9^{12} objects of this scale are the 12-tuples of integers between 1 and 9, and there are $9 \times 12 = 108$ attributes. The derived context thus has nine objects and 108 attributes. Most of these attributes are redundant, and the reduced context (see section 2) has only 19 attributes. This context is shown in figure 11. It has 190 concepts which is slightly too many to obtain a clear line diagram. However, there is a simple and precise way of representing a large concept lattice by *several* line diagrams (cf. [19]), by splitting the attribute set and drawing the concept lattices of the subcontexts separately. Our choice of subcontexts is shown in figure 12, the corresponding concept lattices have 38 and 18 elements, respectively (fig.13 and 14).

As a second plausible attempt we may unfold the Jazz styles data by plain interordinal scaling. There are, however, good reasons to be reluctant: we have already

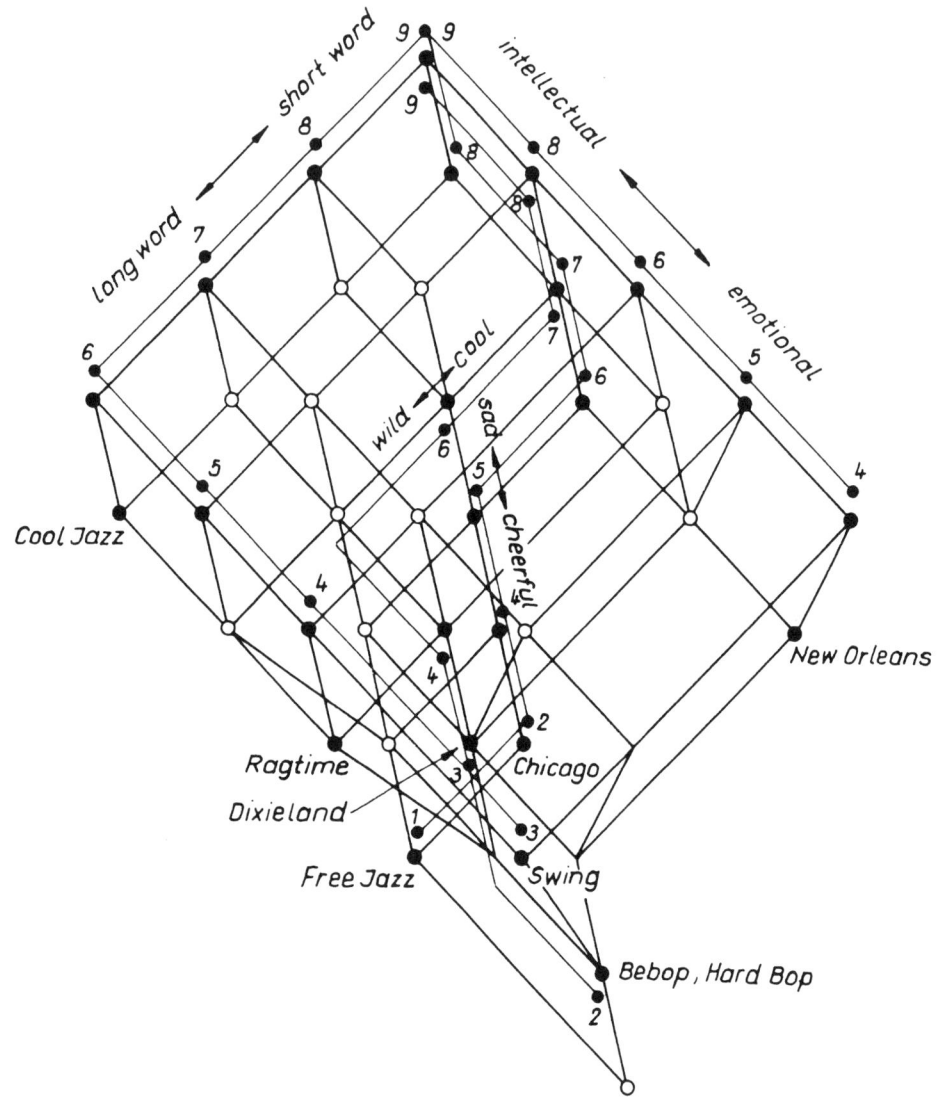

Figure 13: The concept lattice of the first subcontext in figure 12

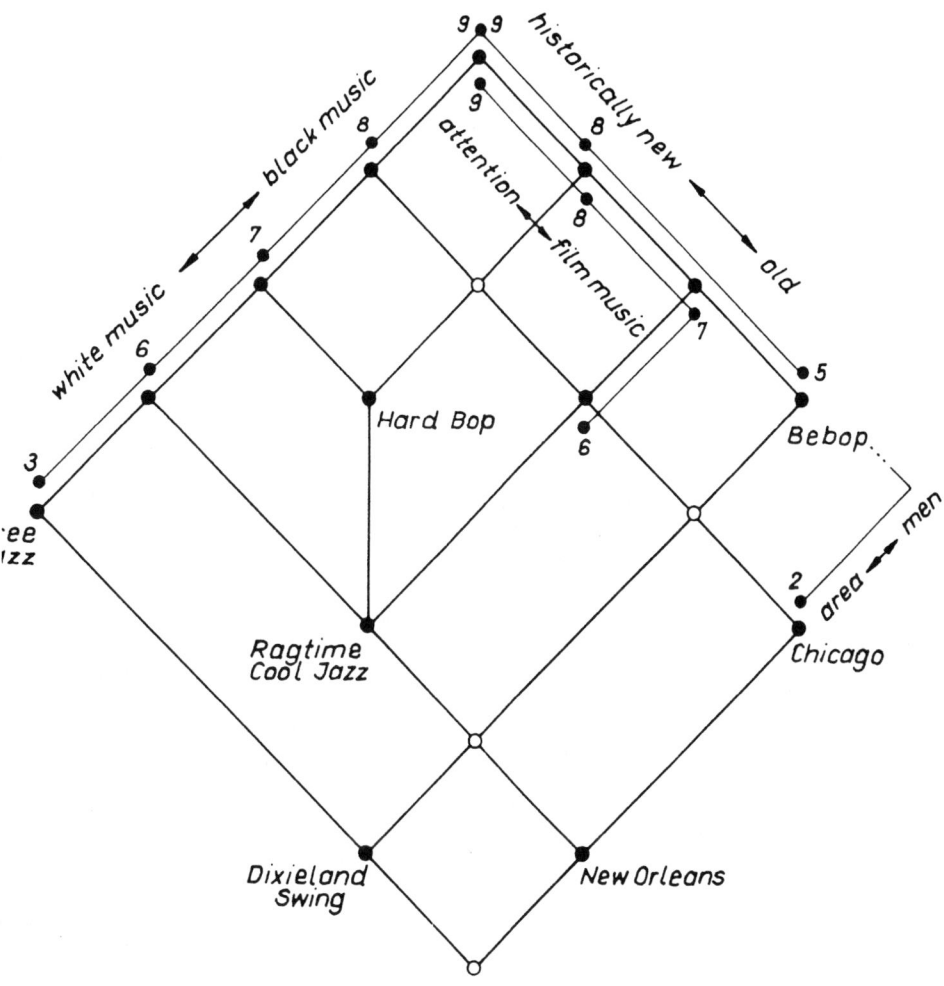

Figure 14: The concept lattice of the second subcontext in figure 12

	0	1
1	×	
2	×	
3	×	
4		
5		
6		
7		×
8		×
9		×

Figure 15: A dichotomic scale for threshold scaling

seen that plain ordinal scaling leads to a rather subtle analysis of the data set, and a comparison with the original data may suggest that we could overinterprete the data by analysing to a higher precision than justified by the data's meaning. The interordinal scale I_9 is finer than the ordinal scale O_9 in the sense that it admits an bijective O_9–measure, and it can be shown that this is inherited to the semiproduct scale. Thus, plain I_9-scaling would lead to a concept lattice which has far more than 190 elements, which is exaggerating. As a second scaling of this many-valued context we therefore discuss a much rougher approach, namely a threshold scaling into the dichotomic scale D^0. This scale has three scale values, while each of our many-valued attributes has nine attribute values. We therefore replace the scale by the equivalent scale shown in figure 15. The use of this scale has the following interpretation: we replace each many-valued attribute m, such as "emotional – intellectual", by two one-valued attributes (which we shall denote as "emotional" and "intellectual"). An object g is given the first attribute in the derived context, if $m(g) \leq 3$, and the second, if $m(g) \geq 7$. The derived context has 24 attributes, it is shown in figure 16. The 42-element concept lattice is given in figure 17.

6 Scaling and Dependency

A general notion of dependency between many-valued attributes has to include scalings of the attribute values. For instance, if we regard the values in fig.10 as unstructured we can only speak of functional dependency (cf. [12]), but in this case we obtain the useless result that each attribute functionally depends on each other attribute. For a meaningful analysis of dependencies we have to consider the ordinal nature of the attributes in fig.10. With the tools of formal concept analysis a general definition of dependency can be given based on the notion of a scaled context. The idea is that an attribute depends on other attributes if its conceptual contribution to the scaled context can already be furnished by the other attributes. Formally this is defined as follows (see [21]): in a scaled context $(\mathbb{K}; \prod_{m \in M} S_m)$ a set Y of attributes *depends* on a set X of attributes if every extent of $(\mathbb{K}; \prod_{m \in X \cup Y} S_m)$ is already an extent of

Figure 16: The derived context, from dichotomic threshold scaling

$(\mathbb{K}; \prod_{m \in X} \mathbb{S}_m)$, i.e.

$$\mathfrak{U}(\mathbb{K}; \prod_{m \in X \cup Y} \mathbb{S}_m) = \mathfrak{U}(\mathbb{K}; \prod_{m \in X} \mathbb{S}_m).$$

A weaker notion of dependency is given by the condition: every extent of $(\mathbb{K}; \prod_{m \in Y} \mathbb{S}_m)$ is an extent of $(\mathbb{K}; \prod_{m \in X} \mathbb{S}_m)$; if this condition is fulfilled, Y is said to be *weakly dependent* on X. Dependency and weak dependency coincide in plain scaled contexts but not generally in many-valued contexts scaled by direct products of scales.

The definition of functional dependency as it is introduced in the theory of relational databases, reads in the language of formal concept analysis as follows: in a complete many-valued context (G, M, W, I) a set Y of attributes *functionally depends* on a set X of attributes if, for all $g, h \in G$, $x(g) = x(h)$ for all $x \in X$ implies $y(g) = y(h)$ for all $y \in Y$, i.e. there exists a function $f : W^X \longrightarrow W^Y$ such that $f(x(g))_{x \in X} = (y(g))_{y \in Y}$ for all $g \in G$. The next proposition shows by a scaling with complementary nominal scales that functional dependency can be understood as a special case of the general dependency defined above.

Proposition 7 *(see [21]) Let $\mathbb{K}$ be a complete many-valued context scaled by $\underset{m \in M}{\times} \mathbb{N}_{n_m}^{c\bar{0}}$. For $X, Y \subseteq M$ the following conditions are equivalent:*

1. *Y functionally depends on X in $\mathbb{K}$.*

2. *Y depends on X in $(\mathbb{K}; \underset{m \in M}{\times} \mathbb{N}_{n_m}^{c\bar{0}})$.*

3. *Y weakly depends on X in $(\mathbb{K}; \underset{m \in M}{\times} \mathbb{N}_{n_m}^{c\bar{0}})$.*

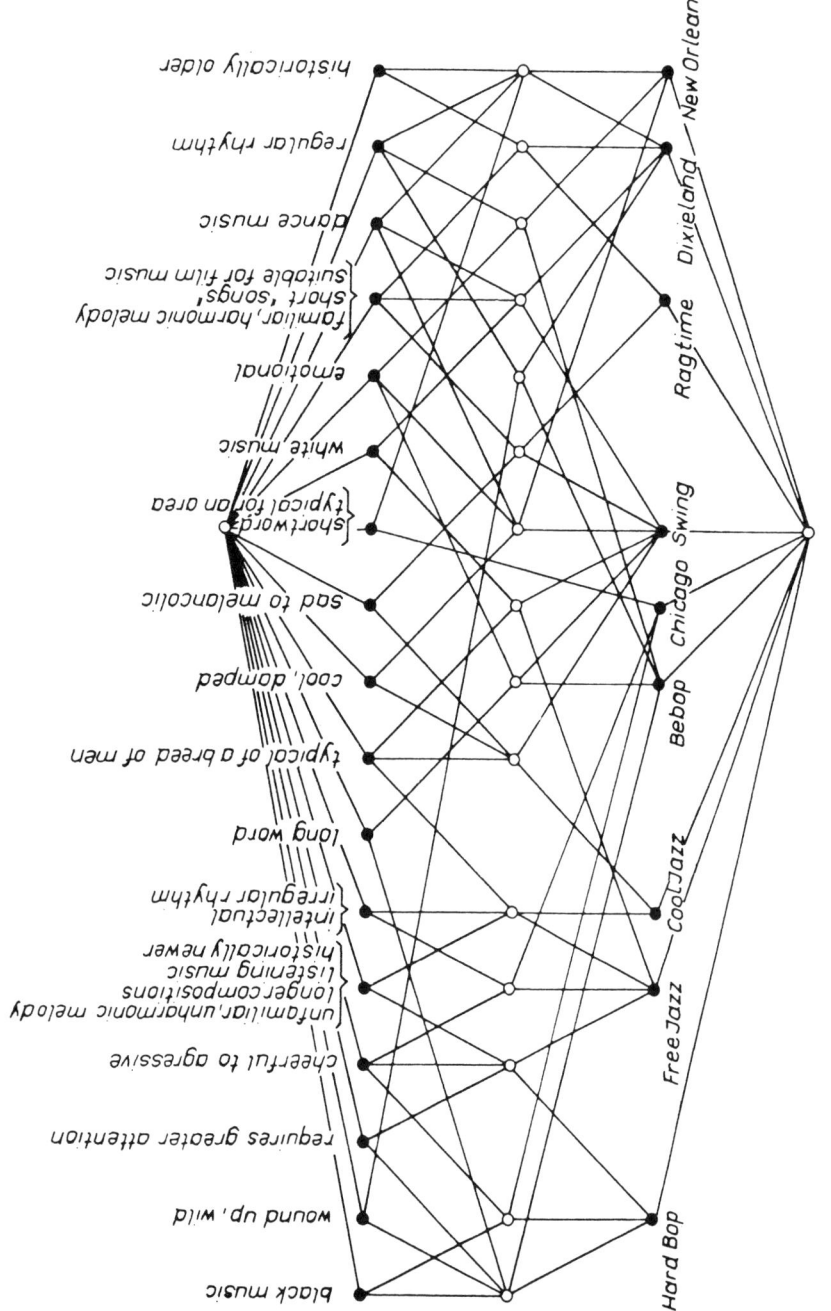

Figure 17: The concept lattice, from dichotomic treshold scaling

If the attribute values carry an order structure it is meaningful to introduce a notion of ordinal dependency (cf. [10]): in a complete many-valued context $(G, M, \mathbf{W}, I)$ with an ordered set $\mathbf{W} := (W, \leq)$ a set Y of attributes *ordinally depends* on a set X of attributes if, for $g, h \in G$, $x(g) \leq x(h)$ for all $x \in X$ implies $y(g) \leq y(h)$ for all $y \in Y$, i.e. there exists an order-preserving function $f : \mathbf{W}^X \longrightarrow \mathbf{W}^Y$ such that $f(x(g))_{x \in X} = (y(g))_{y \in Y}$ for all $g \in G$. It is also useful to introduce interordinal dependency considering the ternary relation *betweeness* which is defined by $[u, v, w] :\Leftrightarrow (u \leq v \leq w$ or $u \geq v \geq w)$. We say that Y **interordinally depends** on X if, for $g, h, k \in G$, $[x(g), x(h), x(k)]$ for all $x \in X$ implies $[y(g), y(h), y(k)]$ for all $y \in Y$, i.e. there exists a betweeness-preserving function $f : \mathbf{W}^X \longrightarrow \mathbf{W}^Y$ such that $f(x(g))_{x \in X} = (y(g))_{y \in Y}$ for all $g \in G$. Ordinal and interordinal dependency can also be understood in the frame of our general definition of dependency; for this we use scalings with contrary ordinal scales and convex ordinal scales, respectively.

Proposition 8 *(see [21]) Let $\mathbf{K}$ be a complete many-valued context scaled by $\underset{m \in M}{\times} \mathbb{O}_{\mathbf{P}_m}^{cd\bar{0}}$ so that the values of each attribute m of $\mathbf{K}$ are ordered by $\mathbf{P}_m$. For $X, Y \subseteq M$ the following conditions are equivalent:*

1. *Y ordinally depends on X in $\mathbf{K}$.*

2. *Y depends on X in $(\mathbf{K}; \underset{m \in M}{\times} \mathbb{O}_{\mathbf{P}_m}^{cd\bar{0}})$.*

3. *Y weakly depends on X in $(\mathbf{K}; \underset{m \in M}{\times} \mathbb{O}_{\mathbf{P}_m}^{cd\bar{0}})$.*

Proposition 9 *Let $\mathbf{K}$ be a complete many-valued context scaled by the apposition $\underset{m \in M}{\times} \mathbb{O}_{\mathbf{P}_m}^{cd\bar{0}} \mid \underset{m \in M}{\times} \mathbb{O}_{\mathbf{P}_m}^{c\bar{1}}$ so that the values of each attribute m of $\mathbf{K}$ are ordered by $\mathbf{P}_m$. For $X, Y \subseteq M$ the following conditions are equivalent:*

1. *Y interordinally depends on X in $\mathbf{K}$.*

2. *Y depends on X in $(\mathbf{K}; \underset{m \in M}{\times} \mathbb{O}_{\mathbf{P}_m}^{cd\bar{0}} \mid \underset{m \in M}{\times} \mathbb{O}_{\mathbf{P}_m}^{c\bar{1}})$.*

3. *Y weakly depends on X in $(\mathbf{K}; \underset{m \in M}{\times} \mathbb{O}_{\mathbf{P}_m}^{cd\bar{0}} \mid \underset{m \in M}{\times} \mathbb{O}_{\mathbf{P}_m}^{c\bar{1}})$.*

Ordinal and interordinal dependency can also be studied in many-valued contexts which are not complete (see [10]), here we restrict our explanations to the case of complete many-valued contexts. A basic question is how to determine and represent the dependencies of a scaled context. A solution for ordinal dependency is described in [10]. The idea is to translate the scaled context into a one-valued context with the same set of attributes so that the implications of the one-valued context are exactly the dependencies of the scaled context. Let us recall that in a (one-valued) context a set X of attributes *implies* a set Y of attributes if $X' \subseteq Y'$. In [8] a computer program is offered to determine functional, ordinal and interordinal dependencies; this program is based on the following proposition (notice that functional dependency is a special case of ordinal dependency).

Proposition 10 *Let $\mathbf{K} := (G, M, \mathbf{W}, I)$ be a complete many-valued context with an ordered set $\mathbf{W} := (W, \leq)$ of attribute values; furthermore, let $\mathbf{K}_o := (G \times G, M, I_o)$ be the context with $(g, h)I_o m \Leftrightarrow m(g) \leq m(h)$ and let $\mathbf{K}_{io} := (G \times G \times G, M, I_{io})$ be the context with $(g, h, k)I_{io} m \Leftrightarrow [m(g), m(h), m(k)]$. Then, for $X, Y \subseteq M$,*

1. $\{7,8,9,11,12\} \Rightarrow \{4\}$
2. $\{11,12\} \Rightarrow \{9\}$
3. $\{7,8,10,12\} \Rightarrow \{6\}$
4. $\{6,10,12\} \Rightarrow \{7,8\}$
5. $\{5,10,12\} \Rightarrow \{2,3\}$
6. $\{5,9,12\} \Rightarrow \{7,8\}$
7. $\{1,9,12\} \Rightarrow \{7,8\}$
8. $\{5,7,8,12\} \Rightarrow \{9\}$
9. $\{1,7,8,12\} \Rightarrow \{9\}$
10. $\{7,12\} \Rightarrow \{8\}$
11. $\{3,6,12\} \Rightarrow \{7,8,9\}$
12. $\{4,5,12\} \Rightarrow \{7,8,9,11\}$
13. $\{1,2,4,12\} \Rightarrow \{7,8,9,11\}$
14. $\{1,3,12\} \Rightarrow \{2,5\}$
15. $\{1,2,4,6,7,8,9,10,11\} \Rightarrow \{12\}$
16. $\{7,9,10,11\} \Rightarrow \{4\}$
17. $\{5,10,11\} \Rightarrow \{3\}$
18. $\{1,10,11\} \Rightarrow \{2\}$
19. $\{2,4,5,6,7,8,9,11\} \Rightarrow \{3\}$
20. $\{3,7,9,11\} \Rightarrow \{4\}$
21. $\{8,11\} \Rightarrow \{7,9\}$
22. $\{7,11\} \Rightarrow \{9\}$
23. $\{6,11\} \Rightarrow \{7,8,9\}$
24. $\{1,3,11\} \Rightarrow \{5\}$
25. $\{5,9,10\} \Rightarrow \{3,4,7,11\}$
26. $\{1,9,10\} \Rightarrow \{2,4,7,11\}$
27. $\{8,10\} \Rightarrow \{7\}$
28. $\{5,7,10\} \Rightarrow \{3,4,9,11\}$
29. $\{1,7,10\} \Rightarrow \{2,4,9,11\}$
30. $\{4,5,10\} \Rightarrow \{3,7,9,11\}$
31. $\{1,5,10\} \Rightarrow \{2\}$
32. $\{1,2,4,10\} \Rightarrow \{7,9,11\}$
33. $\{1,3,10\} \Rightarrow \{2\}$
34. $\{8,9\} \Rightarrow \{7\}$
35. $\{4,5,7,9\} \Rightarrow \{11\}$
36. $\{1,2,4,7,9\} \Rightarrow \{11\}$
37. $\{6,9\} \Rightarrow \{7,8\}$
38. $\{3,5,9\} \Rightarrow \{11\}$
39. $\{4,9\} \Rightarrow \{7\}$
40. $\{1,3,9\} \Rightarrow \{5,11\}$
41. $\{2,9\} \Rightarrow \{4,7\}$
42. $\{5,8\} \Rightarrow \{7\}$
43. $\{4,8\} \Rightarrow \{7\}$
44. $\{3,8\} \Rightarrow \{7,9\}$
45. $\{2,8\} \Rightarrow \{4,7\}$
46. $\{1,8\} \Rightarrow \{7\}$
47. $\{6,7\} \Rightarrow \{8\}$
48. $\{3,7\} \Rightarrow \{9\}$
49. $\{2,7\} \Rightarrow \{4\}$
50. $\{5,6\} \Rightarrow \{7,8\}$
51. $\{4,6\} \Rightarrow \{7,8\}$
52. $\{2,6\} \Rightarrow \{4,7,8\}$
53. $\{1,6\} \Rightarrow \{7,8\}$
54. $\{3,4,5\} \Rightarrow \{7,9,11\}$
55. $\{1,2,3,4\} \Rightarrow \{5,7,9,11\}$
56. $\{1,4\} \Rightarrow \{2\}$

Figure 18: A basis for the ordinal dependencies

1. X ordinally depends on Y if and only if X implies Y in $\mathbf{K}_o$,

2. X interordinally depends on Y if and only if X implies Y in $\mathbf{K}_{io}$.

Let us come back to our example from section 5. Since all rankings in fig.10 are different, each attribute functionally depends on each other attribute; but, by the same reason, no attribute ordinally depends on a single other attribute. All ordinal dependencies valid in our example can be derived from a minimal basis for these dependencies which is listed in fig.18 (cf. [5, 7]). For interpreting this analysis one should explicitly write down the dependencies in verbal form. For instance, dependency 48 of the list means that a Jazz style which is more emotional and more melodically familiar is judged rather to be dance music. Ordinal dependency compares the rankings only in the same direction, but in our example it makes also sense to compare the rankings in opposite directions. This is covered by the interordinal dependencies for which a minimal basis is given in fig.19.

1. $\{7,9,10,11,12\} \Rightarrow \{4\}$
2. $\{3,5,9,10,11,12\} \Rightarrow \{4,7\}$
3. $\{2,10,11,12\} \Rightarrow \{3,5\}$
4. $\{1,3,4,5,6,7,8,9,11,12\} \Rightarrow \{10\}$
5. $\{2,9,11,12\} \Rightarrow \{5\}$
6. $\{1,9,11,12\} \Rightarrow \{5\}$
7. $\{7,11,12\} \Rightarrow \{9\}$
8. $\{6,11,12\} \Rightarrow \{9\}$
9. $\{1,5,11,12\} \Rightarrow \{9\}$
10. $\{1,4,11,12\} \Rightarrow \{3\}$
11. $\{1,3,11,12\} \Rightarrow \{4\}$
12. $\{2,9,10,12\} \Rightarrow \{6,7,8\}$
13. $\{7,10,12\} \Rightarrow \{9\}$
14. $\{3,6,10,12\} \Rightarrow \{1,4,5,7,8,9,11\}$
15. $\{1,6,10,12\} \Rightarrow \{3,4,5,7,8,9,11\}$
16. $\{4,5,10,12\} \Rightarrow \{3\}$
17. $\{2,5,10,12\} \Rightarrow \{3\}$
18. $\{3,4,10,12\} \Rightarrow \{5\}$
19. $\{1,4,10,12\} \Rightarrow \{3,5\}$
20. $\{2,3,10,12\} \Rightarrow \{5\}$
21. $\{1,3,10,12\} \Rightarrow \{5\}$
22. $\{1,2,10,12\} \Rightarrow \{3,4,5,6,7,8,9,11\}$
23. $\{2,5,6,7,8,9,12\} \Rightarrow \{1,3\}$
24. $\{4,6,7,8,9,12\} \Rightarrow \{11\}$
25. $\{2,3,6,7,8,9,12\} \Rightarrow \{1\}$
26. $\{3,5,7,8,9,12\} \Rightarrow \{1\}$
27. $\{1,5,7,8,9,12\} \Rightarrow \{3\}$
28. $\{1,4,7,8,9,12\} \Rightarrow \{3,5,11\}$
29. $\{4,9,12\} \Rightarrow \{7\}$
30. $\{2,3,9,12\} \Rightarrow \{7,8\}$
31. $\{2,5,7,8,12\} \Rightarrow \{9\}$
32. $\{1,2,7,8,12\} \Rightarrow \{9\}$
33. $\{5,6,12\} \Rightarrow \{7,8\}$
34. $\{2,6,12\} \Rightarrow \{7,8,9\}$
35. $\{2,4,5,12\} \Rightarrow \{7,8,9,11\}$
36. $\{1,2,3,5,12\} \Rightarrow \{7,8,9\}$
37. $\{1,2,4,12\} \Rightarrow \{3,11\}$
38. $\{3,4,7,8,9,10,11\} \Rightarrow \{6\}$
39. $\{7,10,11\} \Rightarrow \{9\}$
40. $\{6,10,11\} \Rightarrow \{7,8,9\}$
41. $\{5,10,11\} \Rightarrow \{3\}$
42. $\{4,10,11\} \Rightarrow \{7,9\}$
43. $\{1,2,3,10,11\} \Rightarrow \{5\}$
44. $\{1,10,11\} \Rightarrow \{3\}$
45. $\{2,4,5,6,7,8,9,11\} \Rightarrow \{1\}$
46. $\{2,5,7,8,9,11\} \Rightarrow \{4\}$
47. $\{1,7,8,9,11\} \Rightarrow \{4\}$
48. $\{4,9,11\} \Rightarrow \{7\}$
49. $\{2,3,9,11\} \Rightarrow \{4,7,8\}$
50. $\{2,7,8,11\} \Rightarrow \{9\}$
51. $\{8,11\} \Rightarrow \{7\}$
52. $\{5,7,11\} \Rightarrow \{9\}$
53. $\{3,7,11\} \Rightarrow \{4,9\}$
54. $\{1,7,11\} \Rightarrow \{9\}$
55. $\{5,6,11\} \Rightarrow \{7,8,9\}$
56. $\{3,6,11\} \Rightarrow \{4,7,8,9\}$
57. $\{2,6,11\} \Rightarrow \{7,8,9\}$
58. $\{1,6,11\} \Rightarrow \{4,7,8,9\}$
59. $\{3,4,5,11\} \Rightarrow \{7,9\}$
60. $\{1,4,5,11\} \Rightarrow \{7,8,9\}$
61. $\{1,3,4,7,8,9,10\} \Rightarrow \{6\}$
62. $\{2,7,8,9,10\} \Rightarrow \{6\}$
63. $\{6,9,10\} \Rightarrow \{7,8\}$
64. $\{5,9,10\} \Rightarrow \{3,12\}$
65. $\{4,9,10\} \Rightarrow \{7\}$
66. $\{2,3,9,10\} \Rightarrow \{1,4,6,7,8\}$
67. $\{1,9,10\} \Rightarrow \{3\}$
68. $\{6,7,8,10\} \Rightarrow \{9\}$
69. $\{4,5,7,8,10\} \Rightarrow \{1\}$
70. $\{1,8,10\} \Rightarrow \{4,7\}$
71. $\{5,7,10\} \Rightarrow \{4\}$
72. $\{3,7,10\} \Rightarrow \{4,9\}$
73. $\{1,7,10\} \Rightarrow \{4,8\}$
74. $\{2,6,10\} \Rightarrow \{7,8,9\}$
75. $\{2,4,10\} \Rightarrow \{1,7,8\}$
76. $\{3,5,6,7,8,9\} \Rightarrow \{1\}$
77. $\{1,2,3,5,7,8,9\} \Rightarrow \{12\}$
78. $\{8,9\} \Rightarrow \{7\}$
79. $\{5,6,9\} \Rightarrow \{7,8\}$
80. $\{3,6,9\} \Rightarrow \{7,8\}$
81. $\{2,6,9\} \Rightarrow \{7,8\}$
82. $\{1,6,9\} \Rightarrow \{7,8\}$
83. $\{4,5,9\} \Rightarrow \{7,11\}$
84. $\{2,3,5,9\} \Rightarrow \{7,8\}$
85. $\{1,4,9\} \Rightarrow \{7,8\}$
86. $\{1,2,3,9\} \Rightarrow \{7,8\}$
87. $\{1,2,6,7,8\} \Rightarrow \{9\}$
88. $\{2,3,7,8\} \Rightarrow \{9\}$
89. $\{5,8\} \Rightarrow \{7\}$
90. $\{4,8\} \Rightarrow \{7\}$
91. $\{3,8\} \Rightarrow \{7\}$
92. $\{2,8\} \Rightarrow \{7\}$
93. $\{6,7\} \Rightarrow \{8\}$
94. $\{3,5,7\} \Rightarrow \{9\}$
95. $\{1,4,7\} \Rightarrow \{8\}$
96. $\{1,3,7\} \Rightarrow \{8\}$
97. $\{2,7\} \Rightarrow \{8\}$
98. $\{2,5,6\} \Rightarrow \{7,8\}$
99. $\{4,6\} \Rightarrow \{7,8\}$
100. $\{2,3,4,5\} \Rightarrow \{7,8,9,11\}$
101. $\{1,2,4,5\} \Rightarrow \{7,8\}$

Figure 19: A basis for the interordinal dependencies

References

[1] G. Birkhoff: *Lattice theory.* Third edition. Amer. Math. Soc., Providence / R. I. 1967.

[2] S. Brand (ed.): *The next Whole Earth Catalogue.* Random House, 1980.

[3] P. Burmeister: *Programm zur Formalen Begriffsanalyse einwertiger Kontexte.* TH Darmstadt 1987.

[4] V. Duquenne: *Contextual implications between attributes and some properties of finite lattices.* In: B. Ganter, R. Wille, K.-E. Wolff (eds.): Beiträge zur Begriffsanalyse. B.I.-Wissenschaftsverlag, Mannheim/Wien/Zürich 1987, 213 – 239.

[5] V. Duquenne, J.-L. Guigues: *Familles minimales d'implications informatives resultant d'un tableau de données binaires.* Math. Sci. Hum. **95**(1986), 5 – 18.

[6] M. Erné: *Einführung in die Ordnungstheorie.* B.I.-Wissenschaftsverlag, Mannheim/Wien/Zürich 1982.

[7] B. Ganter: *Algorithmen zur Formalen Begriffsanalyse.* In: B. Ganter, R. Wille, K.-E. Wolff (eds.): Beiträge zur Begriffsanalyse. B.I.-Wissenschaftsverlag, Mannheim/Wien/Zürich 1987, 241 – 254.

[8] B. Ganter: *Programmbibliothek zur Formalen Begriffsanalyse.* TH Darmstadt 1987.

[9] B. Ganter, J. Stahl, R. Wille: *Conceptual measurement and many-valued contexts.* In: W. Gaul, M. Schader (eds.): Classification as a tool of research. North-Holland, Amsterdam 1986, 169 – 176.

[10] B. Ganter, R. Wille: *Implikationen und Abhängigkeiten zwischen Merkmalen.* In: P.O. Degens, H-J. Hermes, O. Opitz (eds.): Die Klassifikation und ihr Umfeld. INDEKS Verlag Frankfurt 1986, 171 – 185.

[11] G. Gigerenzer: *Repertory–Test.* In: H. Bruhn, R. Oerter, H. Rösing (eds.): Musikpsychologie. Urban & Schwarzenberg, München–Wien–Baltimore 1985, 524 – 529.

[12] D. Maier: *The theory of relational databases.* Computer Science Press, Rockville 1983.

[13] F. S. Roberts: *Measurement theory.* Addison–Wesley, Reading/Mass. 1979.

[14] M. Skorsky: *Handbuch für Benutzer und Programmierer des Programmpaketes ANACONDA.* TH Darmstadt, 1986.

[15] S. S. Stevens: *Mathematics, measurement, and psychophysics.* In: S. S. Stevens (ed.): Handbook of experimental psychology. Wiley, New York 1951.

[16] W. S. Torgerson: *Theory and methods of scaling.* Wiley, New York 1958.

[17] H. Wagner: *Begriff.* In: Handbuch philosophischer Grundbegriffe. Kösel, München 1973, 191 – 209.

[18] **R. Wille:** *Restructuring lattice theory: an approach based on hierarchies of concepts.* In: I. Rival (ed.): Ordered sets. Reidel, Dordrecht – Boston 1982, 445 – 470.

[19] **R. Wille:** *Liniendiagramme hierarchischer Begriffssysteme.* In: H. H. Bock (ed.): Anwendungen der Klassifikation: Datenanalyse und numerische Klassifikation. INDEKS Verlag Frankfurt 1984, 32 – 51; engl. translation: Line diagrams of hierarchical concept systems. International Classification 11 (1984), 77 – 86.

[20] **R. Wille:** *Tensorial decompositions of concept lattices.* Order 2 (1985), 81 – 95.

[21] **R. Wille:** *Dependencies of many-valued attributes.* In: H. H. Bock (ed.): Classification and related methods of data analysis. North-Holland, Amsterdam 1988, 581 – 586.

[22] **R. Wille:** *Lattices in data analysis: how to draw them with a computer.* In: I. Rival (ed.): Algorithm and order. Reidel, Dordrecht - Boston (to appear).

THE MICRO-MACRO CONNECTION: EXACT STRUCTURE AND PROCESS*

EUGENE C. JOHNSEN[†]

Abstract. We present an analytic method for determining the model of exact micro- and macrostructures and their set of characterizing microprocesses for a given two-valued social relation in a human group. It is cast in terms which clearly show its applicability to any empirical network representing a relation in a group, human or otherwise, and, in principle, is extendable to networks in which a relation is multi-valued or there is more than one relation. At least for specific cases, such as those discussed in this paper, the method enables us to establish a connection between submodels and the constraining microprocesses which characterize and indirectly generate the micro- and macrostructures in them. A principal result of this method is a clear and exact processual characterization of cliquing and ranking in the macromodel which exactly fits the total empirical sociometric data sets of Davis-Leinhardt and Hallinan. Thus, the method produces a processual solution to some of the substantive structural questions discussed by Homans in The Human Group.

Key words. triadic micromodels, macromodels, submodels, triadic microprocesses

AMS(MOS) subject classifications. 92A20, 05C75

1. Introduction. Over the last several years, social scientists have been paying increasing attention to the relation between social phenomena occurring among individuals, families and other small subgroups (the micro-level), and that occurring within and among larger aggregates of such entities, such as large groups, organizations, nations and institutions (the macro-level). For examples, see Knorr-Cetina & Cicourel 1981; Collins 1981a; Coleman 1986, 1987; and Alexander, Giesen, Münch & Smelser 1987. This has led to various theoretical conceptualizations of how social behavior and attributes at one level might influence or result in social behavior and attributes at the other. For example, institutions and their activities emerging within a society over time may be viewed as resulting from the aggregation and repetition of interpersonal relations and behavior occurring among the individuals and small groups in that society (cf. Collins 1981b). Conversely, emerging interpersonal relations and behavior which can occur among individuals and small groups may be viewed as defined or highly constrained by the already present institutions and their activities (cf. Blau 1987). The relation between social phenomena at the two levels we shall call the *micro-macro connection*. Our distinction between micro and macro is analytic rather than quantitative; we shall typically consider entities at the macro level to consist of collections of entities at the micro level. As discussed by Coleman (1986, 1987), the emergence of macrophenomena from microphenomena is of fundamental sociological interest.

Along this line of approach lies the historical question of how interpersonal relations and activities among humans, during their early social development (possibly

---

*Some of the new results in this paper were presented at the Seventh Annual Social Network Conference in Clearwater Beach, Florida, February 14, 1987. The author wishes to thank the members of the Social Network Seminar at UCSB for their very useful discussions.

[†]Department of Mathematics, University of California, Santa Barbara, California 93106.

going back in time to their evolutionary primate ancestors), developed the kind of social organization and structure that is seen in, e.g., tribes, nations, religious institutions and corporations. For example, the various relations of kinship (especially of close kinship) with all their biological implications, must have had particular effects in the early formation and organization of human groups, some of which later became institutionalized as rights, duties, privileged subgroups, etc. persisting over time.

Since this leads straight to the important endeavor of trying to understand and explain basic social phenomena and the relations between these phenomena at different social levels, we need to begin by formally investigating specific fundamental social micro- and macrophenomena. One such phenomenon of particular importance is the formation of affect, sentiment, or friendship among individuals in a group, for which the micro-macro connection has been studied in an extended line of research in the balance theoretic tradition by Heider 1946; Newcomb 1953, 1961, 1968; Cartwright & Harary 1956; Davis 1963, 1967, 1970; Davis & Leinhardt 1972; Holland and Leinhardt 1970, 1971, 1973, 1976; Hallinan 1974; and Johnsen 1985, 1986.

Affect structure in groups is of considerable natural interest since it is frequently the case that other important interpersonal phenomena, such as personal interaction, information flow, influence, and attitude change tend to follow the network channels of positive sentiment (cf. Davis 1968, p. 549). Next to kinship, affect or friendship would seem to be the most important relation between humans in its effect on social attributes and behavior. The study of affect or friendship formation has led to the discovery of specific micro- and macromodels for the affect structure in empirical social groups (so far, for various sizes from 8 to 79, cf. Johnsen 1985). The macrostructures corresponding to these models can be viewed as ideal structures towards which the empirical structures are tending, although in few, if any, of the empirical structures do the ideal structures appear in their exact form.

In the next section, after some necessary mathematical preliminaries, we summarize as examples the micro- and macromodels for the balance, clustering, ranked clusters of cliques, ranked 2-clusters of cliques, and transitivity models, and in the following section we present the mathematical derivation of the macromodels corresponding to the hierarchical cliques and 39+ models. The first two models stem from balance theory in social psychology, while the latter five are found in the large Davis-Leinhardt data set of empirical sociomatrices (Davis 1970, Johnsen 1985).

A question which naturally arises here concerns the relations between these various models. In section 4 we discuss two types of submodels and supermodels, one of which is the class of submodels and supermodels for which every macrostructure in the submodel can be compatibly combined with every macrostructure in the supermodel by suitably linking all of the members of one with all of the members of the other so as to produce a macrostructure which is still in the supermodel. Such an idea is natural to consider when studying the possible results of the introduction of, say, a minority group into the social environment of a dominant majority group. It is pleasing to find that for every pair of the above models, for which one is a submodel of the other, this property holds.

We now go beyond the question of structure to inquire as to how the structure arises. This can be approached in two ways. In one, we postulate a particular process which may produce the ideal structure previously determined. In the other, we try to discover an incipient process which exactly describes how the structure can be generated. The first way was previously used to investigate affect or friendship formation by Johnsen (1986); we give a brief summary of this approach below. We then give a brief summary of the second way, one of the main objectives of this paper, which will be discussed in section 5.

A process of natural interest in social structure formation is that of how agreement or disagreement (on important issues or other persons) affects friendship or nonfriendship, and vice versa (cf. Lazarsfeld and Merton 1954; Coleman 1957; Newcomb 1961). In Johnsen (1986) we postulated four elementary and eight compound microprocesses by which agreement induces friendship and/or friendship induces agreement and, corresponding to them, their processual contrapositives at equilibrium by which nonfriendship induces disagreement and/or disagreement induces nonfriendship. To each microprocess there corresponds a set of triad types, each of which exhibits a violation of the process at equilibrium, called the *forbidden* triads, and its complementary set of triad types, each of which does not exhibit such a violation, called the *permitted* triads. Either set may be used to characterize in structural terms the outcome of the process proposed and be called the (triadic) *micromodel* of the microprocess. In specific cases the resulting micromodel is used to mathematically derive a corresponding macrostructure or macromodel in which the permitted triads appear but the forbidden ones do not. In each case the resulting macromodel shows the generic macrostructure of the group at equilibrium in terms of, e.g., structurally equivalent subclasses, such as cliques, and the interclass relations between them (cf. White, Boorman & Breiger, 1976).

The elementary microprocesses by which friendship induces agreement (or disagreement induces nonfriendship) produce triadic micromodels which are close approximations to the micromodels in the balance theoretic family arising in the Davis-Leinhardt data set (cf. Davis, 1970), which leads to the conclusion that the affect structure in social groups at or near equilibrium may be accounted for to a significant degree by the social microprocesses by which friendship induces agreement and disagreement induces nonfriendship within the group. This is consistent with and supported by substantive results from earlier empirical and experimental studies (cf. Johnsen 1986, section 6). If a particular microprocess is the predominant one acting in a group and it has progressed a considerable way towards equilibrium, then we should expect the ideal microstructure of the group to correspond exactly or very closely to the micromodel of the process and the ideal macrostructure to be the same as or very close to one of the macrostructures in the macromodel of the process. In the presence of other major processes in the group, however, the ideal macrostructure induced by a particular microprocess may not be highly visible. When two or more strong influences operate simultaneously, it is likely that none of the ideal structures induced by them will appear in its pure form, but it may be possible to account for deviations from one of them, at least in part, by the influence of certain other processes.

An agreement-friendship microprocess in which agreement is with respect to issues or values rather than persons is substantively a different microphenomenon but can also, to some extent, be considered here. Oversimplifying, when an important issue arises in a social group some members become identified with the pro and con sides of the issue, and the members' attitudes towards the issue may transfer more or less faithfully to these members. Then an agreement- friendship microprocess (agreement with respect to persons) may operate to govern the formation of the affect structure of the group. Such a shift, involving multiple issues and values, has been described by Newcomb (1961) for the rooming house groups in his study of the acquaintance process.

In attempting to determine how the generic macrostructure corresponding to a micromodel may have formed, one may sometimes discover incipient microprocesses which exactly describe how the triads in the micromodel could have been generated. This possibility is an attractive alternative to postulating and testing possible microprocesses having either substantive or theoretical appeal, since here there is no need for further analysis to determine how closely the micromodel is generated by the discovered microprocesses.

Recently, the author has developed an analytic procedure for detecting microprocesses from the triads in the micromodel of forbidden triads in question. This procedure has been applied to the empirically based micromodels of forbidden triads obtained for the Davis-Leinhardt data, with rather interesting results. For example, the model corresponding to the total Davis-Leinhardt data is exactly determined by two microprocesses operating concurrently, one inducing hierarchy and the other inducing the formation of cliques by external dissociation, whereby different cliques are separated and delineated by the growth of gaps between their members (to be distinguished from the formation of cliques by internal association, whereby each clique is connected and unified by the growth of strong ties between its members). Thus, the formation of a macrostructure from the hierarchical cliques model, which is the exact model for the total Davis-Leinhardt data set as well as the total Hallinan data set (Hallinan 1974), is generically characterized by two microprocesses for hierarchy formation and clique formation, respectively. We thus obtain a *process* description of how affect relations are structured in social groups: by ranking and cliquing as suggested by Homans (1950), where these are articulated in a manner anticipated by Davis and Leinhardt (1972) and determined by Johnsen (1985).

Further results obtained by this procedure for the other empirical models are also presented. In general, as group size decreases in ranges from $23-38$ to $8-13$ the set of microprocesses needed to characterize the structure of the group expands from two to four. The group size range $39-79$ appears to be an anomaly with three microprocesses required. Intuitively, an increase in number of constraining microprocesses with decreasing group size is consistent with the view that the more intimate the group the greater the opportunity for different interpersonal processes to hold sway.

2. Preliminaries. We formally represent a social group of n persons by the set of n vertices $V = \{v_1, v_2, \ldots, v_n\}$ and the two-valued relation R on the group by the two subsets D^+ and D^- of directed edges, i.e., ordered pairs of distinct vertices from V, where every such $(u,v) \in V \times V$ is labelled either $+$ or $-$, written $R(u,v) = +$ or $R(u,v) = -$ and drawn as a $+$ or $-$ signed arrow from u to v, where $(u,v) \in D^+$ if $R(u,v) = +$ and $(u,v) \in D^-$ if $R(u,v) = -$. We do not specify R for ordered pairs on the same vertex (u,u). Since D^+ and D^- are disjoint sets which together exhaust the ordered pairs of distinct vertices in $V \times V$, we can completely describe this structure by the loop-free digraph (V, D^+). When necessary to avoid clutter we will depict the structure or substructure of a group by an unsigned digraph where $R(u,v) = +$ is represented by an unsigned arrow from u to v and $R(u,v) = -$ by no arrow at all.

Let u, v be distinct vertices in V. If $R(u,v) = +$ and $R(v,u) = +$ then u and v are *mutually* related (connected, linked) and we write uMv. If $R(u,v) = -$ and $R(v,u) = -$ then u and v are *negatively* or null related and we write uNv. And if $R(u,v) = +$ and $R(v,u) = -$ then u is *asymmetrically* related to v and we write uAv. We see that M and N are symmetric relations (connections, links) and that A is an asymmetric relation on ordered pairs of distinct vertices in V. These are extended to full relations on V by taking *as a convention* that uMu and uAu, but not uNu, for all u in V, which makes M and A, but not N, reflexive on V as well. Now, among a set of three distinct vertices from V there are, to within isomorphism, 16 different combinations of M, A and N connections or dyads between the pairs of vertices in the set. Each combination is a *triad type* expressed as an ordered triple of nonnegative integers, $m : a : n$, where m, a and n are the numbers of M, A and N relations, respectively, and $m + a + n = 3$, together with a special letter C, D, T or U standing for "cyclic", "down", "transitive", or "up". The set Θ of the 16 different triad types is given in Figure 1.

Now, a group macrostructure model X can be defined in terms of the subset P_X of Θ, of all triads which appear in the structure, and its complementary subset $P_X^c = \Theta - P_X$ of all triads which do not. Clearly only P_X or P_X^c need be specified, and either one characterizes the generic group structure. For an arbitrary subset P_X it may not be possible to exhibit all of the triads of P_X and none of those in P_X^c in a single general macrostructure. When it is possible the macrostructure is called an *exact macrostructure* corresponding to (or an *exact macrorepresentation* of) P_X and P_X^c and we say that P_X or P_X^c is a (triadic) *micromodel*. The set Ξ_X of all macrostructures which are exact macrorepresentations of P_X is called the *macromodel* for (or corresponding to) P_X. Note that a macromodel may be empty; for example, there is no exact macrorepresentation of $P_Y = \{300, 003\}$, so P_Y is not a micromodel.

In this paper we are interested in how well certain triadic microprocesses account for the known empirically based macrostructures for affect. In the following examples we present the generic exact macrostructures for some prominent micromodels. The first two spring from balance theory in social psychology. The third and fifth were first proposed and the third, fourth and fifth were then discovered

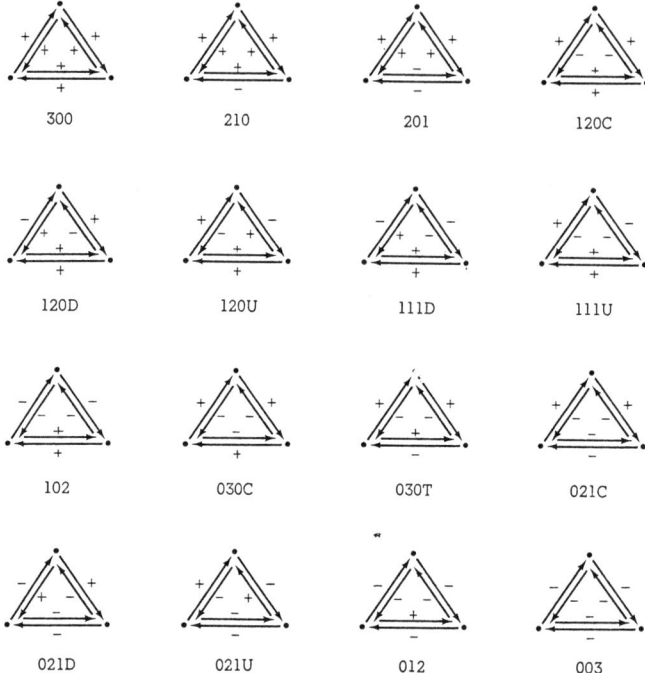

Figure 1. The Sixteen Triad Types

in the Davis-Leinhardt set of 742 empirical sociomatrices (Davis 1970, Davis & Leinhardt 1972, Holland & Leinhardt 1971, Johnsen 1985).

Example 2.1. $P_{BA} = \{300, 102\}$. Here the A dyad does not occur and the M and N dyads can be represented as single undirected signed edges, which reduces the formalization to a complete signed graph $(V, E^\pm)$, where E^+ is the set of all M dyads, E^- is the set of all N dyads, and E^+ and E^- exhaust the set of all unordered pairs of distinct vertices in V. Here Θ reduces to the set of four triads $\{300, 201, 102, 003\}$. In $(V, E^\pm)$ a *cycle* is any set of sequentially adjacent edges $\{\{u_i, u_{i+1}\} | i = 1, 2, \ldots, r; u_{r+1} = u_1\}$, where $u_1, u_2, \ldots, u_r$ are distinct vertices in V and $r \geq 3$; when the cycle consists of r edges and vertices we also call it an *r-cycle*. The analysis of a single triad (3-cycle) representing the attitudes among three persons or two persons and an object (issue, value) arose in the work of Heider (1946), where a psychologically harmonious or nonstressful triad was called *balanced* and a nonharmonious or stressful triad was called *imbalanced*. Oversimplifying Heider's analysis somewhat, it was determined that a balanced triad is one of the two types given in P_{BA} and an imbalanced triad one of the two shown in $\{201, 003\}$. Extending this notion to groups of more than three entities, Cartwright and Harary (1956) defined a signed graph and the group it represents to be *balanced* if every cycle in the graph has an even number of negative edges (N dyads). Thus, every triad in a balanced group and its signed graph is balanced. Flament (1963) showed, conversely, that if every triad is balanced then the group and its signed graph are balanced. From Harary (1954) and Cartwright and Harary (1956) we note that for a complete group and its signed graph to be balanced it is necessary and sufficient that the macrostructure of the group and its graph consists of at most two M-cliques (maximal subsets of vertices completely connected by M links) which are related by N^* (i.e., completely interconnected by N links). When both triad types in P_{BA} appear in the macrostructure then there are exactly two M-cliques and we have the complete *balance model BA* of Cartwright and Harary. See Figure 2.

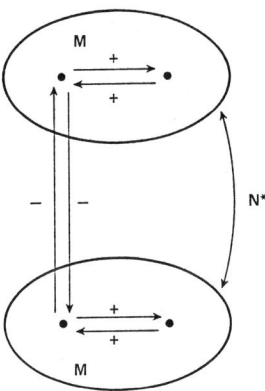

Figure 2. Typical Macrostructure for Complete Balance Model BA (M stands for M-clique)

Example 2.2. $P_{CL} = \{300, 102, 003\}$. As in the balance model above, we represent the group as a complete signed graph, where the only forbidden triad type is 201. Because the structure of the balance model was seen to be too restrictive compared to what is observed in human groups, Davis (1967) proposed to allow any number of M-cliques to be present. Here a group and its signed graph are said to have a *clustering* if its macrostructure consists of any number of M-cliques. Davis showed that for a group and its signed graph to have a clustering it is necessary and sufficient that no cycle in the graph has exactly one negative edge (N dyad) or, equivalently, that no triad in the graph has exactly one negative edge. The resulting exact macrostructure is that of the complete *clustering model CL* proposed by Davis (1967), consisting of three or more M-cliques which are pairwise related by N^*. See Figure 3.

It should be mentioned that in the groups considered by Cartwright and Harary (1956) and Davis (1967) empty edges, as well as positive and negative edges, were allowed between members, which means that the signed graphs representing these groups need not be complete. In this more general context, a cycle has no empty edges and is called *balanced* if it has an even number of negative edges (N dyads), and a group is said to have what we shall call *cycle balance* if every cycle in its signed graph is balanced. A group and its signed graph are said to have what we

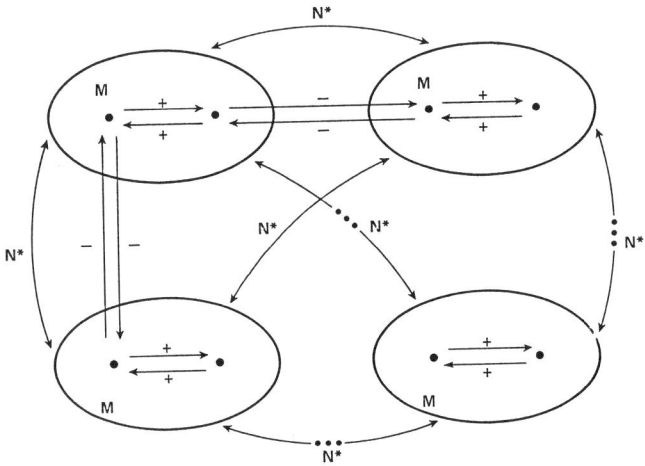

Figure 3. Typical Macrostructure for Complete Clustering Model CL (M stands for M-clique)

shall call *generalized clustering* if the vertices of the graph can be partitioned into components such that each positive edge (M dyad) connects vertices within the same component and each negative edge (N dyad) connects vertices in different components; when the partition consists of exactly s components we shall also call this a *generalized s-clustering*. More generally, Cartwright and Harary showed that a group has cycle balance if and only if the group has a generalized 1- or 2-clustering, and Davis showed that a group has a generalized clustering if and only if no cycle in its signed graph has exactly one negative edge (N dyad).

In this situation we might consider the signed graph of the group to consist of three types of dyads, M, Q, and N, where uQv means that $\{u,v\}$ is an empty edge, in which case every triad in the graph is of one of the 10 triad types $\{300, 210, 201, 120, 111, 102, 030, 021, 012, 003\}$, where type $m:q:n$ represents the triad consisting of m M dyads, q Q dyads, and n N dyads, $m+q+n=3$. Then, in a macrostructure having cycle balance 201 and 003 are forbidden and the other 8 triad types are all permitted, but there are also exact macrostructures meeting these restrictions which do not have cycle balance. Similarly, in a macrostructure having generalized clustering 201 is forbidden and the other 9 triad types are all permitted, but here also there are exact macrostructures meeting these restrictions which do not have generalized clustering. These are examples of families of macrostructures representable by signed graphs and defined in terms of simple graph properties which cannot be characterized simply in terms of their sets of permitted and forbidden triads.

Example 2.3. $P_{RC} = \{300, 102, 003, 120D, 120U, 030T, 021D, 021U\}$. The resulting exact macrostructure is that of the *ranked clusters of M-cliques model RC*, which was proposed by Davis and Leinhardt (1972) as the model to capture cliquing and ranking, two principal features of social structure discussed by Homans (1950). It does not characterize the total Davis- Leinhardt data set of sociomatrices, but it does characterize those in the range of group sizes 8 – 13, assuming triad type 003 is permitted. This macrostructure consists of a single stem hierarchy of cliques at different levels, with cliques at the same level pairwise related by N^* and cliques at different levels pairwise related by A^* (i.e., completely interconnected by the A relation, all A relations going in the same direction) from the lower clique to the higher one. See Figure 4.

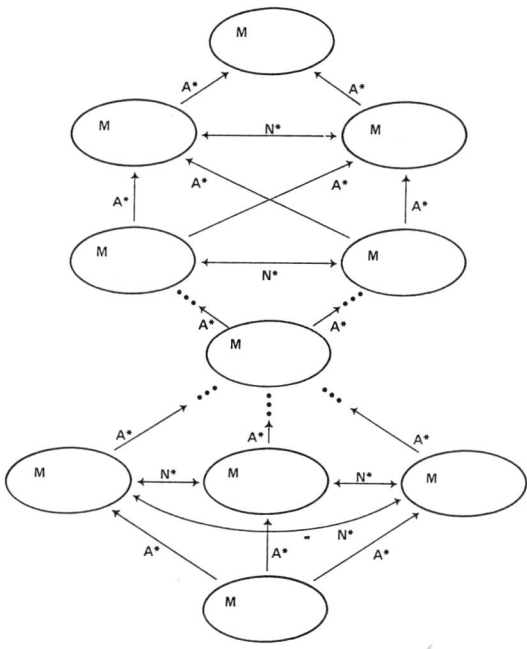

Figure 4. Typical Macrostructure for Ranked Clusters of M-cliques Model RC (M stands for M-clique; all interclique A^* relations implied by transitivity are suppressed)

Example 2.4. A special case of model RC is given by $P_{R2C} = \{300, 102, 120D, 120U, 030T, 021D, 021U\}$. The exact macrostructure here is that of the *ranked 2-clusters of M-cliques model R2C* (Johnsen, 1985) and is the same as that for RC except there are at most two cliques at each level. This model characterizes the sociomatrices in the range of group sizes 8 – 13 from the Davis-Leinhardt data set, when triad type 003 is assumed to be forbidden.

Example 2.5. $P_{TR} = \{300, 102, 003, 120D, 120U, 030T, 021D, 021U, 012\}$. The resulting exact macrostructure is that of the *transitivity model TR* proposed by Holland and Leinhardt (1971) to characterize the total set of sociomatrices in the

Davis-Leinhardt data set. The set P_{TR} comes to within one triad of characterizing this data set exactly, but does characterize exactly those in the range of group sizes 14 – 17. The exact macrostructure here consists of a collection of M-cliques partially ordered by the A^* relation (by convention, every M-clique is in relation A^* to itself) where incomparable M-cliques are pairwise related by N^*. See Figure 5.

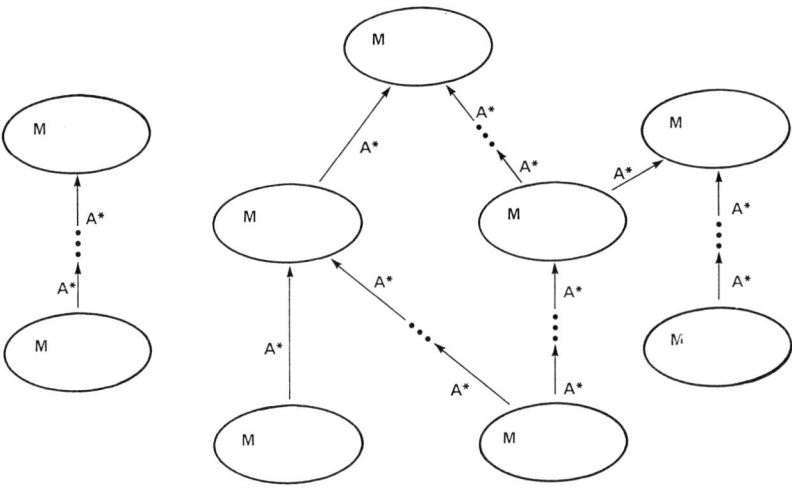

Figure 5. Typical Macrostructure for Transitivity Model TR (M stands for M-clique; all interclique A^* relations implied by transitivitiy are suppressed and all other missing interclique relations are N^*

How are triadic micromodels determined empirically? This is described by Davis and Leinhardt (1972) in testing the ranked clusters of M-cliques model against the empirical data from a sample of a large collection of sociomatrices (the Davis-Leinhardt data set). The statistical details are given by Holland and Leinhardt (1970, 1976). Briefly, for each sociomatrix the census of its triads according to the 16 triad types in Θ is obtained. Using $U|_{MAN}$, the uniform probability distribution over all complete signed digraphs having the same fixed numbers of M, A, and N dyads, one determines for each triad type in each sociomatrix whether that triad type occurs at a frequency less than chance expectation. Then, for various subsets of these sociomatrices one tabulates for each of the 16 triad types the percentage of sociomatrices in which that triad type occurs with less than chance frequency. In the Davis-Leinhardt data, this was done for the total set of sociomatrices and for the subsets of sociomatrices corresponding to group sizes 8 – 13, 14 – 17, 18 – 22, 23 – 38 and 39 – 79. For each of these sets of sociomatrices one obtains an empirical micromodel in terms of the set of triads each of which occurs with less than chance frequency in either (i) less than 50 percent or else (ii) at most 50 percent of the sociomatrices in the set. The sets for group sizes 8-13 and 14-17 yield the permitted triad types for the micromodels in Examples 2.3, 2.4 and 2.5 above. The sets for group sizes $18-22$, $23-38$, $39-79$ and the total data set yield micromodels which will be analyzed in the next section.

It would appear that allowing only the two labels + and − on directed edges is unduly restrictive for realistically representing a directed interpersonal relation, since the absent or zero relation is substantively different from the negative one. If we allow three values, +, 0 and − on directed edges, however, we face a more formidable problem in applying triadic analysis to the determination of macrostructure. With three different values on each directed edge, a simple application of Polya's Enumeration Theorem shows that there are now 138 nonisomorphic triad types. This would appear to be a more suitable set of triad types for analyzing the structure of very large groups, in which it is likely that many dyads will be absent. For statistical purposes we note that, in order to obtain each triad type on average the same number of times in a group with three-valued directed edges as in a group with two-valued directed edges, the group size needs to be approximately doubled.

3. Macromodels for the Larger Group Sizes. Except for perhaps model R2C, the generic macrostructures given in Examples 2.3 - 2.5 were known at the times they were proposed as possible models for the Davis-Leinhardt set of empirical sociomatrices. According to Davis (1979, p. 57), however, the generic macrostructures for the total Davis-Leinhardt set of empirical sociomatrices was not known nor, it would seem, those for the largest group sizes $39-79$. We present here the exact macromodels for these sets of data, as derived in Johnsen (1985).

Before proceeding further, however, we will need the following relation on V.

DEFINITION 3.1. The relation $\widetilde{M}$, called *M-chain-connectedness*, is defined on V as follows:

(a) $u\widetilde{M}u$ for all u in V, and

(b) $u\widetilde{M}v$ for $u \neq v$ in V if and only if for some $r \geq 2$ there exist vertices $x_1 = u, x_2, \ldots, x_r = v$ for which $x_i M x_{i+1}$ for all $i = 1, 2, \ldots, r-1$.

If $u\widetilde{M}v$ we say that u is *M-chain-connected* to v.

Intuitively, u M-chain-connected to v means that there exists a chain of mutual links between u and v. Substantively, when M denotes mutual friendship, it indicates that u and v can reach each other via a chain of pairwise mutual friends. We note that $\widetilde{M}$ is reflexive, symmetric and transitive, which means that $\widetilde{M}$ is an equivalence relation on V. Hence, V is partitioned into equivalence classes under $\widetilde{M}$, called $\widetilde{M}$-*cliques*, where every pair of vertices in the same $\widetilde{M}$-clique are M-chain-connected and every pair in different $\widetilde{M}$-cliques are not. An M-clique is a strong form of $\widetilde{M}$-clique where the shortest M-chain between every pair of vertices in the clique is of length 1 (corresponding to $r = 2$ in the definition). An $\widetilde{M}$-clique will be called an *isolate* if it consists of just one vertex, a *2-clique* if it consists of exactly two vertices, and *proper* if it consists of at least three. An isolate and a 2-clique are both M-cliques. In a macrostructure M_X we can determine when proper $\widetilde{M}$-cliques are M-cliques by the following lemma.

LEMMA 3.2. *Let M_X be an exact macrostructure in the macromodel Ξ_X. If $\{210, 201\} \subset P_X^c$, then every proper $\widetilde{M}$-clique in M_X is an M-clique, and conversely.*

Proof. Let W be a proper $\widetilde{M}$-clique in the exact macrostructure M_X and let u and v be any two distinct vertices in W. Since $u\widetilde{M}v$ there exist vertices $x_1 = u, x_2, \ldots, x_r = v$ in V such that $x_i M x_{i+1}$ for all $i = 1, 2, \ldots, r-1$ for some $r \geq 2$, and since $\widetilde{M}$ is an equivalence relation all of these vertices are in W. Now, if $r = 2$ then uMv. Suppose $r \geq 3$ and assume that $x_1 M x_{k-1}$ for $2 \leq k < r$. Since $x_{k-1} M x_k$ and triad types 210 and 201 are forbidden we must then have $x_1 M x_k$. Thus, by induction on k, we have $x_1 M x_r$ and again uMv. Since u and v were arbitrary, every link in W is M, whence W is an M-clique. Since W was arbitrary, every proper $\widetilde{M}$-clique in M_X is an M-clique. Conversely, suppose every proper $\widetilde{M}$-clique in M_X is an M-clique. Since any triad in M_X having two M dyads must be internal to some proper $\widetilde{M}$-clique, its third dyad must also be an M dyad. Thus, triad types 210 and 201 cannot occur in M_X, whence $\{210, 201\} \subset P_X^c$. □

Before obtaining the main results of this section we need the following

LEMMA 3.3. *Let M_X be an exact macrostructure in the macromodel Ξ_X. If $\{201, 111D, 111U\} \subset P_X^c$, then every link in a proper $\widetilde{M}$-clique of M_X is either an M dyad or an A dyad.*

Proof. Let W be a proper $\widetilde{M}$-clique in the exact macrostructure M_X. Let u and v be any two distinct vertices in W. Since $u\widetilde{M}v$ there exist vertices $x_1 = u, x_2, \ldots, x_r = v$ in W such that $x_i M x_{i+1}$ for all $i = 1, 2, \ldots, r-1$ for some $r \geq 2$. If $r = 2$ then uMv and we are done. Suppose $r = 3$. Then $x_1 M x_2$ and $x_2 M x_3$, and since triad type 201 is forbidden we must have either $x_1 M x_3$, $x_1 A x_3$ or $x_3 A x_1$, and again we are done. Now suppose $r \geq 4$ and, inductively, assume that either $x_1 M x_{r-1}$, $x_1 A x_{r-1}$ or $x_{r-1} A x_1$. Then, since $x_{r-1} M x_r$ and triad types 201, 111D and 111U are forbidden we must have either $x_1 M x_r$, $x_1 A x_r$ or $x_r A x_1$, and we are again done. Thus, by induction on r, we have the lemma. □

If W and W' are two $\widetilde{M}$-cliques for which uAv, respectively uNv, for every pair of vertices u in W and v in W' then W is defined to be in the relation A^*, respectively N^*, to W' and we write WA^*W', respectively WN^*W'.

LEMMA 3.4. *Let M_X be an exact macrostructure in the macromodel Ξ_X. If $\{120C, 111D, 111U\} \subset P_X^c$, then for every pair of distinct $\widetilde{M}$-cliques W and W' in M_X we must have either WA^*W', $W'A^*W$ or WN^*W'.*

Proof. If W and W' are both isolates then the lemma is trivially true. So suppose one of the $\widetilde{M}$-cliques, say W', has at least two vertices (the other $\widetilde{M}$-clique W having one or more vertices). Let u and v be two vertices in W' and let s be a vertex in W. Now there exist r vertices $x_1 = u, x_2, \ldots, x_r = v$ in W' such that $x_i M x_{i+1}$ for all $i = 1, 2, \ldots, r-1$ for some $r \geq 2$. Suppose sAu. Since $x_1 M x_2$ and triad types 120C and 111D are forbidden we must have sAx_2. This argument can be continued inductively to arrive at sAv. Thus, if sAu for some vertex u in W' then sAv for every other vertex v in W'. If W has only the one vertex s then WA^*W'. Now suppose W has another vertex t. Then there exist q vertices $y_1 = s, y_2, \ldots, y_q = t$ in W such that $y_j M y_{j+1}$ for all $j = 1, 2, \ldots, q-1$ for some $q \geq 2$. Since sAu and $y_1 M y_2$ and since triad types 120C and 111U are forbidden we must have $y_2 Au$. This argument can be continued inductively to arrive at tAu. Then by the previous argument we have tAv. Thus sAu implies that tAv for every t in W and v in W' and again we have WA^*W'. Now if, instead of sAu, we assume that uAs, then a similar argument to the above shows that vAt for every v in W' and t in W which means that $W'A^*W$. Finally, if sNu then, since triads 111D and 111U are forbidden, we have inductively that sNx_i and $x_i M x_{i+1}$ imply sNx_{i+1} for $i = 1, 2, \ldots, r-1$ or sNv, and likewise $y_j Nu$ and $y_j M y_{j+1}$ imply $y_{j+1} Nu$ for $j = 1, 2, \ldots, q-1$ or tNu, whence tNv. Thus, sMu implies that tNv for every t in W and v in W' and we have WN^*W'. □

LEMMA 3.5. *Let M_X be an exact macrostructure in the macromodel Ξ_X. If $\{030C, 021C\} \subset P_X^c$, then for every three distinct $\widetilde{M}$-cliques W, W' and W'' in M_X, if WA^*W' and $W'A^*W''$ then WA^*W'', i.e., A^* is transitive on the $\widetilde{M}$-cliques of M_X.*

Proof. Suppose for three distinct $\widetilde{M}$-cliques W, W' and W'' we have WA^*W' and $W'A^*W''$. Let s be any vertex in W and u any vertex in W'', and let t be some vertex in W'. Then sAt and tAu and, since the triad types 030C and 021C are forbidden, we must have sAu. So, for every s in W and u in W'' we have sAu, which means that WA^*W''. Thus $A*$ is transitive on the $\widetilde{M}$-cliques of M_X. □

LEMMA 3.6. *Let M_X be an exact macrostructure in the macromodel Ξ_X. If $\{120C, 030C, 021C\} \subset P_X^c$, then A is transitive on the vertices in M_X. Thus, under the relation A the vertices of M_X form a partially ordered set.*

Proof. Let u, v and w be vertices in M_X where uAv and vAw. Then, since triad types 120C, 030C and 021C are forbidden, we must have uAw. Thus A is transitive on the vertices of M_X. Since A is also reflexive and antisymmetric, the vertices of M_X form a partially ordered set under A. □

a. The total data set. For the total set of sociomatrices and for group sizes 18-22 and 23-38 in the Davis-Leinhardt data set the empirical micromodel to which the data conform is given by the set of forbidden triads

$$P_{HC}^c = \{201, 120C, 111D, 111U, 030C, 021C\}.$$

From the previous lemmas we then have the following result.

COROLLARY 3.7. *Let M_X be an exact macrostructure in the macromodel Ξ_X. If $P_X^c = \{201, 120C, 111D, 111U, 030C, 021C\}$ then*

(a) *under the relation A, the vertices in M_X form a partially ordered set, and*

(b) under the relation $\widetilde{M}$, the vertices in M_X are partitioned into $\widetilde{M}$-cliques such that

(i) every $\widetilde{M}$-clique has only M and A links and is partially ordered with respect to A, and

(ii) the set of all $\widetilde{M}$-cliques form a partially ordered set under the relation A^* induced by A, where $\widetilde{M}$-cliques which are incomparable with respect to A^* are in the relation N^* induced by N.

Proof. By Lemma 3.6 the vertices in M_X form a partially ordered set under the relation A. Under the relation $\widetilde{M}$, the vertices are partitioned into $\widetilde{M}$-cliques. By Lemmas 3.3 and 3.6 each $\widetilde{M}$-clique has only M and A links and is partially ordered with respect to A. By Lemma 3.4 every two $\widetilde{M}$-cliques are in either the A^* relation induced by A or the N^* relation induced by N. By Lemma 3.5 the relation A^* is transitive on the set of all $\widetilde{M}$-cliques and, since A^* is also reflexive (by convention) and antisymmetric, the set of all $\widetilde{M}$-cliques form a partially ordered set under this relation. As a result, $\widetilde{M}$-cliques which are incomparable with respect to A^* are in the relation N^*. □

An exact macrostructure in the model given by Corollary 3.7 is an amalgam of hierarchy and cliquing which we shall call the *hierarchical $\widetilde{M}$-cliques model*, written HC. The set of permitted triads for this model is

$$P_{HC} = \{300, 210, 120D, 120U, 102, 030T, 021D, 021U, 012, 003\}.$$

The generic macrostructure of this model is illustrated in Figure 6a where the structure internal to an $\widetilde{M}$-clique is illustrated in Figures 6b and 6c. We note that each $\widetilde{M}$-clique W contains a unique, possibly empty, M-subclique W_M consisting of all the vertices of W which are in the M relation to every other vertex of W. Thus, from the triad data found by Davis and Leinhardt (cf. Davis, 1970) we have the following important result.

THEOREM 3.8. *In the Davis-Leinhardt data set, the triad data for the empirical sociomatrices in the total set and in the sets for group sizes 18-22 and 23-38 conform exactly to the micromodel P_{HC}. Thus, the ideal macrostructures for these data are in the hierarchical $\widetilde{M}$- cliques macromodel Ξ_{HC}.*

b. The large group sizes. For group sizes 39 − 79 in the Davis-Leinhardt data set the empirical micromodel to which the triad data conform is given by the set of forbidden triads

$$P^c_{39+} = \{201, 111D, 111U, 030C, 021C, 003\}$$

and the set of permitted triads

$$P_{39+} = \{300, 210, 120C, 120D, 120U, 102, 030T, 021D, 021U, 012\}.$$

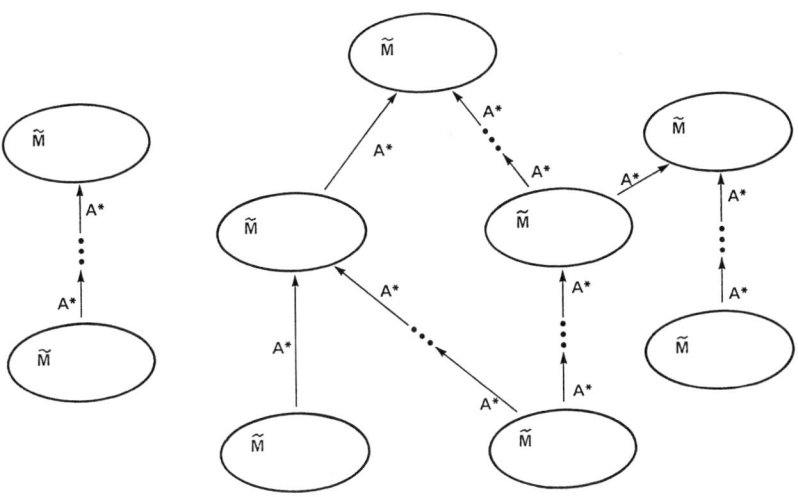

Figure 6(a). Typical Macrostructure for Hierarchical $\widetilde{M}$-cliques Model HC ($\widetilde{M}$ stands for $\widetilde{M}$-clique; all interclique A^* relations implied by transitivity are suppressed and all other missing interclique relations are N^*)

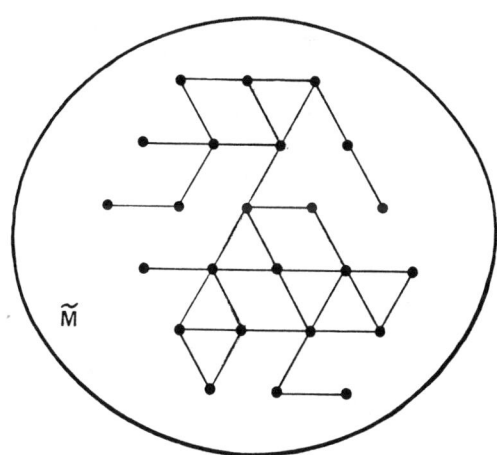

Figure 6(b). Hierarchical $\widetilde{M}$-cliques Model: M-chain-connectivity of an $\widetilde{M}$-clique (M links represented by undirected edges; not all M links shown)

Under the M-chain-connectedness relation $\widetilde{M}$ we again have V partitioned into

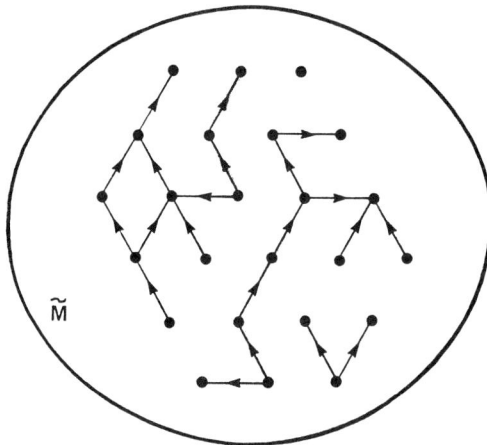

Figure 6(c). Hierarchical $\widetilde{M}$-cliques Model: Partial Ordering under A of an $\widetilde{M}$-clique (A links represented by directed edges; A links implied by transitivity are suppressed)

$\widetilde{M}$-cliques, and by Lemma 3.3 each $\widetilde{M}$-clique has only M and A connections. Furthermore, by Lemma 3.5, A^* is transitive on the set of $\widetilde{M}$-cliques, which means, as above, that the set of all $\widetilde{M}$-cliques forms a partially ordered set under the relation A^*. However, because triad 120C is now permitted, the relation A may not be transitive within $\widetilde{M}$-cliques and so neither V nor any $\widetilde{M}$-clique need be partially ordered by A. In addition, the triad 120C may be present between pairs of $\widetilde{M}$-cliques, which means that if two $\widetilde{M}$-cliques are incomparable with respect to A^* they need not be related by N^*. Nevertheless, since triads 111D ans 111U are forbidden, if any link between two $\widetilde{M}$-cliques is an A relation then every link between them is an A relation. Thus, every two $\widetilde{M}$- cliques W and W' are either completely A-linked (i.e., for every pair of vertices u in W and v in W' either uAv or vAu) or are completely N-linked (i.e., WN^*W').

If $\widetilde{M}$-cliques W and W' are completely A-linked we define them to be in the relation $A^\#$ and write $WA^\#W'$. Note that $A^\#$ is a symmetric relation between $\widetilde{M}$-cliques and that if WA^*W' then $WA^\#W'$. Since WA^*W for all $\widetilde{M}$-cliques W, this implies that $A^\#$ is reflexive on $\widetilde{M}$-cliques. We define the relation $\tilde{A}^\#$, called $A^\#$-chain-connectedness, on the set of all $\widetilde{M}$-cliques by stipulating that $W\tilde{A}^\#W$ for all $\widetilde{M}$-cliques W, and that $W\tilde{A}^\#Y$ for two $\widetilde{M}$-cliques W and Y if and only if for some $r \geq 2$ there are $\widetilde{M}$-cliques $\widetilde{M}_1 = W, \widetilde{M}_2,\ldots,\widetilde{M}_r = Y$ for which $\widetilde{M}_i A^\# \widetilde{M}_{i+1}$ for all $i = 1,2,\ldots,r-1$. If $W\tilde{A}^\#Y$ we say that W is $A^\#$-chain-connected to Y. We note that $\tilde{A}^\#$ is reflexive, symmetric and transitive, which means that $\tilde{A}^\#$ is an equivalence relation on the set $\{\widetilde{M}\}$ of all $\widetilde{M}$-cliques. Hence $\{\widetilde{M}\}$ is partitioned into equivalence classes under $\tilde{A}^\#$, called $\tilde{A}^\#$-superclusters or simply superclusters, where every pair of $\widetilde{M}$-cliques in the same supercluster are $A^\#$-chain-connected and every pair in different superclusters are in the relation N^*.

Now, because the triad 003 is forbidden, a macrostructure M_X here cannot consist of more than two superclusters, and if it consists of two superclusters then every interclique relation within each supercluster must be $A^\#$ (there cannot be any N^* relations). Only in the case of one supercluster is it possible for two $\widetilde{M}$-cliques in a supercluster to be related by N^*. In this case, because triad 003 is forbidden, the $A^\#$-distance between any two $\widetilde{M}$-cliques in the supercluster must be at most 3. In summary, then, from the triad data found by Davis and Leinhardt we have the following result.

THEOREM 3.9. *In the Davis-Leinhardt data set, the triad data for the empirical sociomatrices in the set for group sizes 39-79 conform exactly to the micromodel P_{39+}. Thus, the ideal macrostructures for this data consist of either*

(a) *two superclusters of $\widetilde{M}$-cliques, where each supercluster is a complete graph on its $\widetilde{M}$-cliques with respect to the $A^\#$ relation and the $\widetilde{M}$-cliques of each supercluster are partially ordered with respect to the A^* relation, or else*

(b) *one supercluster of $\widetilde{M}$-cliques, of diameter at most 3 with respect to the $A^\#$-distance between $\widetilde{M}$-cliques, in which the $\widetilde{M}$-cliques are partially ordered with respect to the A^* relation.*

The generic macrostructure for this micromodel is illustrated in Figure 7, where the internal microstructure of the $\widetilde{M}$-cliques is like that shown in Figures 6b and 6c, except that A need not be transitive within the $\widetilde{M}$-cliques.

4. Submodels. Here we investigate some basic ideas concerned with triadic micromodels and their corresponding macrostructures. Let $P_X \subset \Theta$ and $P_Y \subset \Theta$ be micromodels. If $P_X \subset P_Y$ then P_X will be called a *microsubmodel* or *submodel* of P_Y, and P_Y will be called a *microsupermodel* or *supermodel* of P_X. Since subset containment is reflexive and antisymmetric as well as transitive, we see that the micromodels from Θ form a proper subposet of the poset of all subsets of Θ under $\subset$. Corresponding to P_X is the macromodel Ξ_X of all macrostructures which are exact macrorepresentations of P_X. For virtually all of the micromodels of interest to us here the corresponding macromodels will consist of an infinite number of exact macrostructures. For examples, in an exact macrostructure for the complete balance model there are exactly two M-cliques but each may be of any size (at least one ≥ 3), and in an exact macrostructure for the complete clustering model there may be any number ≥ 3 of M-cliques, each of any size (at least one ≥ 3), thus Ξ_{BA} and Ξ_{CL} are both infinite.

Now suppose $P_X \subset P_Y$ and let $M_X \in \Xi_X$ and $M_Y \in \Xi_Y$ be corresponding exact macrostructures. Then, although every triad type which occurs in M_X also occurs in M_Y, it is easily seen that M_X need not be isomorphic to any subgraph (i.e., induced signed subdigraph) of M_Y.

EXAMPLE 4.1. Let $P_X = P_{CL}$ (cf. Example 2.2) and $P_Y = P_{TR}$ (cf. Example 2.5), where $M_X = M_{CL}$ consists of nine M-cliques $U_1, U_2, \ldots, U_9$ of sizes $5, 5, 6, 6, 6, 7, 8, 8, 10$, respectively, which are pairwise related by N^*, and $M_Y = M_{TR}$ consists of five M-cliques $W_1, W_2, \ldots, W_5$ of sizes $3, 3, 3, 4, 4$, respectively, for

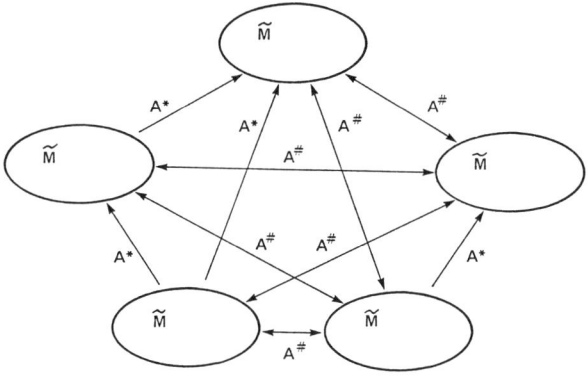

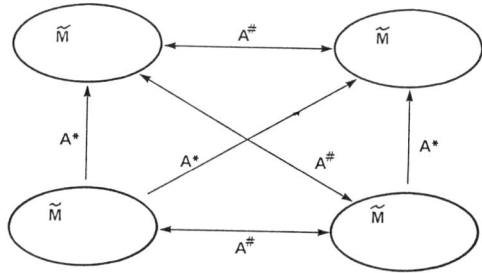

Figure 7. Typical Macrostructure for Group Sizes 39-79 Model 39+ ($\widetilde{M}$ stands for $\widetilde{M}$-clique; all interclique relations between superclusters are N^*; when only one supercluster, some interclique relations may be N^*)

which $W_1A^*W_2, W_1A^*W_3, W_1A^*W_4, W_2A^*W_4, W_3A^*W_4$ and all other interclique relations are N^*. Because M_X has more M-cliques and of larger sizes than does M_Y (as well as having no A^* relations) it is clear that M_X cannot be isomorphic to any subgraph of M_Y.

However, if M_X can be combined with M_Y into a single macrostructure $M'_Y \in \Xi_Y$ by linking all the vertices of M_X to all the vertices of M_Y with suitable choices of M, A and N dyads so that all the newly formed triads are also in P_Y, then M'_Y is a member of Ξ_Y which contains M_X (and also M_Y) as a substructure and we say that M_X is *compatible with* or can be *compatibly combined with* M_Y in Ξ_Y. In general, if M_A and M_B are two macrostructures where M_B contains a subgraph (i.e., induced signed subdigraph) isomorphic to M_A then we say that M_A is (or forms, or is embedded as) a *substructure* of M_B and that M_B is (or forms) a *superstructure* of M_A.

Now, if $P_X \subset P_Y$ then every triad type which appears in all the macrostructures of Ξ_X also appears in all the macrostructures of Ξ_Y and we call Ξ_X a *macrosubmodel* or *submodel* of Ξ_Y and Ξ_Y a *macrosupermodel* or *supermodel* of Ξ_X and denote all this by $\Xi_X < \Xi_Y$. Thus the one-to-one mapping from the set of all micromodels from Θ to their corresponding macromodels induces a partial ordering $<$ on the set of all macromodels so that the resulting poset is isomorphic to the poset of micromodels. If each macrostructure in Ξ_X is a substructure of some macrostructure in Ξ_Y then we say that Ξ_X is a *sharp* submodel of Ξ_Y, that Ξ_Y is a *sharp* supermodel of Ξ_X, and denote all this by $\Xi_X << \Xi_Y$. If every macrostructure in Ξ_X can be compatibly combined with every macrostructure of Ξ_Y in Ξ_Y then we say that Ξ_X is a *totally compatible* submodel of Ξ_Y, that Ξ_Y is a *totally compatible* supermodel of Ξ_X, and denote all this by $\Xi_X \triangleleft \Xi_Y$. We note that if Ξ_X is a totally compatible submodel of Ξ_Y then Ξ_X is also a sharp submodel of Ξ_Y. We also observe that not every macrostructure in Ξ_Y need be a superstructure of some macrostructure in Ξ_X.

EXAMPLE 4.2. P_{RC} and P_{TR} (cf. Examples 2.3 and 2.5) are micromodels for which Ξ_{RC} is a totally compatible submodel of Ξ_{TR}. (If $M_{RC} \in \Xi_{RC}$ and $M_{TR} \in \Xi_{TR}$ then the macrostructure M'_{TR} constructed by forming $M_X N^* M_Y$ is always in Ξ_{TR}.) The M_{TR} in Example 4.1 is a macrostructure in Ξ_{TR}, but is not a superstructure of any macrostructure in Ξ_{RC} because it does not contain a subgraph which exhibits all of the triads in P_{RC} (in particular 030T and 003) but not 012.

Conversely, if Ξ_X and Ξ_Y are macromodels for which P_X is a submodel of P_Y, Ξ_X need not be a sharp (hence not a totally compatible) submodel of Ξ_Y.

EXAMPLE 4.3. $P_{BA} = \{300, 102\} \subset \{300, 102, 120U, 111U\} = P_{1A}$. Here a typical macrostructure M_{1A} in Ξ_{1A} consists of an M-clique W and a vertex z, where W is partitioned into two M-subcliques U and V for which UA^*z and VN^*z. No macrostructure M_{BA} consisting of two M-cliques, each of size ≥ 2, can be a substructure of any macrostructure M_{1A}. Thus, Ξ_{BA} is a submodel but not a sharp submodel of Ξ_{1A}.

In this example we also see that Ξ_{1A} is not a totally compatible submodel of itself, i.e., is not *totally self-compatible*, since the special vertex z which is the target of A relations in a macrostructure of Ξ_{1A} is unique while a combination of two macrostructures of Ξ_{1A} has two such vertices.

Likewise, if Ξ_X and Ξ_Y are macromodels for which P_X is a sharp submodel of P_Y, Ξ_X need not be a totally compatible submodel of Ξ_Y.

EXAMPLE 4.4. $P_{1B} = \{300, 120U\} \subset \{300, 120U, 102, 021U\} = P_{1C}$. A typical macrostructure M_{1B} in Ξ_{1B} consists of one M-clique W and a single vertex z for which WA^*z, and a typical macrostructure M_{1C} in Ξ_{1C} consists of two M-cliques W' and W'' and a single vertex y for which $W'A^*y$, $W''A^*y$ and $W'N^*W''$. It is easy to see that each M_{1B} can be embedded in some M_{1C} by identifying W with

W' and z with y; so we have $\Xi_{1B} \ll \Xi_{1C}$. However, M_{1B} cannot be compatibly combined with an M_{1C} in Ξ_{1C}, since the vertices z and y must remain distinct in the combination while the resulting macrostructure M'_{1C} in Ξ_{1C} can have only one vertex which is the target vertex of an A relation; so we do not have $\Xi_{1B} \triangleleft \Xi_{1C}$.

We note that Ξ_X is a totally compatible, hence sharp, submodel of Ξ_Y in some important cases.

LEMMA 4.5. *Let Ξ_X and Ξ_Y be macromodels for the corresponding micromodels P_X and P_Y, where P_X is a submodel of P_Y and any of the following triad sets is a subset of P_Y:*

(a) $\{102, 012, 003\}$,

(b) $\{120D, 120U, 030T, 021D, 021U\}$,

(c) $\{300, 210, 201\}$.

Then Ξ_X is a totally compatible submodel of Ξ_Y.

Proof. For any $M_X \in \Xi_X$ and any $M_Y \in \Xi_Y$ there are four possible constructions here by which M_X can be compatibly combined with M_Y to form a macrostructure M'_Y in Ξ_Y, namely, completely interconnecting M_X and M_Y by (i) N relations when (a) holds, (ii) A relations from M_X to M_Y or (iii) A relations from M_Y to M_X when (b) holds, and (iv) M relations when (c) holds. In each case all the newly created triads interconnecting M_X and M_Y are in P_Y. Hence $\Xi_X \triangleleft \Xi_Y$. □

LEMMA 4.6. *Let Ξ_X and Ξ_Y be macromodels for the corresponding micromodels P_X and P_Y, where P_X is a submodel of P_Y. Suppose the triads in P_Y either have*

(1) *no M dyads and one of the sets $\{012, 003\}$ or $\{030T, 021D, 021U\}$ is a subset of P_Y, or*

(2) *no A dyads and one of the sets $\{102, 003\}$ or $\{300, 201\}$ is a subset of P_Y, or*

(3) *no N dyads and one of the sets $\{120D, 120U, 030T\}$ or $\{300, 210\}$ is a subset of P_Y.*

Then Ξ_X is a totally compatible submodel of Ξ_Y.

Proof. Similar to that for Lemma 4.5. □

We note that these submodel relations are transitive in certain interesting combinations.

LEMMA 4.7. *Let Ξ_X be a submodel of Ξ_Y and Ξ_Y be a submodel of Ξ_Z.*

(A) *If Ξ_X is a sharp submodel of Ξ_Y and Ξ_Y is a sharp submodel of Ξ_Z then Ξ_X is a sharp submodel of Ξ_Z.*

(B) *If Ξ_X is a sharp submodel of Ξ_Y and Ξ_Y is a totally compatible submodel of Ξ_Z then Ξ_X is a totally compatible submodel of Ξ_Z.*

(C) If Ξ_X is a totally compatible submodel of Ξ_Y and Ξ_Y is a totally compatible submodel of Ξ_Z then Ξ_X is a totally compatible submodel of Ξ_Z.

(D) If Ξ_X is a sharp submodel of Ξ_Y and Ξ_Y is totally self-compatible then Ξ_X is a totally compatible submodel of Ξ_Y.

Proof. If $\Xi_X << \Xi_Y$ and $\Xi_Y << \Xi_Z$ then $\Xi_X << \Xi_Z$, since a substructure of a substructure of an M_Z is a substructure of M_Z, which yields part (A). Now suppose $\Xi_X << \Xi_Y$ and $\Xi_Y \triangleleft \Xi_Z$ and let $M_X \in \Xi_X$ and $M_Z \in \Xi_Z$. Then M_X is a substructure of some $M_Y \in \Xi_Y$ and M_Y can be compatibly combined with M_Z to form an $M'_Z \in \Xi_Z$. Within M'_Z we have M_X compatibly combined with M_Z, forming a substructure $M''_Z \in \Xi_Z$. Since $M_X \in \Xi_X$ and $M_Z \in \Xi_Z$ are arbitrary we have $\Xi_X \triangleleft \Xi_Z$, which yields part (B). Since a totally compatible submodel is also a sharp submodel, part (C) follows from part (B). Finally, setting $Y = Z$ in part (B) we obtain part (D). □

Since the sharp submodel relation $<<$ between macromodels is reflexive and antisymmetric as well as transitive, we see that the set of all macromodels under $<<$ is a partially ordered set. But since Ξ_{1A} is not totally compatible with itself (see remark following Example 4.3), the set of all macromodels under $\triangleleft$ is not reflexive, hence is not a partially ordered set.

It is straightforward to verify that for all the empirical affect models discussed in this paper – $BA, CL, RC, R2C, TR, HC$ and 39+ – whenever we have a not necessarily distinct pair of them (X, Y) where $\Xi_X < \Xi_Y$, we also have $\Xi_X \triangleleft \Xi_Y$. Since $\triangleleft$ is transitive, we only need to verify this property for the pairs of models which have no other model between them. There are 14 such pairs. Since $\{120D, 120U, 030T, 021D, 021U\} \subset P_Y$ for $Y = RC, R2C, TR, HC$ and 39+, we have by Lemma 4.5(b) that $\Xi_X \triangleleft \Xi_Y$ for all (X, Y) pairs $(HC, HC), (TR, HC)$, $(TR, TR), (RC, TR), (RC, RC), (R2C, RC), (R2C, R2C), (R2C, 39+), (39+, 39+)$, (CL, RC) and $(BA, R2C)$, and since $\{102, 003\} \subset P_{CL}$, we have by Lemma 4.6(2) that $\Xi_{CL} \triangleleft \Xi_{CL}$ and $\Xi_{BA} \triangleleft \Xi_{CL}$ as well. Finally, if M_{BA} and M'_{BA} are any two macrostructures in Ξ_{BA}, where M_{BA} consists of two M- cliques W_1 and W_2 with $W_1 N^* W_2$ and M'_{BA} consists of two M-cliques W'_1 and W'_2 with $W'_1 N^* W'_2$, then M_{BA} and M'_{BA} can be compatibly combined in Ξ_{BA} by forming $W_1 M^* W'_1$, $W_2 M^* W'_2$, $W_1 N^* W'_2$ and $W_2 N^* W'_1$; hence $\Xi_{BA} \triangleleft \Xi_{BA}$. This construction shows the only way two pairs of mutually hostile cliques can be compatibly combined, namely, by uniting the two cliques of one hostile pair with separate cliques of the other hostile pair, to form a new and larger structure of two hostile cliques. The total compatibility for some of these pairs (X, Y) can also be established using the criterion in Lemma 4.5(a) for $Y = TR$ and HC.

We interpret these formal ideas and results substantively. We will view the M, A and N relations simply as structural ties, but with the recognition that in an empirical situation they may carry additional substantive importance in that each may be instrumental as a vehicle or a constraint for certain kinds of interpersonal interaction. We interpret M_X and M_Y to be either actual empirical macrostructures or idealizations of actual empirical macrostructures. Corresponding to them are their micromodels P_X and P_Y.

If neither $P_X \subset P_Y$ nor $P_Y \subset P_X$ then neither M_X nor M_Y can be embedded as a substructure in any macrostructure from the macromodel of the other and neither can be compatibly combined with the other within the macromodel of the other. In other words, neither social structure can be successfully embedded in any macrostructure, or transferred into *any* environment, in which the social structure rules and process rules (collectively called the *social rules*) of the other must hold. If M_X were to be **imposed** on M_Y then the two macrostructures in combination would have to undergo a transformation to form a macrostructure in which either the substructure M_X and its corresponding social rules become tolerated by M_Y within the new structure and vice versa, or else the social rules of one or both of the two original macrostructures penetrates the other to produce a new type of macrostructure. In either situation the resulting macrostructure represents a new macromodel for which, in the tolerant case, the corresponding micromodel contains the union of the two original micromodels.

Now suppose $P_X \subset P_Y$. If M_X is not a substructure of any macrostructure from Ξ_Y containing M_Y as a separate substructure (as when Ξ_X is not a totally compatible submodel of Ξ_Y) and is imposed on M_Y then, similar to the situation in the previous discussion, a macrostructure from a new macromodel will be produced. At the outset the only place a problem will arise is among the triads which interconnect M_X with M_Y, since if M_X and M_Y initially stay as substructures of the combined structure then, at first, new triad types not in P_Y must necessarily arise among these interconnecting triads. The remaining possibility is that M_X is embeddable in an M'_Y from Ξ_Y containing an M_Y disjoint from M_X. Here M_X can be embedded or transferred into an environment governed by the social rules of P_Y and Ξ_Y. For example, if M_X combines with M_Y according to (i), (ii), (iii) or (iv) in the proof of Lemma 4.5 (where $\Xi_X \triangleleft \Xi_Y$ is guaranteed) then we have the successful formation of a new macrostructure M'_Y from Ξ_Y which accommodates M_X and is formed within the social rules of P_Y and Ξ_Y. In such a situation $P_X \subset P_Y$ with $P_X \neq P_Y$ means that M_X is being generated and maintained within M'_Y by more restrictive social rules than those which are fully operative within M_Y and M'_Y.

As a hypothetical concrete example, the reader may consider the possibilities and details when a small ethnic group from a foreign locale with social structure M_X immigrates to a neighborhood of a large American city with social structure M_Y, causing negligible displacement. As a brief remark on this example, we note that in many situations $\{102, 012, 003\} \subset P_Y$, which means that an initial arrangement in which each community can ignore or dislike the other will be relatively stable á la alternative (i) in the proof of Lemma 4.5. Here the dyads interconnecting the two communities are initially all N's. Now suppose the inhabitants of this American city are quite gregarious and do not easily tolerate adjacent N relations, making $102, 012$ and 003 "forbidden". What sort of resolution of this situation might occur?

5. Incipient Microprocesses. Here we present an analytic procedure for detecting microprocesses in a micromodel of forbidden triads, P_X^c. As an example allowing a simplified treatment, suppose that $P_X^c = \{120C, 030C, 021C\}$. We want to account for this in terms of some microprocess which acts against the stable formation of these triad types among the members of the group. Now, for any three distinct members a, b, c in the group, if aAb and bAc then we must have aAc since the other three possibilities aMc, cAa and aNc lead, respectively, to the three forbidden triad types listed and are thus ruled out. Assuming that the relations aAb and bAc are relatively stable and not open to near term change, we can interpret this logical implication as a process prescription, namely, aAb and bAc induce aAc, for all a, b, c in V. Considered as a process in time, this prescription can be stated as aAb and bAc at time t induce aAc at time $t+$, for all a, b, c in V, where $t+$ is a point in time shortly after time t. Thus, if this microprocess proceeds until the group is close to social stability and then continues to operate, the three forbidden triads will become relatively rare and we will eventually observe a strong relational tendency towards transitivity of A, whence a strong structural tendency towards a partial ordering of the group with repect to A.

We now formalize this procedure in more detail. Given the binary relation R on V, we define the relation $R^\wedge$ by $xR^\wedge y$ if and only if yRx. Then $M^\wedge = M$, $N^\wedge = N$, $A^\wedge \neq A$ but $(A^\wedge)^\wedge = A$. Consider the triad of vertices $\{a, b, c\}$ for which aJb, bKc, and cLa in cyclic order of the vertices, where J, K, and L are each from the set $\{M, A, A^\wedge, N\}$. We can then represent the *state* of this triad by the three relations in the order JKL. Since there is no preference for one vertex or relation over another, the state of the triad is also represented by KLJ, LJK, $L^\wedge K^\wedge J^\wedge$, $J^\wedge L^\wedge K^\wedge$, and $K^\wedge J^\wedge L^\wedge$. Thus, there are up to six possible ways to represent the state of the triad, although not all six will necessarily be different. As examples, there are only three different representations of the state for MNN, two for AAA, and one for MMM.

To within exchange $\sim$ of the two dyads corresponding to the first two relations of a state representation there are exactly 10 different initial ordered pairs possible in a state representation: $MM, MA, MN, AM, AA, AA^\wedge$, $A^\wedge A, AN, NA$ and NN. (For the other 6 we have $MA^\wedge \sim AM$, $A^\wedge M \sim MA$, $A^\wedge A^\wedge \sim AA, NM \sim MN$, $A^\wedge N \sim NA$, and $NA^\wedge \sim AN$.) This reduction of initial pairs helps organize the analysis when one is interested in finding a triadic microprocess given by the action of the first two relations of a state representation *toward, through,* or *from* the vertex in common to the two dyads having these relations, called the *middle* vertex. For example, in the pair given by aJb and bKc, J and $K^\wedge$ act *toward*, J and K act *through* ($K^\wedge$ and $J^\wedge$ also act *through*), and $J^\wedge$ and K act *from* the middle vertex b. These relational actions are also called, respectively, *nexal, medial,* and *focal*. One may transform any state representation into an equivalent one, having the same middle vertex and its first pair of relations in the list of 10, by the exchange $\sim$, if necessary, of the first two dyads. One may then generate the alternative but equivalent statements of the potential microprocess of interest in its nexal, two medial, and focal forms.

For the micromodel P_X^c we form a 10 × 16 table of symbols, called the *analytic table* for model X, as follows. We label the rows of the table by the 10 ordered pairs mentioned above and the columns by the 16 triads in Θ. For each of the forbidden triads in P_X^c we form all the different state representations of the triad which have their initial ordered pairs among the 10 listed above. Each of these state representations ends with a third symbol. For each such state representation we place the third symbol in the table in the row labelled by the initial ordered pair and the column labelled by the forbidden triad represented. Then, for each row JK of the table we form the set of third symbols appearing in that row, which we call the *forbidden span* of the pair JK in the model X. If the forbidden span of JK is empty this means that this ordered pair can occur in the triads of a macrostructure of the model without restriction.

If the forbidden span of JK is all of $\{M, A, A^\wedge, N\}$ then the pair is forbidden in all triads of every macrostructure of the model. Here, if aJb occurs then bKc and $cK^\wedge b$ cannot. Assuming that the relation aJb is stable, we propose here the microprocess: aJb induces $\neg(bKc)$ (or $\neg(cK^\wedge b)$), for all a, b, c in V, where $\neg(xRy)$ means that xRy does not hold. Considered as a process in time, this is stated as: aJb at time t induces $\neg(bKc)$ (alternatively, $\neg(cK^\wedge b)$) at time $t+$, for all a, b, c in V, where $t+$ is a point in time shortly after time t.

If the forbidden span of a pair JK consists of three of the four relations in $\{M, A, A^\wedge, N\}$, all except L, then the only state representation possible with initial pair JK is JKL. Thus, aJb and bKc imply cLa and $aL^\wedge c$. We propose a microprocess to account for the forbidden triads having state representations JKH with $H \neq L$. Assuming that the relations aJb and bKc are stable, we prescribe from this logical implication the microprocess (med): aJb and bKc induce $aL^\wedge c$ (or cLa), for all a, b, c in V. As a process in time, this can be stated as: aJb and bKc at time t induce $aL^\wedge c$ (or cLa) at time $t+$, for all a, b, c in V, where $t+$ is a point in time shortly after time t. This is a medial form of this process. The corresponding nexal and focal forms are (nex): aJb and $cK^\wedge b$ induce $aL^\wedge c$ (or cLa), and (foc): $bJ^\wedge a$ and bKc induce $aL^\wedge c$ (or cLa), and the other medial form is (med$^\wedge$): $aK^\wedge b$ and $bJ^\wedge c$ induce aLc (or $cL^\wedge a$), for all a, b, c in V.

Now, suppose there are r rows of the analytic table, each row containing at least three of the four relations in $\{M, A, A^\wedge, N\}$, such that their forbidden spans together contain at least one representative from every forbidden triad column of the table. Then there are r microprocesses in time of the types just described which forbid all of the triads in P_X^c and none of those in P_X. These r microprocesses, operating concurrently, induce the micromodel P_X^c and, thus, offer a processual explanation for the formation of macrostructures from the macromodel Ξ_X (such as that for a group which is the object of empirical study). We now apply this analytic procedure to some of the empirically based micromodels of forbidden triads obtained for the Davis-Leinhardt data.

    a. **The hierarchical $\widetilde{M}$-cliques model.** The micromodel for the hierarchical $\widetilde{M}$-cliques model HC is $P_{HC}^c = \{201, 120C, 111D, 111U, 030C, 021C\}$ the analytic table of which is given in Table 1. Since there are 6 forbidden triads and no forbidden span has 4 symbols, a microprocess can act against at most 3 triads; hence, we need at least $r \geq \lceil 6/3 \rceil = 2$ microprocesses. We see that there are, indeed, two incipient microprocesses which, taken together, act against the stable formation of all the forbidden triads of the model. These are given in the result which follows.

Table 1. Analytic Table for the Model HC

initial pair	\multicolumn{16}{c	}{forbidden triads}	permitted triad state (if no. ≤ 1)
	300	210	201	120C	120D	120U	111D	111U	102	030C	030T	021C	021D	021U	012	003	
MM		N															
MA		A						N									
MN		M					A	A^									MNN
AM		A		N													
AA		M								A		N					AAA^
AA^																	
A^A																	
AN								M					A				
NA							M						A				
NN																	

THEOREM 5.1. *The model HC is exactly generated by the following two concurrent microprocesses operating over time : for any a, b, c, in V*

(i.med) : aMb and bNc induce aNc, and

(ii.med) : aAb and bAc induce aAc.

We note that process (ii.med) is required to act against the forbidden triad 030C. At the same time, it acts against the triads 120C and 021C. The remaining forbidden triads 201, 111D, and 111U are cleanly eliminated by the process (i.med), which does not duplicate the effort of process (ii.med) on the other forbidden triads. This yields an elegant process characterization of the hierarchical $\widetilde{M}$-cliques model.

Process (i.med) can be interpreted as *external dissociation*, the process of clique formation by which different $\widetilde{M}$-cliques are separated and delineated by the growth of gaps (N dyads) between their members (to be distinguished from *internal association*, the process by which each clique is connected and unified by the growth of strong ties (M dyads) between its members). It can also be stated as (i.med^): aNb and bMc induce aNc, (i.nex): aMb and cNb induce aNc, and (i.foc): bMa and bNc induce aNc, for all a, b, c in V. As mentioned earlier in this section, process (ii.med) prescribes the formation of hierarchy within the group. Its nexal, focal, and other medial forms do not seem very illuminating, so we do not state them here. Thus, the formation of a macrostructure from the hierarchical $\widetilde{M}$-cliques model (the exact model for the total Davis-Leinhardt data set and for the total Hallinan data set) can be precisely characterized, and hence explained, by the concurrent operation of the two microprocesses (i.med) and (ii.med) inducing the formation of cliques and hierarchy, respectively. This precise process description is a characterization of how affect in social groups is structured by ranking and cliquing as suggested by Homans (1950), articulated by Davis and Leinhardt (1972), and determined by Johnsen (1985).

b. The transitivity model. The micromodel for the transitivity model TR is

$P^c_{TR} = \{210, 201, 120C, 111D, 111U, 030C, 021C\}$. The analytic table for this model appears in Table 2. We see that in this model there are five incipient microprocesses which, taken together, act against the stable formation of the entire set of forbidden triads of the model. On grounds of parsimony, we look for the smallest number of them which together will do this. Since there are 7 forbidden triads and no forbidden span has 4 symbols, a microprocess can act against at most 3 triads; hence, we need at least $r \geq \lceil 7/3 \rceil = 3$ microprocesses. This leads to those given in the result which follows.

Table 2. Analytic Table for the Model TR

initial pair	300	210	201	120C	120D	120U	111D	111U	102	030C	030T	021C	021D	021U	012	003	permitted triad state (if no. ≤ 1)
MM		A,A^	N														MMM
MA		M		A				N									MAA^
MN			M				A	A^									MNN
AM		M		A			N										AMA^
AA			M							A		N					AAA^
AA^																	
A^A																	
AN									M			A					
NA							M					A					
NN																	

THEOREM 5.2. *The model TR is exactly generated by the following three concurrent microprocesses operating over time : for any a, b, c, in V*

(i.med) : aMb and bNc induce aNc,

(ii.med) : aAb and bAc induce aAc, and

(iii.med) : aMb and bMc induce aMc.

Again, process (ii.med) is required to act against the triad $030C$, at the same time acting against triads $120C$ and $021C$, and process (i.med) acts against three of the remaining four forbidden triads, 201, $111D$, and $111U$. The remaining forbidden triad 210 is eliminated by process (iii.med).

As mentioned before, processes (i.med) and (ii.med) express, respectively, clique formation by external dissociation and hierarchy formation. Process (iii.med) expresses clique formation by *internal association*, the process by which a clique is connected and unified by the growth of strong ties (M dyads) between its members. The added effect here of (iii.med), which does not occur as strongly in the HC model, seems natural since the groups in the Davis-Leinhardt data for which the TR model holds (sizes 14-17) are smaller, hence potentially more intimate, than those for which the HC model holds (sizes 18-38). Clique formation in a group for which the transitivity model holds, then, can be viewed as governed by both the centripetal "pull" of association (attraction) and the centrifugal "push" of dissociation (aversion). As with the HC model, the nexal, focal, and other medial forms of (i.med) and the nexal and focal forms of (iii.med) can be stated instead.

c. The two ranked clusters of M-cliques models. The micromodel for the ranked clusters of M-cliques model RC is $P^c_{RC} = \{210, 201, 120C, 111D, 111U, 030C, 021C, 012\}$. Its analytic table is given in Table 3. In this model there are seven incipient microprocesses which, taken together, act against the stable formation of the entire set of forbidden triads of the model. On grounds of parsimony, we again look for the smallest number of microprocesses which together will do this. For 8 forbidden triads and no forbidden span of 4 symbols we need, as before, at least $r \geq \lceil 8/3 \rceil = 3$ microprocesses. Since, in fact, no 3 of them together act against all 8 forbidden triads, we are led to a characterization in terms of 4 microprocesses. On substantive grounds we want to keep this analysis as close as possible to that for models TR and HC; hence, we again use the microprocesses (i.med), (ii.med), and (iii.med), which together act against all forbidden triads of the RC model except 012. Now, this last triad is eliminated by either of the two processes: aAb and bNc induce aAc, or: aNb and bAc induce aAc, for all a, b, c in V. Neither of these two processes, however, appears to have a clear and compelling substantive interpretation as an influence in the formation of affect structure in a group, so we look further for a suitable microprocess. (It may happen that the necessary effect could, instead, be obtained serendipitously via another process acting in the group.) Note that the 012 triad can also be eliminated by the slightly more complicated microprocess obtained from the last row of the analytic table, namely: aNb and bNc induce either aMc or aNc, for all a, b, c in V. We use this process in the statement of the following result.

THEOREM 5.3. *The model RC is exactly generated by the following four concurrent microprocesses operating over time : for any $a, b, c,$ in V*

(i.med) : aMb and bNc induce aNc,

(ii.med) : aAb and bAc induce aAc,

(iii.med) : aMb and bMc induce aMc, and

(iv'.med) : aNb and bNc induce either aMc or aNc.

Table 3. Analytic Table for the Model RC

initial pair	300	210	201	120C	120D	120U	111D	111U	102	030C	030T	021C	021D	021U	012	003	permitted triad state (if no. ≤ 1)
MM		A,A^	N														MMM
MA		M		A			N										MAA^
MN			M				A	A^									MNN
AM		M		A			N										AMA^
AA			M					A		N							AAA^
AA^																	
A^A																	
AN						M				A			N				ANA^
NA					M					A			N				NAA^
NN														A,A^			

The first three processes of Theorem 5.3 have already been discussed. Process (iv′.med) has a simple substantive interpretation as "a mutual nonfriend of a mutual nonfriend will be either a mutual friend or a mutual nonfriend". This is a process which operates *trivially* within any macrostructure of the complete clustering model CL, but only because CL is a context in which A dyads are excluded. Since CL is a submodel of RC, which does not exclude A dyads a priori, (iv′.med) operates *nontrivially* in the context of RC and is thus a substantively appropriate constraining process to occur in any macrostructure from model RC. In fact, (i.med), (iii.med) and (iv.med) taken together indicate the operation of the full set of microprocesses for the clustering model CL within a larger context, a context which also includes the process of hierarchy formation as given by (ii.med). In this way we can formally interpret *in process terms* the Davis-Leinhardt conceptualization of cliquing and ranking suggested from Homans (Davis & Leinhardt 1972; Homans 1950).

We note that model $R2C$ is obtained from RC by additionally forbidding the triad 003. We again need at least 4 microprocesses which act together against the set of 9 forbidden triads. We see that these can be obtained by keeping the first three processes of Theorem 5.3 and modifying the last one to read (iv.med): aNb and bNc induce aMc, for all a, b, c in V. We thus have the following result.

THEOREM 5.4. *The model R2C is exactly generated by the following four concurrent microprocesses operating over time : for any a, b, c, in V*

(i.med) : aMb and bNc induce aNc,

(ii.med) : aAb and bAc induce aAc,

(iii.med) : aMb and bMc induce aMc, and

(iv.med) : aNb and bNc induce aMc.

Process (iv.med) has a simple substantive interpretation as "a mutual nonfriend of a mutual nonfriend will be a mutual friend" which, in its strong form, is the familiar "a (mutual) enemy of a (mutual) enemy will be a (mutual) friend". This process operates *trivially* within a macrostructure of the complete balance model BA, as the context of BA excludes A dyads. Since BA is a submodel of the model $R2C$, in which A dyads are possible, (iv.med) operates *nontrivially* in the context of $R2C$. It is thus a substantively appropriate constraint to have operate in a macrostructure from model $R2C$. So (i.med), (iii.med) and (iv.med) taken together indicate the operation of the full set of microprocesses for the balance model BA within a larger context, a context which also includes the process of hierarchy formation given by (ii.med). Thus, we can formally interpret this model *in process terms* as the more limited Davis-Leinhardt conceptualization of cliquing á la Heider (Heider 1946) with ranking.

   d. **The 39+ (large group) model.** The micromodel for the large group model 39+ is $P^c_{39+} = \{201, 111D, 111U, 030C, 021C, 003\}$ the analytic table of which is given in Table 4. For 6 forbidden triads and no forbidden span of 4 symbols we need at least $r \geq \lceil 6/3 \rceil = 2$ microprocesses. Since no 2 of them together act against all 6 forbidden triads, we are led to a characterization involving at least 3 microprocesses. We again try to keep the analysis as close as possible to that for the previous models; however, the only microprocess which still holds immediately is (i.med), and this only acts against the triads $201, 111D$, and $111U$. By inspection of the table, we see from the fifth row that triads $030C$ and $021C$ are eliminated by the process: aAb and bAc induce aAc or aMc, for all a, b, c in V, and from the last row that triad 003 is eliminated by the process: aNb and bNc induce $\neg(aNc)$, for all a, b, c in V. Using these, we have the following result.

Table 4. Analytic Table for the Model 39+

initial pair	forbidden triads															permitted triad state (if no. ≤ 1)
	300	210	201	120C	120D	120U	111D	111U	102	030C	030T	021C	021D	021U	012	003	
MM		N															
MA							N										
MN		M			A	A^											MNN
AM					N												
AA										A		N					
AA^																	
A^A																	
AN							M					A					
NA							M					A					
NN																N	

   THEOREM 5.5. *The model 39+ is exactly generated by the following three concurrent microprocesses operating over time : for any a, b, c, in V*

   (i.med) : aMb and bNc induce aNc,

   (ii'.med) : aAb and bAc induce either aMc or aAc, and

(iv″.med) : aNb and bNc induce $\neg(aNc)$.

Unfortunately, the required process (ii′.med) involves the A dyad directly, so we cannot take our justification of this process from a context which excludes the A dyad as we did for models RC and $R2C$. And even though process (iv″.med) does not directly involve the A dyad, if we interpret it in a more constrained context free of A dyads it becomes (iv.med), and we will indirectly eliminate the permitted triad 012 if we take this process back to the full context of 39+. A straightforward conclusion from (iv″.med) is that since adjacent null or negative dyads are unstable, some form of gregariousness must be present in the group. Since the groups for which model 39+ holds are large (sizes $39 - 79$ in the Davis-Leinhardt data set), this seems an unlikely phenomenon. A modest interpretation of (ii′.med) is that it induces a *quasitransitivity* of A, which becomes full transitivity when A is aggregated to A^* between $\widetilde{M}$-cliques. Nevertheless, we are still left with trying to satisfactorily justify the total set of microprocesses in Theorem 5.5 substantively. A possible explanation of the difficulty here may be that the original sociometric data collected for the large group sizes in the Davis-Leinhardt data set is too flawed, not sufficiently random, and of insufficient quantity to be used in obtaining, with a high probability, an accurate empirical micromodel of permitted and forbidden triads for affect in such groups.

6. Discussion and problems. In this section we will first make some closing remarks regarding the previous formal mathematical development and social theory, and will then present some mathematical problems which are of interest for their own sake as well as for the development of the social theory.

The presentation in sections 2 and 3 shows the ability of closely reasoned network analysis to contribute both to the formalization of the notion, and to the development of the theory, of social structure. In particular, the analysis allows us to obtain the exact ideal network structure corresponding to the total set of empirical sociomatrices collected by Davis and Leinhardt for the affect relation, the quest for which was earlier attempted but not completed. This analysis shows by example the potential for obtaining rigorous formal connections between microstructure and macrostructure, here in the simple case of one 2-valued directed relation.

In sections 2, 3 and 4 we presented several structural models for affect and developed the notion of a submodel of a model in terms of their permitted and forbidden triads. The definition of a micromodel raises the question of the conditions under which they exist. The definitions of sharp and totally compatible submodels are guided by substantive concerns. The question of embeddability leads to the definition of a sharp submodel. The naturally arising social situation in which one macrostructure is either imposed on or juxtaposed with another leads to that of a totally compatible submodel. We observe that each of the five affect macromodels studied in this paper is a totally compatible, hence also sharp, supermodel of each of its submodels, essentially because we always have $\{120D, 120U, 030T, 021D, 021U\} \subset P_Y$ for $Y = HC, TR, RC, R2C$, and 39+. The analysis and discussion in section 4 should sensitize social scientists to the possibility that submodels need not be totally compatible, nor even self-compatible, that is, micro-compatibility does not necessarily imply macro-compatibility, and that these possibilities become apparent from the formal analysis.

In section 5 we developed an analytic method for finding incipient microprocesseses for the construction of the microstructures (permitted triads) and, as a consequence, the generic formation of the macrostructures in a model. Applying this to the HC model we found that the formation of the macrostructures in this model can be exactly characterized by two microprocesses, one inducing cliquing by dissociation and the other inducing hierarchy. As the set of permitted triads is successively restricted to those of the TR, RC and $R2C$ models we find that other naturally interpretable microprocesses are added to the characterization. By the time we reach model $R2C$ the macrostructures are characterized by the three microprocesses inducing balance and the one inducing hierarchy. The formation of the macrostructures in the 39+ model is not as readily described. However, constant to all five affect models is the microprocess inducing cliquing by dissociation, while the process inducing hierarchy is persistent in all except the anomalous 39+.

We have distinguished three different forms in which the same microprocess can be stated – nexal, medial and focal – to allow for the substantive distinction regarding which members of the triad invoke the first two parts of the process. Suppose a, b, and c are together. In the example of (i.med): aMb and bNc induce aNc, we interpret this to mean that a invokes aMb and b invokes bNc, which results in aNc. This form would occur when a first indicates that b is a mutual friend and b first indicates that c is a mutual enemy. For (i.nex): aMb and cNb induce aNc; this form would occur when a first indicates that b is a mutual friend and c first indicates that b is a mutual enemy. Similarly for (i.foc). In all three cases the eventual result aNc is the same. A more pleasant process is the one for mutual friendship formation, which we elaborate a bit more. For (iii.med): aMb and bMc induce aMc; this form would occur when a first indicates that b is a mutual friend (by an arm around the shoulders) and b first indicates that c is a mutual friend (by warmly inviting him over and speaking favorably of him). Similarly for (iii.nex). Finally, for (iii.foc): bMa and bMc induce aMc; this form would occur when b first indicates (by inviting both a and c over and warmly recommending them to each other) that a is a mutual friend and c is a mutual friend.

In summary, we have showed, in the case of the fundamental relation of affect, that a rigorous formal connection between micro- and macro- social phenomena is possible in terms of both structure and process. As a result, this study should be informative for research into other fundamental dyadic relations besides affect. As an exercise in applied mathematics, this is an attempt to satisfactorily address an urgent major question in the social sciences. We hope this is construed to be in the spirit of the points made by Fred Roberts in his introductory remarks to this Workshop.

In closing, we mention several interesting mathematical and methodological problems which have arisen in this study:

(1) Which subsets of Θ are micromodels, i.e., have exact macrorepresentations?
(2) If P_X and P_Y are micromodels, when are $P_X \cap P_Y$ and $P_X \cup P_Y$ also micromodels?
(3) More generally, what is the poset of micromodels (and macromodels)?
(4) What other results are there regarding the total compatibility of a submodel in a supermodel?
(5) More generally, what is the structure of the set of all macromodels under the relation of total compatibility?
(6) What other results are there regarding when a model is a sharp submodel of a supermodel?
(7) More generally, what is the poset of macromodels under the sharp relation?
(8) Can the triadic analysis of structures having $+/0/-$ valued directed edges be finessed by a carefully chosen set of analyses involving only $+/-$ directed edges?

REFERENCES

[1] ALEXANDER, JEFFREY C., BERNHARD GIESEN, RICHARD MÜNCH AND NEIL J. SMELSER (EDS.), *The Micro-Macro Link*, University of California Press, Berkeley, 1987.
[2] BLAU, PETER M., *Contrasting theoretical perspectives*, in Jeffrey C. Alexander, Bernhard Giesen, Richard Münch and Neil J. Smelser (eds.), *The Micro-Macro Link*, University of California Press, Berkeley, 1987, pp. 71–85.
[3] CARTWRIGHT, DORWIN AND FRANK HARARY, *Structural balance: A generalization of Heider's theory*, Psychological Review, 63 (1956), pp. 277–293.
[4] COLEMAN, JAMES S., *Community Conflict*, The Free Press, Glencoe Illinois, 1957.
[5] COLEMAN, JAMES S., *Micro foundations and macrosocial theory*, in Siegwart Lindenberg, James S. Coleman and Stefan Nowak (eds.), *Approaches to Social Theory*, Russell Sage Foundation, New York, 1986, pp. 345–363.
[6] COLEMAN, JAMES S., *Microfoundations and macrosocial behavior*, in Jeffrey C. Alexander, Bernhard Giesen, Richard Münch and Neil J. Smelser (eds.), *The Micro- Macro Link*, University of California Press, Berkeley, 1987, pp. 153–173.
[7] COLLINS, RANDALL, *On the microfoundations of macrosociology*, American Journal of Sociology, 86 (1981a), pp. 984–1014.
[8] COLLINS, RANDALL, *Micro-translation as a theory-building strategy*, in K. Knorr-Cetina and A.V. Cicourel (eds.), *Advances in Social Theory and Methodology: Toward an Integration of Micro- and Macro-Sociologies*, Routledge & Kegan Paul, Boston, 1981b, pp. 81–108.
[9] DAVIS, JAMES A., *Structural balance, mechanical solidarity, and interpersonal relations*, American Journal of Sociology, 68 (1963), pp. 444–462.
[10] DAVIS, JAMES A., *Clustering and structural balance in graphs*, Human Relations, 20 (1967), pp. 181–187.
[11] DAVIS, JAMES A., *Social structures and cognitive structures*, in Robert P. Abelson et al. (eds.), *Theories of Cognitive Consistency: A Sourcebook*, Rand-McNally, Chicago,, 1968, pp. 544–550.
[12] DAVIS, JAMES A., *Clustering and hierarchy in interpersonal relations: Testing two graph theoretical models on 742 sociomatrices*, American Sociological Review, 35 (1970), pp. 843–851.
[13] DAVIS, JAMES A. AND SAMUEL LEINHARDT, *The structure of positive interpersonal relations in small groups*, in Joseph Berger et al (eds.), *Sociological Theories in Progress Vol. 2*, Houghton Mifflin, Boston, 1972, pp. 218–251.
[14] FLAMENT, CLAUDE, *Applications of Graph Theory to Group Structure*, Prentice-Hall, Englewood Cliffs, New Jersey, 1963.
[15] HALLINAN, MAUREEN T., *The Structure of Positive Sentiment*, Elsevier Scientific, New York, 1974.

[16] HARARY, FRANK, *On the notion of balance of a signed graph*, Michigan Mathematical Journal, 2 (1954), pp. 143–146.
[17] HEIDER, FRITZ, *Attitudes and cognitive organization*, Journal of Psychology, 21 (1946), pp. 107–112.
[18] HOLLAND, PAUL W. AND SAMUEL LEINHARDT, *A method for detecting structure in sociometric data*, American Journal of Sociology, 70 (1970), pp. 492–513.
[19] HOLLAND, PAUL W. AND SAMUEL LEINHARDT, *Transitivity in structural models of small groups*, Comparative Group Studies, 2 (1971), pp. 107–124.
[20] HOLLAND, PAUL W. AND SAMUEL LEINHARDT, *The structural implications of measurement error in sociometry*, Journal of Mathematical Sociology, 3 (1973), pp. 85–111.
[21] HOLLAND, PAUL W. AND SAMUEL LEINHARDT, *Local structure in social networks*, in D. Heise (ed.) *Sociological Methodology 1976*, Jossey-Bass, San Francisco, 1976, pp. 1–45.
[22] HOMANS, GEORGE C., *The Human Group*, Harcourt, Brace and World, New York, 1950.
[23] JOHNSEN, EUGENE C., *Network macrostructure models for the Davis-Leinhardt set of empirical sociomatrices*, Social Networks, 7 (1985), pp. 203–224.
[24] JOHNSEN, EUGENE C., *Structure and Process: Agreement Models for Friendship Formation*, Social Networks, 8 (1986), pp. 257-306.
[25] KNORR-CETINA, K. AND A.V. CICOUREL (EDS.), *Advances in Social Theory and Methodology: Toward an Integration of Micro- and Macro-Sociologies*, Routledge & Kegan Paul, Boston, 1981.
[26] LAZARSFELD, PAUL F. AND ROBERT K. MERTON, *Friendship as social process: A substantive and methodological analysis (1954)*, reprinted, in Patricia L. Kendall (ed.), *The Varied Sociology of Paul F. Lazarsfeld*, Columbia University Press, New York 1982, pp. 298–348.
[27] NEWCOMB, THEODORE M., *An approach to the study of communicative acts*, Psychological Review, 60 (1953), pp. 393–404.
[28] NEWCOMB, THEODORE M., *The Acquaintance Process*, Holt, Rinehart and Winston, New York, 1961.
[29] NEWCOMB, THEODORE M., *Interpersonal balance*, in Robert P. Abelson et al. (eds.), *Theories of Cognitive Consistency: A Sourcebook*, Rand-McNally, Chicago, (1968), pp. 28–51.
[30] WHITE, HARRISON C., SCOTT A. BOORMAN AND RONALD L. BREIGER, *Social structure from multiple networks I. Blockmodels of roles and positions*, American Journal of Sociology, 81 (1976), pp. 730–780.

SIGN–PATTERNS AND STABILITY*

VICTOR KLEE[†]

Abstract. This is a survey of recent results and problems concerning the relationship between the stability properties of a linear system and the sign–pattern of its coefficients.

Key words. sign–pattern, stable, semistable, quasistable, Lyapunov function, graph, digraph, cycle, coloring, matching, algorithm, computational complexity.

1. Introduction. Our primary concern here is the simplest sort of dynamical system — one associated with a system of first–order linear differential equations with constant coefficients. For an $n \times n$ real coefficient matrix $A = [a_{ij}]$, the system equation is

$$\dot{x} = Ax, \tag{1}$$

where $x \in \mathbf{R}^n$ and $\dot{}$ indicates the derivative with respect to time; equivalently,

$$\frac{dx_i}{dt}(t) = \sum_{j=1}^{n} a_{ij} x_j(t) \qquad (1 \leq i \leq n). \tag{2}$$

For each point $x_0 \in \mathbf{R}^n$, there is a unique function

$$x = (x_1, \ldots, x_n)^T : \mathbf{R} \longrightarrow \mathbf{R}^n$$

that satisfies the system equation (1) and has $x(0) = x_0$. Each such function x is a *trajectory* of the system, and its restriction to the interval $[0, \infty[$ is a *positive trajectory*.

We are interested in the asymptotic behavior of the system's positive trajectories. Is the system *stable* in the sense that for each positive trajectory x, $x(t)$ converges to the origin as $t \longrightarrow \infty$? If not, is the system at least *quasistable* in the sense that each positive trajectory is bounded? And if not that, is the system at least *semistable* in the sense that whenever a positive trajectory is unbounded, it "diverges to infinity" at a polynomial rather than an exponential rate? Such questions naturally involve A's eigenvalues and hence its characteristic polynomial. They also involve A's *minimum polynomial*, which is the unique monic polynomial p of smallest degree such that $p(A) = 0$. The following well–known results can be proved by considering the real canonical form of A (a real analogue of the Jordan normal form).

---

*Preparation of this report was supported in part by a grant from the National Science Foundation.

[†]Department of Mathematics, University of Washington, GN-50, Seattle, WA 98195

1.1 Solutions of Linear systems. If A is an $n \times n$ real matrix and the function $x = (x_1, \ldots, x_n)^T : \mathbf{R} \longrightarrow \mathbf{R}^n$ is such that $\dot{x} = Ax$, then each component $x_i(t)$ of $x(t)$ is a linear combination of functions of the forms

(3) $$t^k e^{t\alpha} \cos \beta t \quad \text{or} \quad t^\ell e^{t\alpha} \sin \beta t$$

where $\lambda = \alpha + \iota\beta$ is an eigenvalue of A and k and ℓ are nonnegative integers less than λ's multiplicity as a root of A's minimum polynomial.

1.2 Eigenvalues and Stability. The system $\dot{x} = Ax$ is —

stable if and only if each eigenvalue of A has negative real part;

quasistable if and only if each eigenvalue has nonpositive real part and each zero or pure imaginary eigenvalue is a simple root of A's minimum polynomial;

semistable if and only if each eigenvalue has nonpositive real part.

It may be instructive to think of Theorems 1.1 and 1.2 in terms of a decomposition of the trajectories of the system $\dot{x} = Ax$. Let $F_<$, F_δ, F_ϵ, and $F_>$ denote the set of all linear combinations of functions of the form (3) with $\alpha < 0$, $\alpha = 0 \neq \beta$, $\alpha = 0 = \beta$, and $\alpha > 0$ respectively. Then each trajectory x admits a unique decomposition

$$x = x_< + x_\delta + x_\epsilon + x_>$$

with $x_* \in F_*^n$ for $* \in \{<, \delta, \epsilon, >\}$. Since $\dot{x}_* = Ax_*$ in each case, the equation $\dot{x} = Ax$ splits naturally into four parts corresponding to the four choices of $*$. For each positive trajectory x, it is true that —

x converges to the origin if and only if x_δ, x_ϵ and $x_>$ are all identically zero, so that x is equal to its *transient part* $x_<$;

x is bounded if and only if each component of x_δ is a linear combination of sines and cosines of multiples of t, x_ϵ is constant, and $x_>$ is identically zero;

x does not "diverge to infinity" at an exponential rate if and only if $x_>$ is identically zero.

From these facts it is clear that the "if" parts of (1.2) follow immediately from (1.1). Detailed proofs of (1.1) and (1.2) can be found in books by Hahn [Ha], Hirsch and Smale [HS], and elsewhere.

Under the circumstances described in (1.2), the terms *stable*, *quasistable*, and *semistable* are applied to the matrix A as well as to the associated linear system. (Varying terminology is used by other authors.) However, our primary concern here is not with these notions but with related notions that take into account only the *sign–pattern* of the matrix A. With each $n \times n$ matrix A we associate the convex cone $Q(A)$ consisting of all $n \times n$ matrices $\tilde{A}$ that have the same sign–pattern as A — that is,

$$\text{sgn } \tilde{a}_{ij} = \text{sgn } a_{ij} \quad \text{for all } i \text{ and } j.$$

The matrix A and the associated linear system are said to be *sign–stable* if every matrix $\tilde{A} \in Q(A)$ is stable. (In terminology suggested by Johnson [Jo2], A is a sign–stable if its sign–pattern *requires* stability and *potentially stable* if its sign–pattern *allows* stability.) Analogous definitions in terms of A's sign–pattern are applied to quasistability and semistability. In discussing systems such as (2), the matrix entries a_{ij} are in some contexts called *interaction coefficients*. Initial interest in sign–stability and its relatives was stimulated by the 1947 observation of Samuelson [Sa] that in the mathematical modeling of problems from economics, it often happens that many of the interaction coefficients are known only qualitatively; typically, the sign–pattern is known, but quantitative estimates for some or all of the coefficients may be extremely rough or unavailable. This led to questions concerning the extent to which important properties of a system — for example, various sorts of stability — can be deduced from the sign–pattern alone. These questions were studied by mathematical economists in the 1960's. An influential paper by Quirk and Ruppert [QR], characterizing sign–semistability in finitely computable terms, appeared in 1965, and in 1969 Maybee and Quirk [MQ] recognized the essentially graph–theoretic nature of much of the analysis. In the 1970's, some mathematical biologists became interested in the area because of their recognition that the qualitative nature of many economic models is also often found in models of ecological systems — involving, for example, interactions among the population levels of various species (May [Ma'1,2], Jeffries [Je1], Svirežev and Logofet [SL]). The notion of sign–stability attracted some chemists in the 1970's (Clarke [Cl1, 2], Tyson [Ty]), and sociologists in the 1980's (Shirakura [Sh]).

In 1974, Jeffries [Je1] published an important paper that gave the first correct, finitely computable characterization of sign–stability. Since then, the investigation has broadened, and by now the phenomena of sign–semistability, sign–quasistability and sign–stability are all well understood. For each of these notions, there are characterizations in terms of graphs and digraphs associated with the matrix A. The characterizations are mathematically illuminating and they lead to extremely efficient recognition algorithms, of time–complexity

$$0(n+ \text{ number of nonzero entires of } A).$$

On the other hand, the results concerning potential stability and some other qualitative notions of stability are still quite primitive, and there are many opportunities for further research concerning these notions.

In the present survey, Section 2 discusses the main results on sign–semistability [QR] and sign–stability (Jeffries, Klee and van den Driessche [JKV1,2]). Section 3 covers sign–quasistability [JKV2]. Section 4 deals with a variety of results and problems, due to various researchers, concerning other qualitative stability notions. There are some indications of methods and of the intuitive ideas behind some of the arguments, but detailed proofs are not included. In each case, the primary purpose is to state the main definitions and theorems, to supply references to the proofs, and (in Section 4) to state some unsolved problems. I am indebted to Pauline van den Driessche and Clark Jeffries for some helpful comments.

The main results reported here may be regarded as a very special sort of interval analysis of stability problems. In a completely general interval analysis, for each pair (i,j) of indices between 1 and n, it would be known that the entry a_{ij} of the matrix A belongs to a certain interval I_{ij} of real numbers, and one would seek to determine whether this information guarantees some sort of stability for A. In classical discussions of stability, each I_{ij} consists of a single number, so that the matrix A is completely specified. In the present discussion, each interaction coefficient is classified only as negative, zero or positive, so that only the intervals $]-\infty, 0[$, $\{0\}$ and $]0, \infty[$ are involved. Since this is a very crude classification, the results on sign–stability and its relatives are limited so far as direct application is concerned (see Rader [Ra]). Some possible extensions that could increase applicability are mentioned in Section 4.

2. Sign–semistability and sign–stability. We begin with the simplest notion, sign–semistability. The characterization is due to Quirk and Ruppert [QR].

2.1 Characterization of Sign–Semistability. *An $n \times n$ real matrix $A = [a_{ij}]$ is sign–semistable if and only if it satisfies the following three conditions:*

(α) *for each i, $a_{ii} \leq 0$;*

(β) *for all $i \neq j$, $a_{ij}a_{ji} \leq 0$;*

(γ) *for each sequence of $k \geq 3$ distinct indices $i(1), \ldots, i(k)$ in $\{1, \ldots, n\}$,*

$$a_{i(1)i(2)}a_{i(2)i(3)} \cdots a_{i(k-1)i(k)}a_{i(k)i(1)} = 0.$$

To see that the conditions are necessary, suppose that one of them is violated, fix at ± 1 the entries of $\tilde{A} \in Q(A)$ that correspond to the violation, and note that as the remaining entries of $\tilde{A}$ converge to 0, $\tilde{A}$'s eigenvalues converge to the zeros of the polynomial $(\lambda - 1)\lambda^{n-1}$, or $(\lambda^2 - 1)\lambda^{n-2}$, or $(x^k \pm 1)x^{n-k}$. In each case there is an eigenvalue with positive real part when the remaining entries are sufficiently small in magnitude.

The *digraph* $D(A)$ of the matrix A is useful in connection with condition (γ) and essential in dealing with some conditions to be encountered later. This digraph has $\{1, \ldots, n\}$ as its node–set, and has its arc–set the set of ordered pairs

$$\{(j, i) : a_{ij} \neq 0\}.$$

For most purposes, it is more common to take $\{(i, j) : a_{ij} \neq 0\}$ as the arc-set. However, we want to think of the arc (j, i) as represented by an arrow from j to i, and that is appropriate in the present context when $a_{ij} \neq 0$, for then, in the system $\dot{x} = Ax$, the level of the variable x_j influences the level of the variable x_i.

A *k–cycle* in a digraph is a (simple) cycle that is formed from precisely k arcs. In terms of the digraph $D(A)$, condition (γ) says that $D(A)$ has no k-cycle for $k \geq 3$. As noted by [KV], this can be tested in time

$$0(n + \text{number of nonzero entries of } A)$$

by an algorithm based on depth–first search of $D(A)$. Of course, this estimate does not apply when A is presented explicitly as an $n \times n$ array, but does apply when a presentation of $D(A)$ in terms of adjacency lists is available.

If, whenever $a_{ij} \neq 0$, one associates the sign of a_{ij} with the edge (j,i), thus turning the digraph $D(A)$ into the *signed digraph* $SD(A)$, then condition (α) says that each 1–cycle in $SD(A)$ consists of a negative arc and condition (β) says that each 2–cycle in $SD(A)$ consists of one negative arc and one positive arc. Thus conditions $(\alpha)-(\gamma)$ are called the *cycle conditions*. When the interactions given by A are those of the population levels in an ecological system, condition (α) may be interpreted as saying that each self–interacting species is in fact self–regulating, and condition (β) as saying that each pair of mutually interacting species is in a prey–predator relationship. It seems remarkable that when conditions (α) and (β) are assumed to be satisfied, the signs of the entries do not play any further role in the analysis of sign–semistability, sign–quasistability, or sign–stability. In particular, when $a_{ij}a_{ji} < 0$ it is not necessary to know which of a_{ij} and a_{ji} is positive and which is negative.

In this section and the next one, it is assumed henceforth that the cycle conditions are satisfied by the matrix A.

Thus when $a_{ii} \neq 0$, we know that in fact $a_{ii} < 0$. This is recorded by making the node i *distinguished* rather than using an arc (i,i) and recording the sign explicitly. Similarly, since each 2–cycle consists of one negative arc and one positive arc, and since it turns out not to matter which is which, each 2–cycle in $D(A)$ can be collapsed into a single undirected edge. Thus the arcs of $D(A)$ can be represented by a mixture of one–way directed *arcs* (ordered pairs (j,i) for which $a_{ij} \neq 0 = a_{ji}$) and two–way undirected *edges* (unordered pairs $\{j,i\}$ for which $i \neq j$ and $a_{ij}a_{ji} < 0$). The resulting *mixed* graph, with its distinguished nodes, will be denoted by $M(A)$, and the *undirected* graph that results from discarding the one–way arcs (but keeping the distinguished nodes) will be denoted by $G(A)$.

In the mixed graph M of Figure 1 (from [JKV2]), there are 3 distinguished nodes. For each of the 15! labelings of the nodes, M represents 2^{18} different sign–patterns that satisfy the cycle conditions. The components of the underlying graph G are trees, as they must be when condition (γ) holds. The one–way arcs organize these trees into an acyclic digraph, which is also shown.

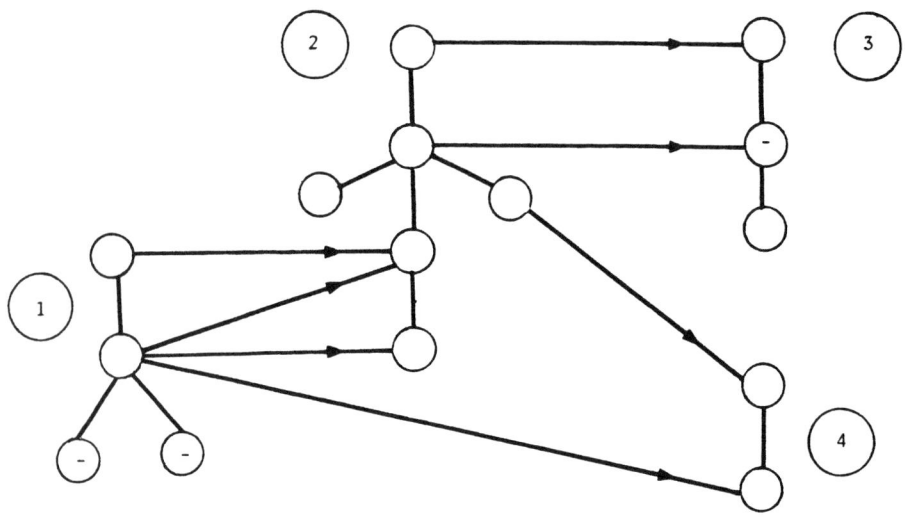

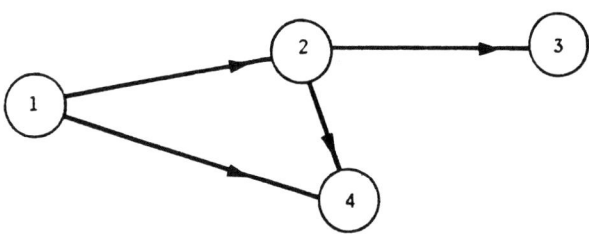

Fig. 1

When the cycle conditions are satisfied, A's characteristic polynomial is the same as the characteristic polynomial of the matrix $\overline{A}$ that is formed from A by replacing with 0 all the entries a_{ij} for which $a_{ij} \neq 0 = a_{ji}$. Hence the one–way arcs are not relevant to the study of sign–semistability or sign–stability, though they turn out to play a crucial role in connection with the sign–quasistability of reducible matrices.

We may assume without loss of generality that $\overline{A}$ is in block–diagonal form, so that its characteristic polynomial is just the product of the characteristic polynomials of the *components* of $\overline{A}$ (the submatrices associated with the blocks or, equivalently, with the components of $G(A)$). Without loss of generality, we may focus on a $k \times k$ block of $\overline{A}$, hence on a k-node component of $G(A)$, say with k nodes $1,\ldots,k$. By means of a simple inductive procedure, it is possible to produce positive constants $\lambda_1 = 1, \lambda_2, \ldots, \lambda_k$ such that $\lambda_i \overline{a}_{ij} = -\lambda_j \overline{a}_{ji}$ for all $i \neq j$ in $\{1,\ldots,k\}$. The construction relies heavily on condition (β). It is described in detail in [JKV1], but may in essence go back at least to Volterra [Vo]. The subsystem

$$\dot{x}_i = \sum_{i=1}^{k} \overline{a}_{ij} x_j \qquad (i = 1,\ldots k)$$

then admits the positive definite *Lyapunov function*

$$\Lambda(x) = \sum_{i=1}^{k} \lambda_i x_i^2,$$

whose derivative

$$\dot{\Lambda}(x) = \sum_{i=1}^{k} 2\lambda_i x_i \dot{x}_i = 2 \sum_{i=1}^{k} \sum_{j=1}^{k} \lambda_i x_i \overline{a}_{ij} x_j = 2 \sum_{i=1}^{k} \lambda_1 \overline{a}_{ii} x_i^2$$

is negative semidefinite by condition (α). The level sets of Λ are ellipsoids centered at the origin, and from the fact that $\dot{\Lambda} \leq 0$ it follows that whenever a positive trajectory of the system hits one of these ellipsoids, it stays from then on in the convex body bounded by the ellipsoids. That proves the sufficiency of the cycle conditions in Theorem 2.1 and also in the following result of [JKV1], where the matrix A is called *combinatorially symmetric* [Ma″1] if $A = \overline{A}$.

2.2. Sign–Quasistability of Combinatorially Symmetric Matrices. *A combinatorially symmetric matrix is sign–quasistable if and only if it satisfies the cycle conditions* (α), (β) *and* (γ).

Theorem 2.1 can also be proved by a more direct argument [QR] that does not involve Lyapunov functions. However, the function Λ plays an essential role in dealing with sign–stability and sign–quasistability.

The remaining conditions for sign–stability involve only $G(A)$, a graph with a set of distinguished nodes. For any such graph G, let us define a *δ-coloring* as a partition of all the nodes into two sets, black and white, such that

(i) each distinguished node is black;

(ii) no black node has exactly one white neighbor;

(iii$_\delta$) each white node has at least one white neighbor.

An ϵ-*coloring* is defined by conditions (i), (ii), and

(iii$_\epsilon$) no white node has a white neighbor.

A coloring is said to be *trivial* if all nodes are black. In terms of these notions, the characterization of sign–stability is as follows [JKV2].

2.3. Characterization of Sign–Stability. *An $n \times n$ real matrix A is sign–stable if and only if A's signed digraph $SD(A)$ satisfies the cycle conditions (α) and (β), A's digraph $D(A)$ satisfies the cycle condition (γ), and A's graph $G(A)$ satisfies the following two color conditions:*

(δ) *each δ-coloring of $G(A)$ is trivial;*

(ϵ) *each ϵ-coloring of $G(A)$ is trivial.*

Now suppose that $G(A)$ and $G(B)$ are simple paths with 5 and 3 nodes respectively, each having the center node as its sole distinguished node. If this node is colored black and the remaining nodes are colored white, there results a nontrivial δ-coloring in the case of A and a nontrivial ϵ-coloring in the case of B. Since the cycle conditions are satisfied, these colorings correspond to the facts that A has a pure imaginary eigenvalue and 0 is an eigenvalue of B. The facts that each ϵ-coloring of $G(A)$ is trivial and each δ-coloring of $G(B)$ is trivial correspond respectively to the fact that for each $\tilde{A} \in Q(A)\langle$ resp. $\tilde{B} \in Q(B)\rangle$ the only positive trajectories of the associated linear system that do not converge to the origin are sinusoidal $\langle$ resp. constant$\rangle$.

In [JKV2], the Lyapunov function Λ is used to prove that when the cycle conditions are satisfied, the *color conditions* (δ) and (ϵ) imply sign–stability. By means of an interplay between the system equation $\dot{x} = Ax$ and the equations $\lambda_i a_{ij} = -\lambda_j a_{ji}$, it is shown that if A has a pure imaginary eigenvalue then $G(A)$ admits a nontrivial δ-coloring, while if 0 is an eigenvalue of A then $G(A)$ admits a nontrivial ϵ-coloring. In the reverse direction, it is shown that when $G(A)$ is connected, the failure of condition (δ) implies the existence of a sinusoidal trajectory (and hence a pure imaginary eigenvalue) for *some* member of $Q(A)$, while the failure of condition (ϵ) corresponds to the existence of a nonzero constant trajectory and (hence a zero eigenvalue) for *every* member of $Q(A)$.

For a graph G with a set of distinguished nodes, let us define the $\delta-rim\langle\epsilon-rim\rangle$ of G as the set of all nodes i such that i is white in some δ-coloring $\langle$ resp. ϵ-coloring$\rangle$ of G. The δ-core and the ϵ-core are the complements of the respective cores. (In particular, each distinguished node belongs to both cores.) In testing for the sign–stability of A, it suffices to know whether the rims of $G(A)$ are empty. However, as we shall see in Section 3, in testing for sign–quasistability it is necessary to know exactly which nodes belong to the respective rims and hence to identify the cores.

When a graph G with distinguished nodes is presented by means of adjacency lists, its δ-core can be determined in time

$$0(\text{number of nodes} + \text{number of edges})$$

[KV], and the same is true of the ϵ-core when G is a forest [JKV2]. For easy understanding of the procedure for the δ-core, imagine that initially all the distinguished nodes are black and all the remaining nodes are white. Then let blackness spread in such a way that a white node turns black if it has no white neighbor or is the sole white neighbor of a black node. After the spread has been completed according to these rules, the black nodes are exactly those in the δ-core. As is shown in [KV], the spreading process can be carried out in the specified time by maintaining a stack which lists the nodes that may play a role in spreading the blackness. When G is a forest, the ϵ-core can be produced by a somewhat similar procedure in which a crucial role is played by the nodes of valence ≤ 1 [JKV2].

The original characterization of Jeffries [Je1] (established in detail in [JKV1]; see also Logofet and Ul'yanov [LU1] [LU2]), uses a matching condition in place of condition (ϵ). For a graph G with a set of distinguished nodes, define the *value* of a matching as the number of undistinguished nodes that it covers, and call a matching *perfect* if it covers all undistinguished nodes. Then the condition is that $G(A)$ should admit a perfect matching. It is easy to see [KV] that in any forest with distinguished nodes, a matching of maximum value can be found in time

$$0\,(\text{number of nodes} + \text{number of edges}).$$

And it turns out [JKV2] that the ϵ-core is equal to the set of all nodes that are distinguished or are covered by every matching of maximum value.

3. Sign–quasistability. In dealing with sign–quasistability, the one–way arcs play an essential role and the analysis is much more complicated than for sign–semistability and sign–stability. The results reported in this section are all from [JKV2].

As an aid to understanding, we take some pains to explain the underlying intuitive idea. Think of each δ-rim $\langle$ resp. ϵ-rim$\rangle$ node of the mixed graph $M(A)$ as being capable of emitting a nonzero sinusoidal $\langle$ resp. nonzero constant$\rangle$ signal that can be propagated through $M(A)$ according to certain rules. When such a signal can be propagated so as to "drive" another node of the same sort, the two nodes can combine forces to produce (for some choice of $\tilde{A} \in Q(A)$) an unbounded signal and hence an unbounded positive trajectory. That is the correct general idea, but of course it remains to find the right definitions of driving and to work out the details of the argument.

The simplest sort of driving is that which is *direct*. Node j *directly drives* node i if the ordered pair (j,i) is an arc of $M(A)$. According to our definition of $M(A)$, this implies that $a_{ij} \neq 0 = a_{ji}$ and hence, since condition (γ) is assumed to hold, i and j are in different components of $M(A)$.

Figure 2 (like Fig. 1) is taken from [JKV2]. It satisfies the cycle conditions but not the color conditions. On A's main diagonal there are four nonzero (hence negative) entries, and they produce the four distinguished nodes labeled with a minus sign in the figure. Each of the other nodes belongs to the δ-rim or the ϵ-rim, but not to both. Rim nodes are directly driven by other rim nodes, but when the broken edges are absent (i.e., when the related entries of A are zero), no rim node is driven by another rim node of the same sort and the resulting matrix A is sign–quasistable. However, if either broken edge is present, then A is not sign–quasistable. This is true regardless of the signs (negative or positive) associated with any arc.

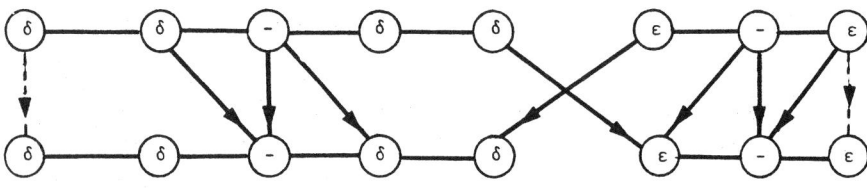

Fig. 2

The key to finding a finitely computable characterization of sign–quasistability lies in formulating the appropriate notions of δ-driving and ϵ-driving, each being an extension of direct driving. In terms of these formulations, the characterization is as follows.

3.1. Characterization of Sign–Quasistability. *For the sign–quasistability of a sign–semistable A, it is necessary and sufficient that in the mixed graph $M(A)$, no δ-rim node is δ-driven by another δ-rim node and no ϵ-rim node is ϵ-driven by another ϵ-rim node.*

The proof of Theorem 3.1 is long and involved, and even the definition of ϵ-driving is very complicated. However, the following corollary of 3.1 is easy to understand and to prove.

3.2. An Easy Sufficient Condition. *For the sign–quasistability of a sign–semistable matrix A, it is sufficient that no path in $M(A)$ should lead from a rim node to another rim node of the same sort that lies in a different component of $M(A)$.*

In seeking the proper definitions of δ-driving and ϵ-driving, one must consider the ways in which nonzero sinusoidal or nonzero constant signals can propagate though $M(A)$ without losing their strength. Only rim nodes can initiate such signals, but under certain circumstances other nodes that are "downstream" from rim nodes can serve as relay stations that pass the signals on. What are the circumstances? How can a node i of $M(A)$ become a δ-driver, meaning that there exists $\tilde{A} \in Q(A)$ and a trajectory x with $\dot{x} = \tilde{A}x$ such that $x_i \neq 0$ and $x_i(t)$ is a linear combination of $\sin t$ and $\cos t$? Or an ϵ-driver, in the sense that for some $\tilde{A} \in Q(A)$ and some trajectory x with $\dot{x} = \tilde{A}x$, x_i is a nonzero constant? The answer to the first question turns out to be fairly simple.

3.3. Characterization of δ-Drivers. *A node i of a sign–semistable matrix A is a δ-driver if and only if either*

(i) i belongs to A's δ-rim,

or (ii) i is pushed by a member j of the δ-rim, in the sense that there are nodes $i =$

$i_0, \ldots, i_k = j$ such that (i_k, i_{k-1}) is an arc of $M(A)$, and for $1 \leq h < k$, $(i_h, i_{h=1})$ is an arc of $M(A)$ or $\{i_h, i_{h-1}\}$ is an edge of $M(A)$.

The notion of *pushing* is an extension of direct driving, and it leads at once to the desired notion of δ-driving. A node j is said to δ-*drive* a node i if j directly drives i or j pushes a node that directly drives A. In Figure 3, for example, the nodes directly driven by node 1 are 4 and 5. All nodes except 1, 2 and 3 are pushed by 1, and the nodes that are δ-driven by 1 are 4, 5, 8, 9 and 10.

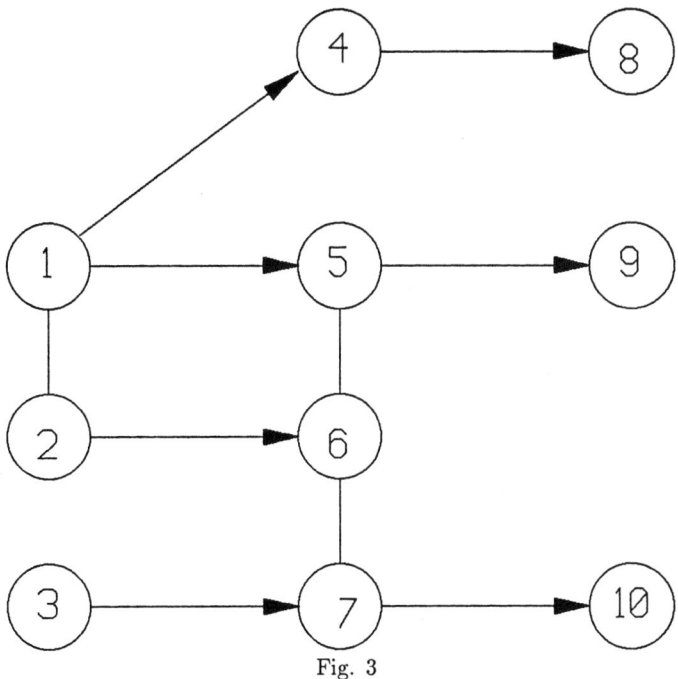

Fig. 3

Here are the two main results in the analysis of sign–quasistability.

3.4. δ-Driving and Imaginary Eigenvalues. *For a sign–semistable A, the following conditions (i), (ii) and (iii) are equivalent and they imply condition (iv):*

(i) *some pure imaginary number is a multiple root of the minimum polynomial of some member of $Q(A)$;*

(ii) *A has a δ-driver that directly drives a node of the δ-rim;*

(iii) *there is a δ-rim node of A that δ-drives another such node;*

(iv) *there is a path in $M(A)$ that leads from a δ-rim node to another δ-rim node in a different component.*

3.4. ϵ-Driving and Zero Eigenvalues. *For a sign–semistable A, the following conditions (i), (ii) and (iii) are equivalent and they imply condition (iv):*

(i) *zero is a multiple root of the minimum polynomial of some member of $Q(A)$;*

(ii) A has an ϵ-driver that directly drives a node of the ϵ-rim;

(iii) there is an ϵ-rim node of A that ϵ-drives another such node;

(iv) there is a path in $M(A)$ that leads from an ϵ-rim node to another ϵ-rim node in a different component.

When Theorems 3.3 and 3.4 are combined, they yield 3.1 as a corollary. The parallelism between 3.3 and 3.4 may appear to be complete, but there are two important differences between them. The most important difference is that while δ-driving is easily described, there seems to be no similarly simple description of ϵ-driving. The definition in [JKV2] involves a coloring–and–recoloring algorithm that is quite complicated. However, it is computationally very satisfactory, since for ϵ-driving as for δ-driving, it can be determined in time

$$0\,(n+ \text{ number of nonzero entries})$$

whether one rim node drives another.

Another difference is in the behavior of the drivers. When A is sign–semistable, both of the following statements are true:

(i) for each δ-driver i there exist $\tilde{A} \in Q(A)$ and a trajectory x with $\dot{x} = \tilde{A}x$ such that $x_i \ne 0$ and *all* components of x are sinusoidal (some may be zero);

(ii) there exist $\tilde{A} \in Q(A)$ and a trajectory x with $\dot{x} = \tilde{A}x$ such that x_i is nonzero and sinusoidal for *all δ-drivers i simultaneously*.

However, in the case of ϵ-drivers, the analogues of statements (i) and (ii) both fail for the matrix $A = \begin{bmatrix} 0 & 1 \\ 0 & 0 \end{bmatrix}$.

4. Other qualitative stability notions. The aim of the preceding two sections was to explain the main results of the papers [QR], [Je1], [JKV1], [KV] and [JKV2]. These provide a good understanding, from both mathematical and algorithmic viewpoints, of the notions of sign–semistability, sign–quasistability, and sign–stability. The present section describes some other directions of research in the qualitative study of stability, still concentrating on sign–patterns or closely related notions.

Qualitative approach to stability of discrete systems. In a context related to difference equations or to Markov processes, the term *stable* is often applied to a matrix all of whose eigenvalues are of magnitude less than 1. Is there, for this notion of stability, an analogue of the theory of sign–stability? The sign–pattern would not be an appropriate tool here, but perhaps something of interest would emerge if comparison of the entries to 0 (and their resulting classification as negative, zero or positive) were replaced by comparison of the entries' magnitudes to 1.

Possible refinement of sign–stability. If one could extend the analysis of sign–stability to cover the situation in which the interactions (i.e., the entries a_{ij} of the matrix A) are classified not only as to sign but also roughly as to magnitude, the range of applicability might be greatly extended. For example, one might classify the interaction coefficients as "negative large" (normalized to mean ≤ -1), "negative small" (between -1 and 0), zero, "positive small" (between 0 and 1), and "positive large" (≥ 1). Two matrices A and $\tilde{A}$ would then be regarded as equivalent if, for each i and j, the entries a_{ij} and $\tilde{a}_{ij}$ lie together in one of the five intervals $]-\infty,-1]$, $]-1,0[$, $\{0\}$, $]0,1[$ and $[1,\infty[$. The analogue of recognizing sign–stability would be to recognize those matrices A for which every matrix equivalent to A is stable. In applications, this would be especially appropriate when all of the x_i's are measured in the same units (e.g., when each x_i represents the biomass of some species). But (as pointed out by A. Nijenhuis) this approach might be inappropriate when the various variables x_i represent quite different sorts of units, because an interaction coefficient a_{ij} could be changed from "small" to "large", or vice–versa, simply by rescaling of x_i or x_j.

Sign–inertia. For each $n \times n$ matrix A, let $N(A) \langle$ resp. $P(A) \rangle$ denote the number of A's eigenvalues having negative $\langle$ resp. positive $\rangle$ real part, where the eigenvalues are counted according to their multiplicities as roots of A's characteristic polynomial. Then of course $0 \leq N(A) + P(A) \leq n$. It would be of interest, for each pair of nonnegative integers (r,s) having $0 \leq r+s \leq n$, to be able to recognize which $n \times n$ sign–patterns guarantee that $(N(A), P(A)) = (r,s)$. The case $(0,0)$ is trivial, and the cases $(n,0)$ and $0,n)$ are settled by the results on sign–stability. Partial results on some other cases appear in papers by Ishida, Adachi and Tokumaru [IAT] and Jeffries and Johnson [JJ].

A hierarchy of qualitative notions of stability. For two $n \times n$ matrices A and B, let $A \circ B$ denote the Hadamard product of A and B (the (i,j) entry of $A \circ B$ is $a_{ij}b_{ij}$). For $1 \leq k \leq n$, let $D_{n,k}$ denote the set of all $n \times n$ real matrices A such that the matrix $A \circ B$ is stable whenever B is a matrix of rank $\leq k$ all of whose entries are positive. Then $D_{n,n}$ is the class of all $n \times n$ sign–stable matrices, and

$$D_{n,n} \subseteq D_{n,n-1} \subseteq \cdots \subseteq D_{n,2} \subseteq D_{n,1}.$$

As noted by Johnson and van den Driessche [JV], $D_{n,1}$ is another class of matrices that has been studied extensively — the *D-stable* matrices. (See [MQ], [Qu2], Johnson [Jo1], Carlson, Datta and Johnson [CDJ], and Berman and Hershkowitz [BH1] for some results on D-stability.) Thus the classes $D_{n,k}$ interpolate between the sign–stable and the D-stable matrices. Some basic results concerning the classes $D_{n,k}$ are established in [JV], and several open questions are raised. In particular, it is shown that for $n \geq 3$, $D_{n,2}$ is a proper subset of $D_{n,1}$, but it is not known whether the other inclusions are strict.

Potential stability. A square matrix (or sign–pattern) A is called *potentially stable* [Qu1] if the sign–pattern allows stability (i.e., some member of $Q(A)$ is stable), and the lack of potential stability is called *structural instability* [Ha]. From augmentation of the well–known Routh–Hurwitz stability criterion by the inequalities expressing the sign–pattern, there results a finite system of polynomial inequalities whose consistency is equivalent to potential stability of the sign–pattern. Hence, as noted by Bone [Bo2], it follows from the general decision method of Tarski [Ta] that there is a finite algorithm for testing potential stability. It seems probable, however, that the problem of recognizing potential stability is NP–complete. If this is true (and if $P \neq NP$), then in contrast to the case of sign–stability, there is no polynomial–time algorithm for recognizing potential stability. For certain sign–patterns, potential stability or the lack thereof has been established by Campbell [Ca], Bone [Bo1, 2], Jeffries and Johnson [JJ], and Johnson and Summers [JS].

Viable systems. A linear algebraic system $Ax = b$ is said to be *positively sign–solvable* if for each $\tilde{A} \in Q(A)$ and $\tilde{b} \in Q(b)$, the system $\tilde{A}x = \tilde{b}$ is solvable and each solution x has exclusively positive components. (For basic results on sign–solvability, see Maybee [Ma″2], Manber [Ma], Klee, Ladner and Manber [KLM], Klee [K1] and their references.) Sign–solvability and sign–stability are combined in a paper of Bone, Jeffries and Klee [BJK], where the linear system $\dot{x} = Ax + b$ is said to be *viable* if the algebraic system $Ax = -b$ is positively sign–solvable and the matrix A is sign–stable; equivalently, the sign–pattern of (A, b) guarantees that the system $\dot{x} = Ax + b$ has an asymptotically stable constant equilibrium in the first orthant. [BJK] provide a general recusive method for constructing viable systems, and also an

$$0\,(n+ \text{number of nonzero entries})$$

algorithm for recognizing such systems. (For recognition alone, the two properties of sign–stability and positive sign–solvability could of course be tested separately — [K1] has an $0(n^2)$ algorithm for the latter.)

Nonlinear systems. In dealing with linear dynamical systems, it is not necessary to distinguish between local and global stability, because the two notions coincide. For nonlinear systems, some insight can often be obtained by considering linear approximations to the system, but this insight is usually only local in nature. However, it turns out that when the matrix A is sign–semistable, there are unusually close global relationships between the asymptotic behavior of the linear system

$$\dot{x}_i = b_i + \sum_{j=1}^{n} a_{ij} x_j \qquad (1 \leq i \leq n)$$

and the associated nonlinear Lotka–Volterra system

$$\dot{x}_i = x_i \left(b_i + \sum_{j=1}^{n} a_{ij} x_j \right) \qquad (1 \leq i \leq n).$$

Such relationships have been established under various supplementary hypotheses by [BJK] and by Redheffer and Zhou [RZ3], but some basic questions are still open. In particular, Redheffer has asked whether, assuming only sign–semistability of A, the boundedness of all solutions of the linear system implies that of the associated Lotka–Volterra system.

In the case of the Lotka–Volterra system, there is special interest in the existence of a global attractor trajectory that lies entirely in the positive orthant, for this corresponds to persistence of species in the ecological model. For some other approaches to the matter of persistence, see Gard [Ga], Hutson and Vickers [HV], Butler and Waltman [BW], and some of their references.

Other graph–theoretic approaches. We have here discussed several ways of using graph–theoretical methods in studying the stability behavior of dynamical systems. Other examples are provided by Berman and Hershkowitz [BH2], Yamada [Ya], and Jeffries and van den Driessche [JV]. And we want in particular to mention the extensive work of Krikorian [Kr], Redheffer [Re1,2,3,4], Redheffer and Zhou [RZ1, 2, 3], and Redheffer and Walter [RW1, 2].

Higher–dimensional complexes. Since a graph may be regarded as a 1-dimensional simplicial complex, and graph–theoretic methods have proved to be so fruitful in dealing with asymptotic properties of certain sorts of dynamical systems, it is natural to consider the use of higher–dimensional simplicial complexes in studying stability properties of more complicated, nonlinear systems. An example of a successful analysis of this sort (dealing with attractor regions rather than attractor trajectories) is provided by Jeffries [Je2].

REFERENCES

[BH1] A. BERMAN AND D. HERSHKOWITZ, *Characterization of acyclic D–stable matrices*, Linear Algebra Appl. 58 (1984) 17–31.

[BH2] A. BERMAN AND D. HERSHKOWITZ, *Graph theoretical methods in studying stability*, Contemporary Math. 47 (1985) 1–6.

[Bo1] T. BONE, *Positive feedback may sometimes promote stability*, Linear Algebra Appl. 51 (1983) 143–151.

[Bo2] T. BONE, *Qualitative stability properties of matrices*, Ph.D. dissertation, Math. Dept., Univ. of Washington, Seattle, 1985.

[BJK] T. BONE, C. JEFFRIES AND V. KLEE, *A qualitative analysis of $\dot{x} = Ax + b$*, Discrete Appl. Math. 20 (1988) 9–30.

[BW] G.J. BUTLER AND P. WALTMAN, *Persistence in dynamical systems*, J. Differential Equations 63 (1986) 255–263.

[Ca] R.C. CAMPBELL, *Some new results on potentially stable matrices*, Ph.D. dissertation, Math. Dept., Univ. of Colorado, Boulder, 1975.

[CDJ] D. CARLSON, B. DATTA AND C.R. JOHNSON, *A semi–definite Lyapunov theorem and the characterization of tridiagonal D–stable matrices*, SIAM J. Alg. Discrete Methods 3 (1982) 293–304.

[Cl1] B.L. CLARKE, *Theorems on chemical network stability*, J. Chemical Physics 62 (1975) 773–775.

[Cl2] B.L. CLARKE, *Stability of topologically similar chemical networks*, J. Chemical Physics 62 (1975) 3726–3738.

[Ga] T.C. GARD, *Persistence in food chains with general interactions*, Math. Biosciences 51 (1980) 165–174.

[Ha] W. HAHN, *Stability of Motion*, Springer–Verlag, New York, 1967.

[HS] M. HIRSCH AND S. SMALE, *Differential Equations, Dynamical Systems and Linear Algebra*, Academic Press, New York, 1974.

[HV]   V. HUTSON AND G.T. VICKERS, *A criterion for permanent coexistence of species, with an application to a two–prey one–predator system*, Math. Biosciences 63 (1983) 253–269.

[IAT]   Y. ISHIDA, N. ADACHI AND H. TOKUMARU, *Some results on the qualitative theory of matrix*, Trans. Soc. Instrument and Control Engineers 17 (1981) 49–55.

[Je1]   C. JEFFRIES, *Qualitative stability and digraphs in model ecosystems*, Ecology 55 (1974) 1415–1419.

[Je2]   C. JEFFRIES, *Qualitative stability of certain nonlinear systems*, Linear Algebra Appl. 75 (1986) 133–144.

[JJ]   C. JEFFRIES AND C.R. JOHNSON, *Some sign patterns that preclude matrix stability*, SIAM J. Matrix Anal. Appl. 9 (1988) 19–25.

[JKV1] C. JEFFRIES, V. KLEE AND P. VAN DEN DRIESSCHE, *When is a matrix sign stable?*, Canad. J. Math. 29 (1977) 315–326.

[JKV2] C. JEFFRIES, V. KLEE AND P. VAN DEN DRIESSCHE, *Qualitative stability of linear systems*, Linear Algebra Appl. 87 (1987) 1–48.

[JV]   C. JEFFRIES AND P. VAN DEN DRIESSCHE, *Eigenvalues of matrices with tree graphs*, Linear Algebra Appl., 101 (1988) 109–120.

[J1]   C. R. JOHNSON, *Sufficient conditions for D–stability*, J. Economic Theory 9 (1974) 53–62.

[Jo2]   C.R. JOHNSON, *Combinatorial matrix analysis: An overview*, Linear Algebra Appl. 108 (1988) (to appear).

[JS]   C.R. JOHNSON AND T. A. SUMMERS, *The potentially stable tree sign patterns for dimensions less than five*, Research report, Math. Dept., Clemson Univ., 1987.

[J'V]   C.R. JOHNSON AND P. VAN DEN DRIESSCHE, *Interpolation of D–stability and sign stability*, Linear and Multilinear Algebra, 23 (1988) 363–368.

[K1]   V. KLEE, *Recursive structure of S^*-matrices, and on an $0(m^2)$ algorithm for recognizing strong sign–solvability*, Linear Algebra Appl. 96 (1987) 233–247.

[KLM]   V. KLEE, R. LADNER AND R. MANBER, *Signsolvability revisited*, Linear Algebra Appl. 59 (1984) 131–157.

[KV]   V. KLEE AND P. VAN DEN DRIESSCHE, *Linear algorithms for testing the sign–stability of a matrix and for finding Z–maximum matchings in acyclic graphs*, Numer. Math. 28 (1977) 273–285.

[Kr]   N. KRIKORIAN, *The Volterra model for three species predator–prey systems: boundedness and stability*, J. Math. Biology 7 (1979) 117–132.

[LU1]   D.O. LOGOFET AND N.B. UL'YANOV, *Necessary and sufficient conditions for sign stability of matrices (in Russian)*, Dokl. Akad. Nauk. SSSR 264 (1982) 542–546. (English transl. in Soviet Math. Dokl. 25 (1982) 676–680).

[LU2]   D.O. LOGOFET AND N.B. UL'YANOV, *Sign stability in model ecosystems: A complete class of sign–stable patterns*, Ecological Modelling 16 (1982) 173–189.

[Ma]   R. MANBER, *Graph theoretical approach to qualitative solvability of linear systems*, Linear Algebra Appl. 48 (1982) 457–470.

[Ma'1] R.M. MAY, *Qualitative stability in model ecosystems*, Ecology 54 (1973) 638–641.

[Ma'2] R.M. MAY, *Stability and Complexity in Model Ecosystems*, Princeton Univ. Press, 1973.

[Ma"1] J. MAYBEE, *Combinatorially symmetric matrices*, Linear Algebra Appl. 8 (1974) 529–537.

[Ma"2] J. MAYBEE, *Sign solvability*, in Computer–Assisted Analysis and Model Simplification (H. Greenberg and J. Maybee, eds.), Academic Press, New York, 1981, pp. 544–563.

[MQ]   J. MAYBEE AND J. QUIRK, *Qualitative problems in matrix theory*, SIAM Rev. 11 (1969) 30–51.

[Qu1]   J. QUIRK, *The correspondence principle: a macroeconomic application*, Internat. Econom. Rev. 9 (1968) 294–306.

[Qu2]   J. QUIRK, *Qualitative stability of matrices and economic theory: A survey article*, in Computer–Assisted Analysis and Model Simplification (H. Greenberg and J. Maybee, eds.), Academic Press, New York, 1981, pp. 113–164.

[QR]   J. QUIRK AND J. RUPPERT, *Qualitative economics and the stability of equilibrium*, Rev. Economic Studies 32 (1965) 311–326.

[Ra] T. RADER, *On the impossibility of qualitative economics: Excessively strong correspondence principles in production-exchange economics*, Zeitschr. f. Nationalökonomie 32 (1972) 397–416.

[Re1] R. REDHEFFER, *Labeled graphs and a class of generalized Volterra systems*, in Trends in Theory and Practice of Nonlinear Differential Equations (V. Lakshmikantham, ed.), 1984, Marcel Dekker, New York, pp. 495–503.

[Re2] R. REDHEFFER, *Volterra multipliers I, II*, SIAM J. Alg. Discrete Methods 6 (1985) 592–563.

[Re3] R. REDHEFFER, *Volterra multipliers III*, (to appear).

[Re4] R. REDHEFFER, *A new class of Volterra differential equations for which the solutions are globally asymptotically stable*, (to appear).

[RW1] R. REDHEFFER AND W. WALTER, *On parabolic systems of the Volterra prey–predator type*, Nonlinear Analysis 7 (1983) 333–347.

[RW2] R. REDHEFFER AND W. WALTER, *Solution of the stability problem for generalized Volterra prey–predator systems*, J. Differential Equations 52 (1984) 245–263.

[RZ1] R. REDHEFFER AND Z. ZHOU, *Global asymptotic stability for a class of many-variable Volterra prey–predator systems*, Nonlinear Analysis 5 (1981) 1303–1309.

[RZ2] R. REDHEFFER AND Z. ZHOU, *A class of matrices connected with Volterra prey–predator equations*, SIAM J. Discrete Methods 3 (1982) 122–134.

[RZ3] R. REDHEFFER AND Z. ZHOU, *Sign semistability and global asymptotic stability*, (to appear).

[Sa] P.A. SAMUELSON, *Foundations of Economic Theory*, Harvard Univ. Press, 1947 (repub. by Atheneum, N.Y., 1971).

[Sh] Y. SHIRAKURA, *Jeffries' colour point method and Simons–Homan model (in Japanese)*, Sociological Theory and Methods 1 (1986) 57–70.

[SL] YU M. SVIREŽEV AND D.O. LOGOFET, *Stability of Biological Relations (in Russian)*, "Nauka", Moscow, 1978.

[Ta] A. TARSKI, *A Decision Method for Elementary Algebra and Geometry*, Univ. of California Press, 1951.

[Ty] J.J. TYSON, *Classification of instabilities in chemical reaction systems*, J. Chemical Physics 62 (1975) 1010–1015.

[Vo] V. VOLTERRA, *Lecons sur la théorie mathématique de la lutte pour la vie*, Gauthier-Villars, Paris, 1931.

[Ya] T. YAMADA, *Generic matrix sign-stability*, Canad. Math. Bull. 30 (1986) 370–376.

FOOD WEBS, COMPETITION GRAPHS, COMPETITION-COMMON ENEMY GRAPHS, AND NICHE GRAPHS

J. RICHARD LUNDGREN*

Abstract. This paper surveys the recent work on competition graphs of food webs and some new graphs related to competition graphs, namely, competition-common enemy graphs and niche graphs. Also investigated are digraphs having interval competition graphs, and a partial solution to this problem for a class of (i,j)-competition graphs is given. Several open problems related to these graphs as well as generalized competition graphs are mentioned.

1. Introduction. Cohen [5] introduced competition graphs associated with food web models of an ecosystem as a means of determining the dimension of ecological phase space. Surprisingly, as we shall see later, Cohen [4] found that most food webs have competition graphs that are interval graphs. Roberts [34] established several fundamental properties of competition graphs. Research in this area has varied from statistical analysis of data on food web models to development of a stochastic theory of food web models to a graph theoretic analysis of competition graphs and several natural generalizations and extensions. In this survey paper we will focus on the graph theoretic analysis, most of which has occurred in the last ten years. For details of the other research areas, see Cohen [4,6], Sugihara [38], and the series of papers on stochastic theory by Cohen, Briand, and Newman [7,8,9,29]. In a digraph model of a food web, the vertices represent species, and there is an arc from species x to species y if x preys on y. These digraphs are generally assumed to be acyclic. Figure 1 shows a food web from Burnett, Fisher, and Zim [2], Harary [20], and Roberts [33].

The *competition graph* of an acyclic digraph $D = (V, A)$ is the graph $C(D) = (V, E)$ where $xy \in E$ if and only if $x \neq y$ and for some $z \in V$ both xz and $yz \in A$; i.e., if and only if x and y have a common prey. Analogously, in the *common enemy* graph $CE(D)$, xy is an edge if and only if x and y have a common predator. $C(D)$ and $CE(D)$ for the digraph in Figure 1 are shown in Figure 2.

Another interpretation of these graphs will be useful in understanding some of the theorems. If we let $M = A(D)$ be the adjacency matrix of D, then we see that $G = C(D)$ is simply the "row graph" of M studied by Greenberg, Lundgren, and Maybee [17,18]. In the *row graph* $G = RG(M)$, the rows of M are the vertices of G, and two rows are adjacent if and only if they have a nonzero entry in the same column of M. Similarly, we can define the *column graph* of M, $CG(M)$, which is

---

*Department of Mathematics, University of Colorado at Denver, Denver, CO 80204. This research was partially supported by ONR Research Contract N00014-88-K-0087.

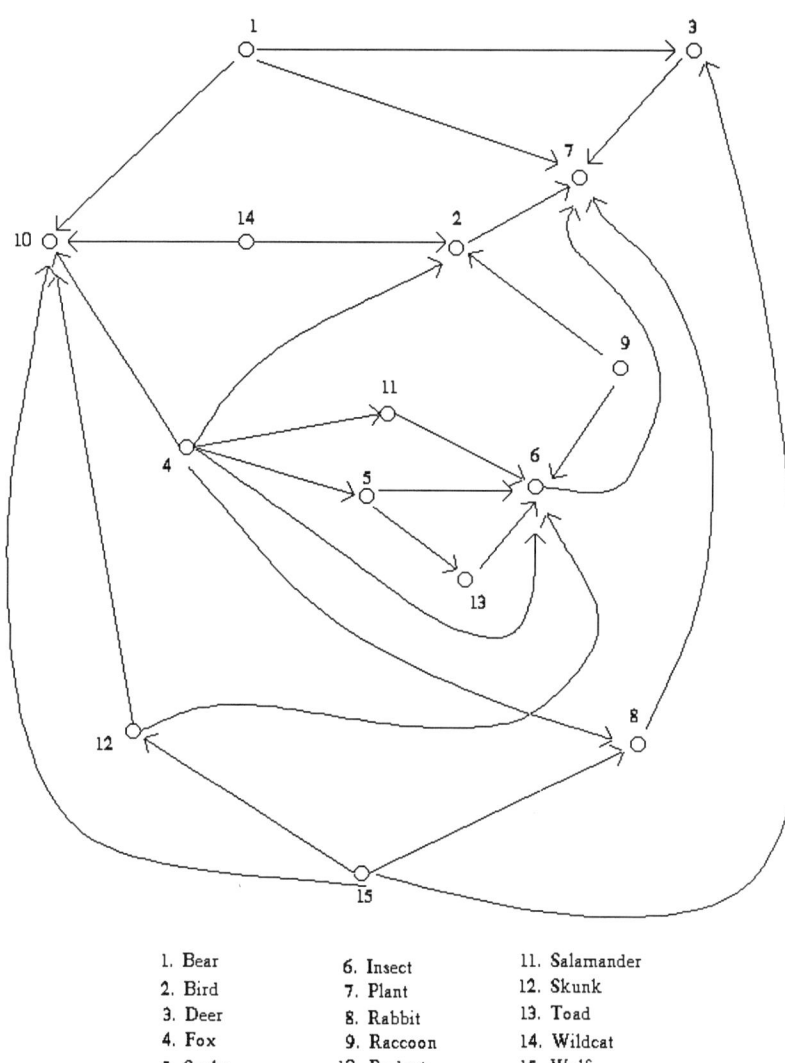

1. Bear	6. Insect	11. Salamander
2. Bird	7. Plant	12. Skunk
3. Deer	8. Rabbit	13. Toad
4. Fox	9. Raccoon	14. Wildcat
5. Snake	10. Rodent	15. Wolf

Figure 1.
Illustration by James Gordon Irving from *Zoology: A Golden Guide*.
© 1958 Western Publishing Company, Inc. Used by permission.

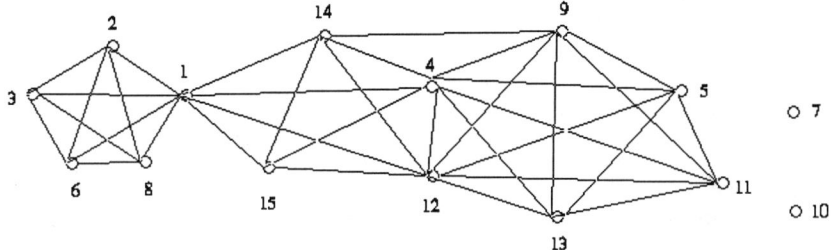

C(D)

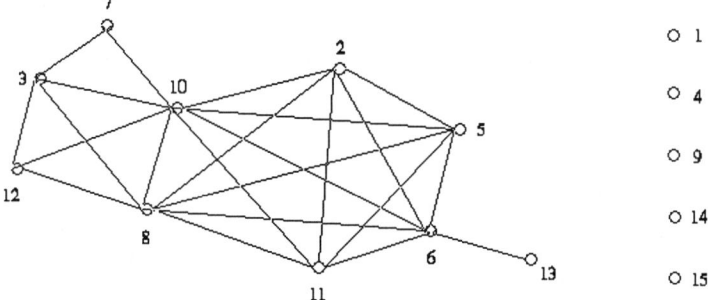

CE(D)

Figure 2.

equal to $CE(D)$. Particularly useful is the fact that by Theorem 10.1 of Harary, Norman, and Cartwright [21], since D is acyclic, the vertices of D can be labeled so that $M = A(D)$ is strictly lower triangular.

To graph theorists and ecologists, much of the initial interest in this problem came from the fact that most food webs have interval competition graphs. A graph G is an interval graph if G is the intersection graph of a set of intervals of real numbers. For example, it is not hard to check that both $C(D)$ and $CE(D)$ in Figure 2 are interval graphs. Roberts [33,34] posed the following fundamental questions:
1. Which acyclic digraphs have interval competition graphs?
2. What graphs are competition graphs of acyclic digraphs?

While the second question has been answered, the first remains elusive, with only partial results. The research on these problems has led to several new problems and applications that we will discuss. Section 2 covers some basic results on competition graphs. In Section 3 we consider question one about interval competition graphs. Some new graphs, competition-common enemy graphs and niche graphs, are discussed in Sections 4 and 5. In Section 6 we return to a special case of question one with a solution for some (i,j)-competition graphs. Finally, in Section 7 we conclude with a brief survey of some of the recent results, applications, and open problems associated with generalized competition graphs.

2. Characterizations of competition graphs. Ecologists generally assume that digraph models of food webs are acyclic. We say that a graph G is a *competition graph* if G is the competition graph of some acyclic digraph D. Since D is acyclic, it has at least one vertex with no outgoing arcs, so G has at least one isolated vertex. Roberts [34] showed that every graph G together with enough isolated vertices is a competition graph. He defined the *competition number* of a graph, $k(G)$, to be the least integer k such that $G \cup I_k$ is a competition graph, where I_k is the graph of k isolated vertices.

If $G = (V, E)$ has e edges, we see that $k(G) \leq e$ as follows. Let D have the same vertices as G plus e additional vertices determined by the edges of G. If $xy \in E$, add vertex α_{xy} to D with arcs in D from x and y to α_{xy}. Then $G \cup I_e$ is the competition graph of D. So if G is connected, we have that $1 \leq k(G) \leq e$.

Opsut [30] showed that computation of $k(G)$ is an np-complete problem. However, if G is triangle free, it is easy to calculate $k(G)$ by the following theorem.

THEOREM 2.1. *(Roberts [34])* If G is a connected, triangle-free graph with n vertices and e edges, then $k(G) = e - n + 2$.

From this result we see that trees have competition number one and cycles Z_n, $n \geq 4$, have competition number two. Roberts [34] used this result to show that for every m, there is a graph G with $k(G) > m$.

We turn now to characterizations of competition graphs. We will call a collection $\mathcal{C}$ of sets of vertices from G an *edge cover* of G if each set in $\mathcal{C}$ is either a clique or the empty set, and every edge in G is in at least one of the cliques in $\mathcal{C}$. By a clique we mean a (not necessarily maximal) complete subgraph K_m of G. The following characterization was obtained independently by Dutton and Brigham [14] and Lundgren and Maybee [26].

THEOREM 2.2. *A graph G with n vertices is a competition graph if and only if the vertices of G can be labeled so that G has an edge cover* $C = \{C_1, \ldots, C_n\}$ *such that if* $v_i \in C_j$, *then* $i > j$.

The idea of the proof is illustrated in the following example. Let G be the graph in Figure 3.

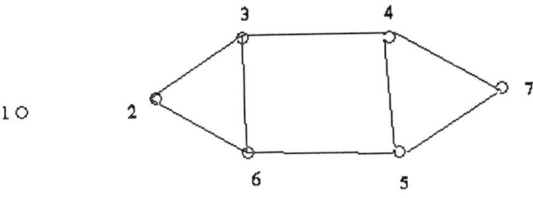

Figure 3.

Then an edge covering C is obtained as follows. Let $C_1 = \{2, 3, 6\}$, $C_2 = \{3, 4\}$, $C_3 = \{4, 5, 7\}$, $C_4 = \{5, 6\}$, $C_5 = \{7\}$, $C_6 = C_7 = \{\emptyset\}$. Clearly C satisfies the conditions of the theorem. We find the acyclic digraph D satisfying $C(D) = G$ by first constructing its adjacency matrix $M = A(D)$ from C. M has a 1 in the i, j-entry if and only if $i \in C_j$.

$$M = \begin{bmatrix} 0 & 0 & 0 & 0 & 0 & 0 & 0 \\ 1 & 0 & 0 & 0 & 0 & 0 & 0 \\ 1 & 1 & 0 & 0 & 0 & 0 & 0 \\ 0 & 1 & 1 & 0 & 0 & 0 & 0 \\ 0 & 0 & 1 & 1 & 0 & 0 & 0 \\ 1 & 0 & 0 & 1 & 0 & 0 & 0 \\ 0 & 0 & 1 & 0 & 1 & 0 & 0 \end{bmatrix}.$$

Since M is strictly lower triangular, D, which is shown in Figure 4, is acyclic.

A graph G is called a *rigid circuit graph* if it does not contain Z_n, $n \geq 4$, as a generated subgraph. By a result of Dirac [13], every rigid circuit graph has a simplicial vertex, a vertex whose neighborhood is a clique. If G is a rigid circuit graph with an isolated vertex, we can find an edge cover satisfying Theorem 2.2 by using a sequence of simplicial vertices. So Theorem 2.2 has as a corollary the following result of Roberts [34].

COROLLARY 2.3. *If G is a rigid circuit graph with an isolated vertex, then G is a competition graph.*

Since interval graphs are rigid circuit graphs, connected interval graphs have competition number one. The following result of Lundgren and Maybee [26] gives an upper bound on the competition number of a graph (this is a corrected version of the theorem provided by Kim [personal communication]).

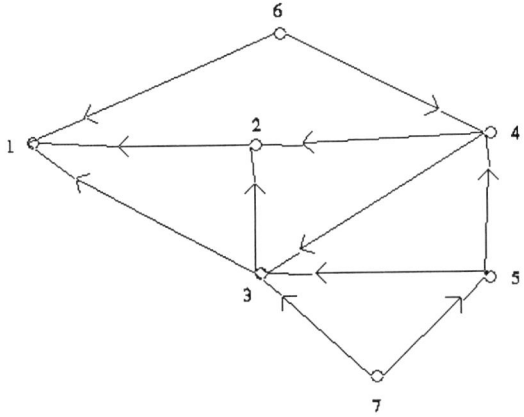

Figure 4.

THEOREM 2.4. *If G is a graph with n vertices, and $m \leq n$, then $k(G) \leq m$ if and only if G has an edge cover $C = \{C_1, \ldots, C_{n+m-2}\}$ and a sequence of distinct points $a_1, \ldots, a_n$ such that if $a_i \in C_j$, then $i \geq j - m + 1$.*

An interesting question is for graphs on n vertices: What is the largest $k(G)$? Let $f(n)$ denote this number. Kim [24] shows that this maximum is attained by the complete bipartite graph $K_{\lfloor \frac{n}{2} \rfloor, \lceil \frac{n}{2} \rceil}$. So $f(n) = \lfloor \frac{n^2}{4} \rfloor - n + 2$.

Now suppose we drop the requirement that D be acyclic and characterize those digraphs that are competition graphs of arbitrary digraphs. The relationship between the columns of the adjacency matrix M of D and the cliques of $RG(M)$ lead to the following result of Dutton and Brigham [14].

THEOREM 2.5. *A graph G with n vertices is the competition graph of an arbitrary digraph (loops allowed) if and only if G has an edge covering of at most n cliques.*

Surprisingly, if we require no loops in D, almost the same result holds.

THEOREM 2.6. *(Roberts and Steif [35])   A graph G with n vertices is the competition graph of an arbitrary digraph without loops if and only if G is not K_2 and G has an edge covering of at most n cliques.*

These results illustrate that a lot is known about competition graphs. In the next section, we consider what digraphs have interval competition graphs.

3. Digraphs with interval competition graphs. We return now to the observation by Cohen [4] that most food webs have interval competition graphs. To understand the significance of this to ecologists, we need some background on the origin of the problem.

Think of a species as determined by certain parameters which characterize its usual healthy environment (temperature, pH, size of prey, etc.). Each of these k parameters determines a dimension in Euclidean k-space. If on every such "ecological dimension," the species' normal healthy environment determines a range of values, the corresponding intervals together define a box in Euclidean k-space, called the species' ecological niche. A basic principle of ecology is that two species compete if and only if their corresponding ecological niches overlap. Cohen [5] asked the following question: If we start with an independent notion of competition, what is the minimum dimension k so that we can find ecological niches in k-space so that competition corresponds to niche overlap? This corresponds to finding the boxicity of the competition graph.

The *boxicity* of a graph G is defined to be the smallest k so that G can be represented as an intersection graph of boxes in k-space. One can show that a graph on n vertices is the intersection graph of a family of boxes in n-space, so boxicity is well-defined. Of course, if G has boxicity one, then G is an *interval graph*. Since, as discussed in Section 2, any graph together with enough isolated vertices is a competition graph, Cohen's empirical discovery that most competition graphs are interval is a surprising phenomenon. It raises the important question of characterizing those acyclic digraphs with interval competition graphs. Such digraphs will be called *interval digraphs*.

Cozzens [10] and Yannakakis [39] showed that computation of the boxicity of a graph is an NP-complete problem. While several good characterizations of interval graphs exist (Gilmore and Hoffman [16], Fulkerson and Gross [15], Lekkerkerker and Boland [25]), graphs with boxicity two have not been characterized. Z_4 is an example of such a graph.

In trying to characterize interval digraphs, it seems reasonable to look for a forbidden subgraph characterization. Steif [37] showed that this is not possible. The digraph D in Figure 5 is acyclic, and its competition graph, $K_6 \cup I_1$, is interval. However, the generated subgraph $D - \{g\}$ has competition graph $Z_4 \cup I_2$, which is not interval.

Another approach to finding forbidden characterizations for acyclic digraphs with interval competition graphs is to use the *sink* and *source* food webs introduced by Cohen [4]. A *sink-induced subdigraph* H of a digraph D is a generated subdigraph with the following additional property: if $x \in H$ and (x,y) is an arc in D, then $y \in H$. A *source-induced subdigraph* H of a digraph D is a generated subdigraph with the following additional property: if $y \in H$ and (x,y) is an arc in D, then $x \in H$. Cohen [4] obtained the following characterization of digraphs with interval competition graphs.

THEOREM 3.1. *An acyclic digraph has a competition graph that is interval if and only if every sink-induced subdigraph does.*

It is possible for a digraph to have an interval competition graph while some source-induced subdigraph contained in it has a competition graph which is not an interval graph (see Roberts [34] for an example). However, source-induced subdigraphs provide the dual to Theorem 3.1 for common enemy graphs.

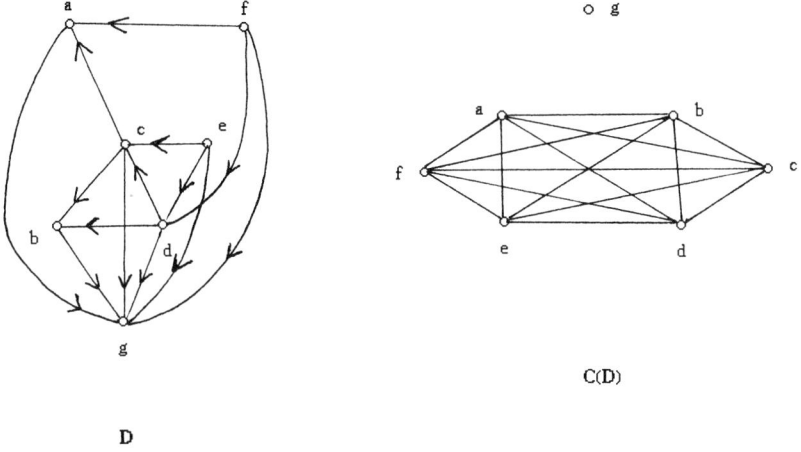

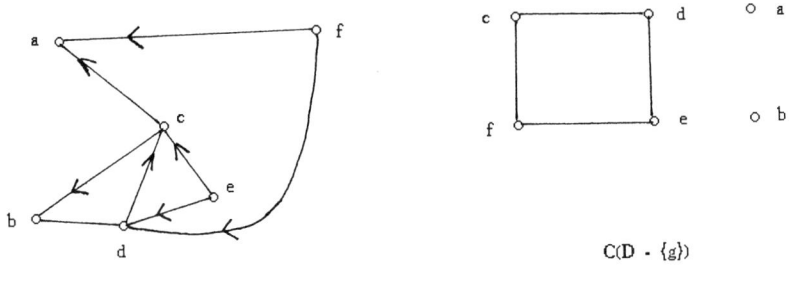

Figure 5.

THEOREM 3.2. *(Lundgren and Maybee [27])* An acyclic digraph D has a common enemy graph that is interval if and only if every source-induced subdigraph does.

From the above discussion we see that, for both the competition and common enemy graphs to be interval, the conditions of both Theorems 3.1 and 3.2 have to be satisfied. These theorems are not useful from a practical standpoint. However, we give an example in Figure 6 to illustrate the case where neither graph is interval. Here H is a sink-induced subdigraph of D with competition graph G_1, which is not an interval graph; so $G = C(D)$ is not an interval graph, and F is a source-induced subdigraph of D with common enemy graph E_1 which is not an interval graph, so $E = CE(D)$ is not an interval graph. Observe that G_1 and E_1 are induced subgraphs of G and E, respectively.

Steif [37] has pursued these ideas as follows. A *forbidden sink-induced subdigraph list* for property P is a list L of digraphs such that a digraph D has property P if and only if D does not contain any element of L as a sink-induced subdigraph. Steif [37] shows that such a list exists for interval digraphs. However, to find such a list appears to be a difficult problem.

THEOREM 3.3. *(Steif [37])* There exists a forbidden sink-induced subdigraph list for (acyclic) digraphs with interval competition graphs.

Another approach to characterizing interval digraphs is to find sets in D corresponding to maximal cliques in $C(D)$. A set $S = \{C_1, \ldots, C_r\}$ of vertices of D will be called a *competition cover* of D if the following conditions are satisfied:

(1) $i, j \in C_m$ implies that there exists a vertex k such that (i, k) and (j, k) are arcs,

(2) if (i, k) and (j, k) are arcs in D, then $i, j \in C_m$ for some m.

The idea of the proof of the following theorem is that the sets C_i correspond to cliques in G, so we can use the Fulkerson and Gross [15] characterization of interval graphs. We say that a ranking $C_1, \ldots, C_r$ of sets of vertices of D is consecutive if whenever a vertex u is in C_i and C_j for $i < j$, then for all $i < k < j$, $u \in C_k$.

THEOREM 3.4. *(Lundgren and Maybee [27])* D is an interval digraph if and only if D has a competition cover S which has a ranking that is consecutive.

While all of these results indicate some progress in characterizing interval digraphs, we are looking for a better characterization in terms of properties of D. Some recent results in this direction are discussed in Section 6 where restrictions are placed on vertex degrees in D.

A different approach to explaining the high frequency of interval competition graphs is taken by Sugihara [38]. First he illustrates that for some food webs, the common enemy graph (Sugihara calls it the resource graph) provides more information than the competition graph. An example is the food web of the Knysna Estuary that was studied by Day [11] and is shown in Figure 7. In the graphs $C(D)$ and $CE(D)$ for the Knysna Estuary, there are cliques of prey in $CE(D)$ corresponding to an individual predator in $C(D)$.

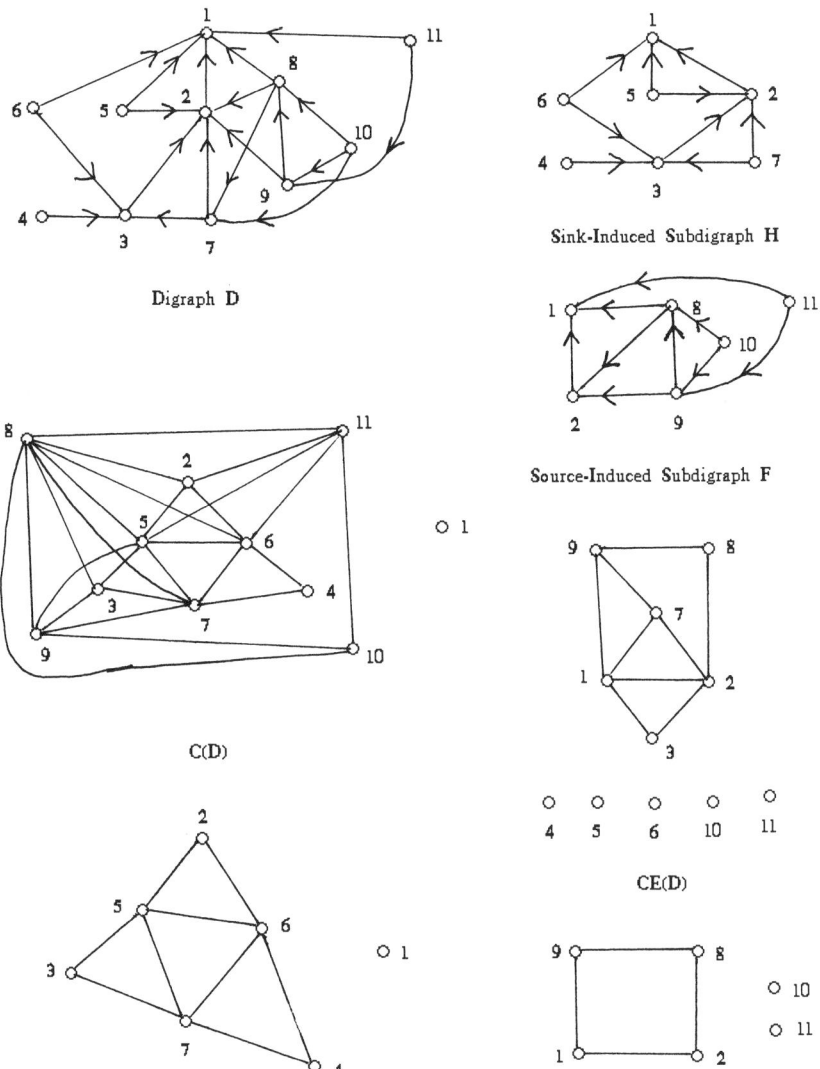

Figure 6.

Simplified Food Web Matrix

	2	5	6	7	8	9	10	11	13	14
12			1	1						
13				1	1	1	1	1		
14	1				1	1		1		
15		1	1						1	1

Legend--Knysna Estuary

2. attached plants
5. Hyporhamphus
6. Mugil
7. Upogebia
8. Lamya
9. Solen
10. Arenicola
11. Hymenosoma
12. Johnius
13. Lithognathus
14. Rhabdosargus
15. Hypacanthus

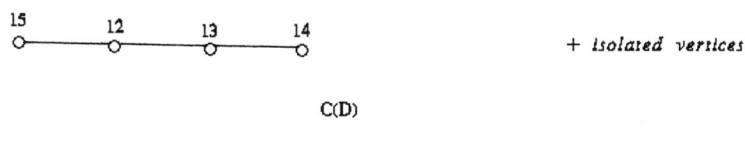

C(D)   + isolated vertices

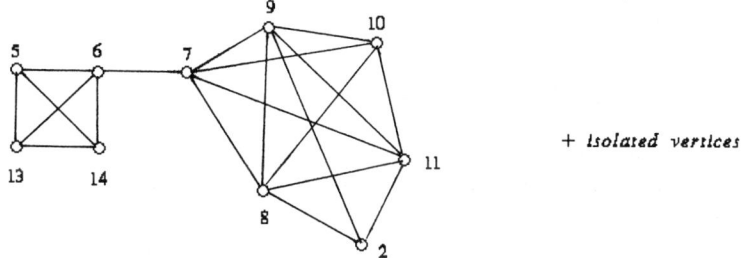

CE(D)   + isolated vertices

Figure 7.

Sugihara [38] gives an ecological explanation of why common enemy graphs are frequently rigid circuit graphs. A simplified version of this explanation can be applied to the digraph D in Figure 8. The fact that $CE(D)$ is frequently a rigid circuit graph follows from the implication that resources (prey) in the environment are ordered or correlated with each other by various means (e.g., spatially, taxonomically, by size) and that this ordering is perceived similarly by all the species involved. For example, in Figure 8 we assume that the predators 1–4 are numbered from smallest to largest (say various-sized lizards), and the prey from smallest to largest (say various-sized insects). If the largest predator, 4, also ate the smallest insect, 5, thereby creating a 4-cycle, then it is reasonable to assume that 4 would also eat the middle-sized insects 6 and 7, so $CE(D)$ would be a rigid circuit graph. For more details on why $C(D)$ as well as $CE(D)$ are usually rigid circuit for actual food webs, see Sugihara [38].

Figure 8.

It was also shown statistically by Sugihara [38] that the frequency of interval food webs could be accounted for by requiring that the competition graph be a rigid circuit graph. This interesting approach deserves more graph-theoretic analysis. For example, what are the structural properties of the digraphs of actual food webs, and how is this related to Sugihara's statistical observations? What acyclic digraphs have competition graphs that are rigid circuit graphs? We will return to this question in Section 6.

4. Competition-common enemy graphs. A natural extension of competition and common enemy graphs is the competition-common enemy graph recently introduced by Scott [36]. The *competition-common enemy* graph of an acyclic digraph $D = (V, A)$ is the graph $CCE(D) = (V, E)$ where $xy \in E$ if and only if $x \neq y$ and for some $w, z \in V$, the arcs $wx, wy, xz, yz \in A$. Figure 9 gives an example of an acyclic digraph D and $CCE(D)$.

Since D is acyclic, it follows that $CCE(D)$ always has at least two isolated vertices. Scott [36] defined the *double competition number* $dk(G)$ of a graph G to be the smallest integer k so that $G \cup I_k$ is the CCE graph of some acyclic digraph.

If $G = (V, E)$ has e edges, we see that $dk(G) \leq 2e$ as follows. Let D have the same vertices as G plus $2e$ additional vertices determined by the edges of G. If $xy \in E$, add vertices x_α and y_α to D with arcs in D from x_α to x and y, and from x

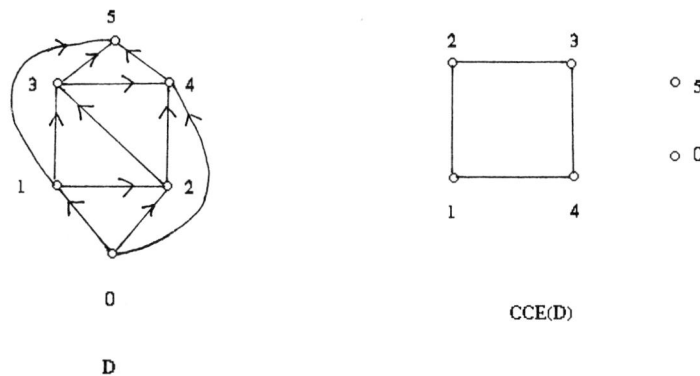

Figure 9.

and y to y_α. Then $G \cup I_{2e} = CCE(D)$. Scott [36] found the following relationship between $dk(G)$ and $k(G)$.

THEOREM 4.1. $dk(G) \leq k(G) + 1$.

If $G = P_n$, $n \geq 2$, then $dk(G) = 2 = k(G) + 1$, while if $G = C_n$, $n \geq 4$, $dk(G) = 2 < k(G) + 1 = 3$. Scott [36] was unable to find any graphs with $dk(G) > 2$ even though she considered several triangle-free graphs with large competition number. So, she raised the question of whether any such graph existed.

Let $K(n,n,n)$ be the complete 3-partite graph on $3n$ vertices. Jones, Lundgren, Roberts, and Seager [23] proved the following theorem.

THEOREM 4.2. $dk(K(n,n,n)) \geq 2\sqrt{n}$.

They also calculate the exact values for $n = 2, 3,$ and 4 ($3, 4,$ and 4, respectively), but the exact value of $dk(K(n,n,n))$ is unknown for $n \geq 5$.

Triangle-free graphs can be found with large competition number, but the only known triangle-free graph G with $dk(G) > 2$ is the graph in Figure 10. Jones et al. [23] show by exhaustive computer search that this graph has $dk(G) = 3$. It remains an unsolved problem to find a class of triangle-free graphs with $dk(G) > 2$.

Since these graphs have just been introduced, several other interesting questions remain unanswered. Scott [36] gave a characterization of competition-common enemy graphs, but it is not particularly useful, so a good characterization is needed. It is very difficult to calculate $dk(G)$, even for small graphs. Lower bounds on $dk(G)$ in terms of other graph parameters would be particularly useful. It would be interesting to explore possible applications, with or without D acyclic.

5. Niche graphs. Since the CCE graph of a digraph D is the intersection of $C(D)$ and $CE(D)$, it is natural to also consider the union of $C(D)$ and $CE(D)$, introduced by Cable, Jones, Lundgren, and Seager [3]. The *niche graph* of an acyclic

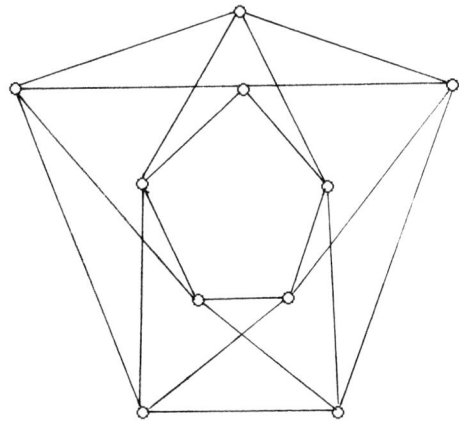

Figure 10.

digraph $D = (V, A)$ is the graph $N(D) = (V, E)$ where $xy \in E$ if and only if $x \neq y$ and for some $z \in V$, the arcs $xz, yz \in A$ or the arcs $zx, zy \in A$.

Some of the relationships between these graphs are evident from Figure 11. In particular, $CCE(D) \subseteq C(D) \subseteq N(D)$ and $CCE(D) \subseteq CE(D) \subseteq N(D)$. Following the pattern established for competition graphs and CCE graphs, the *niche number* $n(G)$ is defined to be the smallest k such that $G \cup I_k$ is the niche graph of an acyclic digraph. From Figure 11, we see that $n(G) = 0$ is possible for a connected graph. However, it also turns out that there are classes of graphs that cannot be made into niche graphs by the addition of isolated vertices. For such graphs we say that the niche number is infinite and write $n(G) = \infty$. A good exercise in finding niche numbers is to convince yourself that $n(K_{1,3}) = \infty$.

This fact gives hope to a characterization of niche graphs with $K_{1,3}$ as a forbidden subgraph. However, Figure 12 gives an example of a niche graph with a generated $K_{1,3}$.

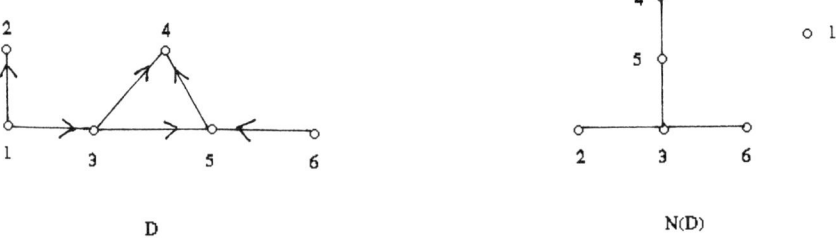

Figure 12.

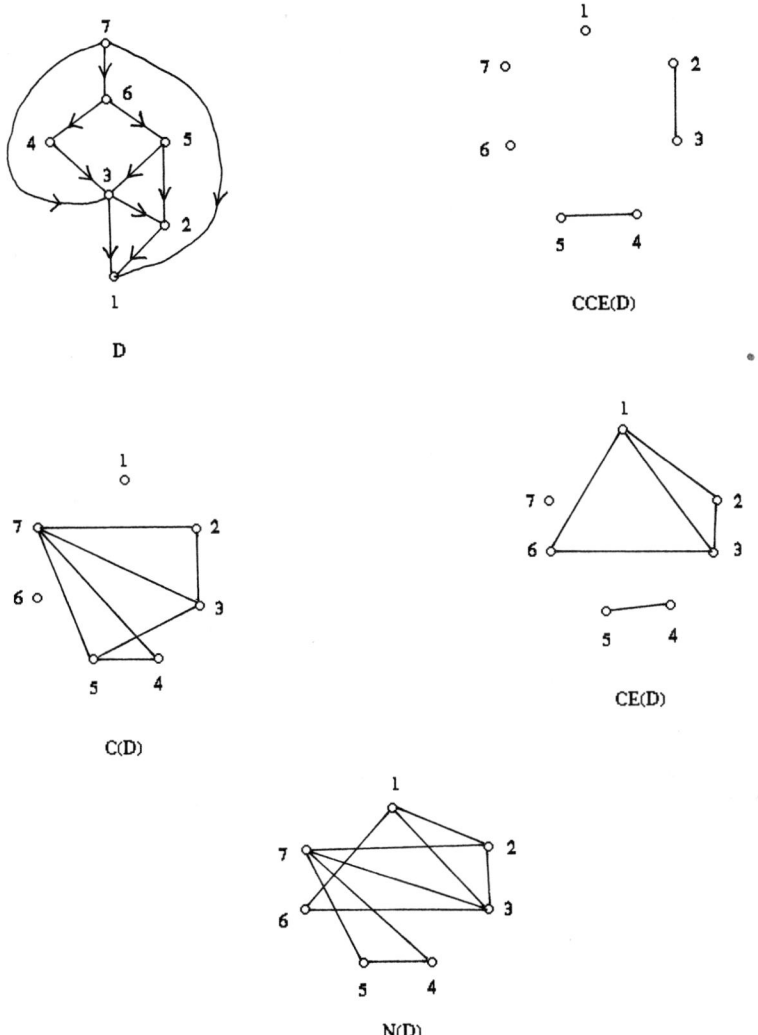

Figure 11.

A *nova* is a graph obtained by replacing each edge of the star $K_{1,n}$, where $n \geq 3$, by a clique on at least 2 vertices. The graph in Figure 13 is a nova.

THEOREM 5.1. *(Cable et al. [3])* If G is a nova, then $n(G) = \infty$.

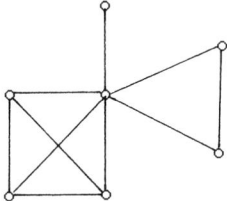

Figure 13.

If G has components $G_1, \ldots, G_m$, then it is easy to see that

$$n(G) \leq n(G_1) + \cdots + n(G_m).$$

In particular, if all of the components of G are niche graphs, then so is G. But the converse can fail quite spectacularly, as illustrated in Figure 14. Here we see that $n(K_{1,3} \cup K_{1,3}) = 0$ even though $n(K_{1,3}) = \infty$.

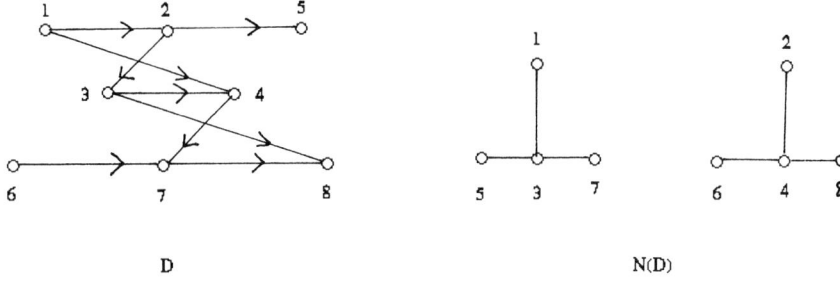

Figure 14.

Cable et al. [3] calculate $n(G)$ for several classes of graphs. We give some of these results in the following theorem.

THEOREM 5.2. $n(K_n) = 1$, $n \geq 2$. $n(P_2) = 1$, $n(P_n) = 0$, $n \geq 3$.

$$\begin{aligned} n(C_n) &= 0 && \text{for } n = 7, \ n \geq 9; \\ n(C_n) &= 1 && \text{for } n = 3 \text{ and } 8; \\ n(C_n) &= 2 && \text{for } n = 4, \ 5, \text{ and } 6. \end{aligned}$$

In general, calculating $n(G)$ is a difficult problem, so any results providing bounds on $n(G)$ or showing that $n(G) = \infty$ are useful. The following result is particularly useful in finding graphs with $n(G) = \infty$. For example, it follows from the following theorem that if G is a tree with a vertex of degree ≥ 5, then $n(G) = \infty$.

THEOREM 5.3. *(Cable et al. [3])* If G is K_{m+1}-free and $n(G) < \infty$, then G has maximum degree at most $2m(m-1)$.

So far, no one has found a graph with $2 < n(G) < \infty$. It is also not clear that such a graph exists, so this is a particularly interesting problem. There is a wealth of other interesting problems on niche graphs. A good characterization of graphs with $n(G) < \infty$ is needed, as are lower or upper bounds for $n(G)$. An investigation of the structure of niche graphs for known food webs would be useful, particularly determining which food webs have interval niche graphs. It would be worthwhile to consider possible applications, including the most general case where M is a $0, 1$-matrix and $G = RG(M) \cup CG(M)$.

6. (i, j) competition graphs. In this section we return to the problem of characterizing interval digraphs. Since the general problem has proved to be so difficult, here we will consider the problem with constraints put on the indegrees and outdegrees in D. All of the results in this section can be found in a recent paper by Hefner, Jones, Kim, Lundgren, and Roberts [22]. Suppose that D is an acyclic digraph, and let $id(x)$ denote the indegree of x and $od(x)$ the outdegree of x.

D is an	if for every vertex x
(i, j) digraph	$id(x) \leq i$ and $od(x) \leq j$;
$(\bar{i}, \bar{j})$ digraph	$id(x) = 0$ or i and $od(x) = 0$ or j;
$(\bar{i}, j)$ digraph	$id(x) = 0$ or i and $od(x) \leq j$;
$(i, \bar{j})$ digraph	$id(x) \leq i$ and $od(x) = 0$ or j.

We say that a graph G is a (u, v) competition graph, where $u = i$ or $\bar{i}$ and $v = j$ or $\bar{j}$, if it is the competition graph of a (u, v) digraph. We call the digraph in Figure 15 $P(2, 2)$. Note that forbidding $P(2, 2)$ as a subgraph says that any two species can have at most one common prey. If D has no $P(2, 2)$ subgraph, we say D is *irredundant*.

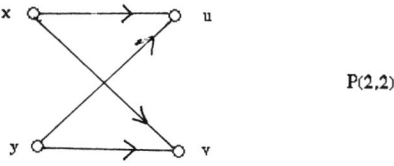

P(2,2)

Figure 15.

We will consider the following three problems:

(1) Characterize the (u, v) competition graphs;
(2) Characterize the (u, v) interval competition graphs;
(3) Characterize the (u, v) interval digraphs.

All three problems are solved by Hefner et al. [22] in the $(2, 2)$ case as well as problems (2) and (3) in the $(\bar{i}, \bar{j})$ irredundant case. For all other cases, all three problems remain open. It is interesting that in the cases where the problems have been solved, interval can be changed to rigid circuit with the same results holding. Here we will summarize some of the main results and indicate open problems.

The $(2, 2)$ case is greatly simplified by establishing the following two facts about $(2, 2)$ competition graphs G:

(i) G has no generated $K_{1,3}$;
(ii) If G has a triangle, it is a component.

The next three results of Hefner et al. [22] solve the three basic problems in the $(2, 2)$ case.

THEOREM 6.1. *A graph is a $(2, 2)$ competition graph if and only if each component is an isolated vertex, a path, or a cycle, and the number of isolated vertices is at least 2 if every component is a cycle of length > 3 and at least 1 otherwise.*

COROLLARY 6.2. *A graph is a $(2, 2)$ interval competition graph if and only if each component is an isolated vertex, a path, or a triangle, and the number of isolated vertices is at least 1.*

A similar characterization of $(\bar{2}, \bar{2})$ competition graphs exists. It is also possible to find a forbidden subgraph characterization of $(2, 2)$ interval digraphs, but it is not particularly useful.

THEOREM 6.3. *Suppose D is a $(2, 2)$ digraph. Then D is an interval digraph if and only if every $(\bar{2}, \bar{2})$ irredundant subgraph of D with at least one arc contains one of the three digraphs S, T, and U in Figure 16 as a generated subgraph.*

Recalling Sugihara's [38] observation about rigid circuit graphs, it is interesting that we can replace interval with rigid circuit in Corollary 6.2 and Theorem 6.3 and the conclusions stay the same.

Now consider the three questions for $(\bar{i}, \bar{j})$ *irredundant competition graphs* ($C(D)$ is irredundant if D is irredundant). Existence of these graphs will depend on the existence of certain block designs. To save a page of definitions, the reader is referred to Dembrowski [12] or Hall [19] for all definitions related to block designs.

To lay the groundwork for the theory, consider the $(\bar{3}, \bar{2})$ digraph and its competition graph given in Figure 17. Existence of the digraph corresponds to existence of a $(6, 4, 3, 2, 1)$-design. Notice that the nontrivial part of $C(D)$ is 4-regular, and since it contains a generated 4-cycle, it is neither a rigid circuit graph nor an interval graph. In the general $(\bar{i}, \bar{j})$ case, existence will depend on existence of certain block designs and the nontrivial components of $C(D)$ will be $j(i-1)$-regular. It then follows that for $C(D)$ to be an interval graph, the nontrivial components must be complete.

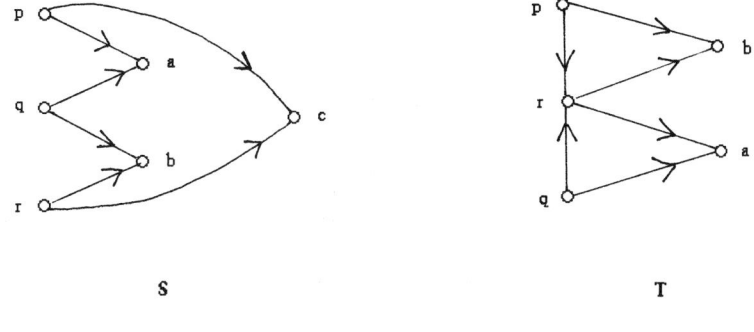

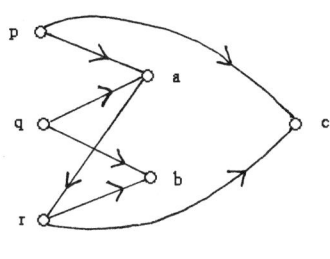

Figure 16.

THEOREM 6.4. *(Hefner et al. [22])* Suppose $i, j \geq 2$. Then there is a nontrivial $(\overline{i}, \overline{j})$ irredundant competition graph if and only if there is a (b, v, r, k) mixed 2-design with parameters $r = j$ and $k = i$.

Note that this theorem does not characterize $(\overline{i}, \overline{j})$ irredundant competition graphs, but from it we can deduce information about the structure of such graphs such as that observed from the example in Figure 17. In particular, we can show that for these graphs to be interval, every component must be complete. Question two is answered by the following result of Hefner et al. [22].

THEOREM 6.5. Suppose $i, j \geq 2$ and G is a nontrivial graph. Then G plus sufficiently many isolated vertices is an $(\overline{i}, \overline{j})$ irredundant interval competition graph if and only if every nontrivial component of G is $K_{[j(i-1)+1]}$ and there is a $(b, v, r, k, 1)$-design with $r = j$ and $k = i$.

As a consequence of this theorem, it follows from design theory that there are no nontrivial $(\overline{i}, \overline{j})$ irredundant interval competition graphs when $i > j$. However, as

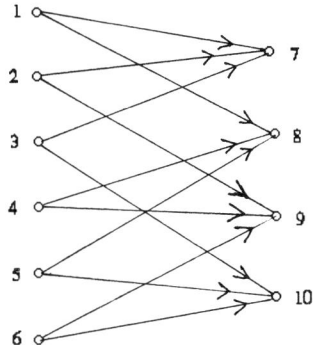

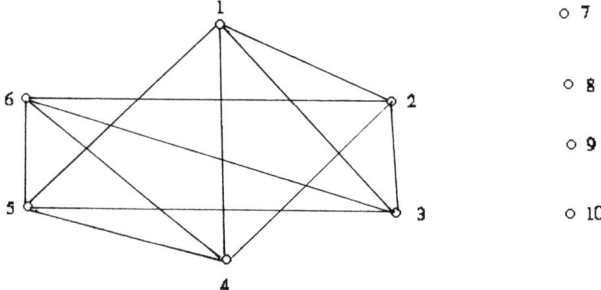

Figure 17.

illustrated in Figure 17, there are such competition graphs that are not interval.

We now turn to the question of what the $(\bar{i},\bar{j})$ irredundant interval digraphs look like. Let us say that a vertex x in a digraph D is *di-simplicial* if whenever there are arcs (x,u), (y,u), (x,v), and (z,v), then there are arcs (y,w) and (z,w). It follows that x is di-simplicial in D if and only if x is simplicial in $C(D)$. The next result of Hefner et al. [22] gives the characterization.

THEOREM 6.6. *Suppose $i,j \geq 2$. Then an $(\bar{i},\bar{j})$ irredundant digraph D is interval if and only if every vertex of D is di-simplicial.*

It turns out that in each of the last two theorems, interval can be replaced by rigid circuit with the same characterization holding. This is because those $(\bar{i},\bar{j})$ irredundant competition graphs that are not interval have no simplicial vertices in their nontrivial components.

Several questions remain open. The three basic problems have not been solved in the $(\bar{i},\bar{j})$ case if we remove the assumption that digraphs must be irredundant. They have not been solved with or without this assumption in the general (i,j) case or the mixed cases $(\bar{i},j)$ and $(i,\bar{j})$. Whether or not there is a forbidden subgraph characterization remains open in all but the (2, 2) case. It would also be interesting to solve the same problem for other restrictions on the digraphs. Appropriate restrictions could be determined by investigating the known food webs.

7. Generalized competition graphs. A brief discussion of generalized competition graphs completes this survey paper. A more thorough treatment is given in Roberts [32] and Raychaudhuri and Roberts [31]. Here we will consider some of the problems discussed earlier in this more general setting.

Suppose that $D = (V, A)$ is a digraph and B and C are (not necessarily disjoint) sets of vertices in D. The *generalized competition graph* $G(D, B, C)$ corresponding to D, B, and C is an undirected graph with vertex set equal to the set B and with an edge between two distinct vertices x and y of B if and only if for some z in C, (x, z) and (y, z) are both arcs of D.

These graphs have numerous applications, including confusion graphs associated with communnication over a noisy channel, conflict graphs associated with radio or television transmission, and row and column graphs used in analyzing large scale computer models. In this last example, signed graphs have also been used.

Most of the research in this area has focused on the problem of recognizing the generalized competition graphs arising in certain circumstances or applications. That is, given restrictions on D, B, and C, characterize the generalized competition graphs under these restrictions. Several interesting open problems remain and are summarized along with known results in Raychaudhuri and Roberts [31].

The fact that every graph is a generalized competition graph of some digraph has been discovered by several authors in different contexts (see Roberts [32]). The graph inversion method of Greenberg, Lundgren, and Maybee [18] shows how to construct all the digraphs having a given graph as its competition graph.

New problems of interest include studying competition-common enemy graphs and niche graphs in this more general setting. Important applications of these graphs may arise where the restriction of D being acyclic is removed. Characterizations of these graphs should be found under appropriate restrictions on D, B, and C. For example, with $B = C$ but no restrictions on D, $K_{1,3}$ is still not a niche graph. Another problem would be to investigate the interval graph question with different restrictions on D, B, and C.

Finally, another type of problem, only recently investigated, is as follows. Given a pair of graphs G and H, when does there exist a digraph D such that G is the competition graph and H the common enemy graph of D? Bergstrand and Jones [1] and Lundgren, Maybee, and McMorris [28] have some results when D is either acyclic or the digraph of a poset. Other possibilities for D should be considered.

REFERENCES

1. D. J. Bergstrand and K. F. Jones, *On upper bound graphs of partially ordered sets*, mimeographed, Department of Mathematics, University of Colorado, Denver, CO, 1988; submitted for publication,.
2. R. W. Burnett, H. I. Fisher, and H. S. Zim, "Zoology: An Introduction to the Animal Kingdom," Golden Press, New York, 1958.
3. C. Cable, K. F. Jones, J. R. Lundgren, and S. Seager, *Niche graphs*, (1988) to appear in Disc. Appl. Math.
4. J. E. Cohen, "Food Webs and Niche Space," Princeton University Press, Princeton, NJ, 1978.
5. _____, *Interval graphs and food webs: A finding and a problem*, Rand Corporation Document 17696-PR, Santa Monica, CA, 1968.
6. _____, *Recent progress and problems in food web theory*, in "Current Trends in Food Web Theory," (D. F. DeAngelis, W. M. Post, and G. Sugihara, eds.), Tech. Rept. ORNL/TM-8643, Oak Ridge National Laboratory, Oak Ridge, TN, 1983.
7. J. E. Cohen, F. Briand, and C. M. Newman, *A stochastic theory of community food webs. III. Predicted and observed lengths of food chains*, Proc. R. Soc. Lond. **B 228** (1986), 317–353.
8. J. E. Cohen and C. M. Newman, *A stochastic theory of community food webs. I. Models and aggregated data*, Proc. R. Soc. Lond. **B 224** (1985), 421–448.
9. J. E. Cohen, C. M. Newman, and F. Briand, *A stochastic theory of community food webs. II. Individual webs*, Proc. R. Soc. Lond. **B 224** (1985), 449–461.
10. M. B. Cozzens, *Higher and multi-dimensional analogues of interval graphs*, Ph.D. Thesis, Department of Mathematics, Rutgers University, New Brunswick, NJ, 1981.
11. J. H. Day, *The biology of the Knysna Estuary, South Africa*, in "Estuaries," (G. H. Lauff, ed.) (AAAS publication 83), Washington, DC, 1967, pp. 397–404.
12. P. Dembrowski, "Finite Geometries," Springer-Verlag, Berlin, 1965.
13. G. A. Dirac, *On rigid circuit graphs*, Abhandlungen Mathematischen Seminar Universität Hamburg **25** (1961), 71–76.
14. R. D. Dutton and R. C. Brigham, *A characterization of competition graphs*, Discr. Appl. Math. **6** (1983), 315–317.
15. D. R. Fulkerson and O. A. Gross, *Incidence matrices and interval graphs*, Pacific J. Math. **15** (1965), 835–855.
16. P. C. Gilmore and A. J. Hoffman, *A characterization of comparability graphs and of interval graphs*, Canad. J. Math. **16** (1964), 539–548.
17. H. J. Greenberg, J. R. Lundgren, and J. S. Maybee, *Graph theoretic methods for the qualitative analysis of rectangular matrices*, SIAM J. Algebraic Discrete Methods **2** (1981), 227–239.
18. _____, *Inverting graphs of rectangular matrices*, Discr. Appl. Math. **8** (1984), 255–265.
19. M. Hall, Jr., "Combinatorial Theory," second ed., Wiley, New York, 1986.
20. F. Harary, *Who eats whom*, General Systems **6** (1961), 41–44.
21. F. Harary, R. Z. Norman, and D. Cartwright, "Structural Models: An Introduction to the Theory of Directed Graphs," John Wiley & Sons, Inc., New York, 1965.
22. K. A. S. Hefner, K. F. Jones, S. Kim, J. R. Lundgren, and F. S. Roberts, (i,j) *competition graphs*, mimeographed, Department of Mathematics, Rutgers University, 1988; submitted for publication.
23. K. F. Jones, J. R. Lundgren, F. S. Roberts, and S. Seager, *Some remarks on the double competition number of a graph*, Congressus Numerantium **60** (1987), 17–24.
24. S. Kim, Ph.D. Thesis, Rutgers University, 1988.
25. C. B. Lekkerkerker and J. Ch. Boland, *Representation of a finite graph by a set of intervals on the real line*, Fund. Math. **51** (1962), 45–64.
26. J. R. Lundgren and J. S. Maybee, *A characterization of graphs with competition number m*, Discr. Appl. Math. **6** (1983), 319–322.

27. _____, *Food webs with interval competition graphs*, in "Graphs and Applications: Proceedings of the First Colorado Symposium on Graph Theory," Wiley, New York, 1984, pp. 231–244.
28. J. R. Lundgren, J. S. Maybee, F. R. McMorris, *Two-graph inversion of competition graphs and bound graphs*, mimeographed, Department of Mathematics, University of Colorado, Denver, CO, 1988; submitted for publication.
29. C. M. Newmann and J. E. Cohen, *A stochastic theory of community food webs. IV. Theory of food chain lengths in large webs*, Proc. R. S. Lond. **B 228** (1986), 355–377.
30. R. J. Opsut, *On the computation of the competition number of a graph*, SIAM J. Alg. Disc. Meth. **3** (1982), 420–428.
31. A. Raychaudhuri and F. S. Roberts, *Generalized competition graphs and their applications*, in "Methods of Operations Research," (P. Brucker and R. Pauly, eds.), 49, Anton Hain, Konigstein, West Germany, 1985, pp. 295–311.
32. F. S. Roberts, *Applications of edge coverings by cliques*, Discr. Appl. Math. **10** (1985), 93–109.
33. _____, "Discrete Mathematical Models, with Applications to Social, Biological, and Environmental Problems," Prentice-Hall, Englewood Cliffs, NJ, 1976.
34. _____, *Food webs, competition graphs, and the boxicity of ecological phase space*, in "Theory and Applications of Graphs," (Y. Alavi and D. Lick, eds.), Springer-Verlag, New York, 1978, pp. 477–490.
35. F. S. Roberts and J. E. Steif, *A characterization of competition graphs of arbitrary digraphs*, Discr. Appl. Math. **6** (1983), 323–326.
36. D. D. Scott, *The competition-common enemy graph of a digraph*, Discr. Appl. Math. **17** (1987), 269–280.
37. J. E. Steif, *Frame dimension generalized competition graphs, and forbidden sublist characterizations*, Henry Rutgers Thesis, Dept. of Math., Rutgers University, New Brunswick, NJ, 1982.
38. G. Sugihara, *Graph theory, homology, and food webs*, in "Population Biology," (S. A. Levin, ed.), 30, American Math. Society, Providence, RI, 1983.
39. M. Yannakakis, *The complexity of the partial order dimension problem*, SIAM J. Alg. & Disc. Meth. **3** (1982), 351–358.

QUALITATIVELY STABLE MATRICES AND CONVERGENT MATRICES

JOHN S. MAYBEE*

Abstract. We review the basic known facts about stable matrices. From these and other, related, results we show some of the relationships that exist between qualitatively (sign) stable matrices and Hicksian stable matrices. We also show that the relationship between stable matrices and convergent matrices can be viewed essentially as a problem on inverses. Finally we derive a variety of results about inverses of sign stable matrices and use these to obtain information about the properties of corresponding convergent matrices.

1. Introduction. The classical definition of a stable matrix is derived from the theory of ordinary differential equations with constant coefficients, namely, the problem of solving the system

$$\dot{x} = Ax \tag{1}$$

for A an $n \times n$ matrix whose elements are real (or complex) numbers. Stability for such a system has been interpreted in the sense of all solutions tending to zero as $t \to \infty$. For this to happen it is necessary and sufficient that all eigenvalues of A have negative real parts, i.e., the spectrum, $\sigma(A)$, of A lies in the open left half of the complex plane. Thus we say that A is a *stable matrix* if $Re\lambda < 0$ for each $\lambda \in \sigma(A)$.

A large amount of literature has accumulated over the past 150 years on the subject of stable matrices. We will not attempt to summarize or even reference this literature here.

In the late 1930's and 1940's two outstanding mathematical economists identified stability problems which occur in economic theory. Historically, the first of these problems was raised by Sir John Hicks [H39] who called a real matrix *perfectly stable* if each principal minor of order p has sign $(-1)^p$ for $p = 1, 2, \ldots, n$. In subsequent years this concept has been identified as *Hicksian stability* by economists and we will use this terminology. The second stability problem proposed by Paul Samuelson [S47], is called the qualitative stability problem or sign stability problem. To define it properly suppose A is a real $n \times n$ matrix and let $Q(A)$ be the set of all real $n \times n$ matrices $B = [b_{ij}]$ satisfying sign b_{ij} = sign a_{ij} for all i and j. The class $Q(A)$ consists of all matrices having the same sign pattern as A. The matrix A is called *qualitatively stable* (or sign stable) if every element $B \in Q(A)$ is a stable matrix. In other words, a matrix A is qualitatively stable if, on the basis of sign pattern information alone, we can conclude that it is a stable matrix.

---

*Department of Mathematics, University of Colorado at Boulder, Boulder, Colorado 80309-0426. The author's research was supported in part by the Office of Naval Research under contract N00014-88-K-0087.

The author wishes to express his thanks to the College of William and Mary, Williamsburg, Virginia 23185, where he is visiting while on sabbatical leave, for their support.

Next, we recall that in numerical analysis it is customary to call the real matrix A *convergent* if for any vector x the sequence $\{A^n x\}$ tends to zero as $n \to \infty$. It is known that the matrix A is convergent if and only if $|\lambda| < 1$ for all $\lambda \in \sigma(A)$. We can see that this is also a stability problem by recognizing that every solution of the difference equation

(2) $$x_{n+1} = A x_n$$

tends to zero as $n \to \infty$ if and only if A is convergent.

As a final concept we call A *potentially stable* if there exists at least one $B \in Q(A)$ which is a stable matrix.

It is our intention here to point out some of the relationships holding between these concepts and, at the same time, to point out some of the open problems. Most of the results presented below concerning the inverses of sign stable matrices and the relations existing between sign stable matrices and the corresponding class of convergent matrices are new. We do not have much that is new on the potential stability problem which seems to be inherently very difficult. On the other hand there are several observations made about the relationship between sign stability and Hicksian stability. These seem to be new but not particularly exciting.

2. Some basic facts. First we observe that, if A is stable or Hicksian stable, then A^{-1} exists and is stable, respectively, Hicksian stable. For A stable the result follows from the facts that $\lambda \neq 0$ for $\lambda \in \sigma(A)$ and $\dfrac{1}{\lambda} = \dfrac{\bar{\lambda}}{|\lambda|^2}$ ($\bar\lambda$ is the complex conjugate of λ), hence has negative real part. For A Hicksian stable the result follows from the Jacobi identities

Obviously a convergent matrix need not be nonsingular and, if it is, the inverse is certainly not convergent.

Next we observe that if the matrix A has negative principal diagonal, $a_{ii} < 0$, $i = 1, 2, \ldots, n$ then A is potentially stable. This result follows at once from the fact that the eigenvalues of A depend continuously on the elements of A. It in turn implies that every Hicksian stable matrix is potentially stable.

To investigate the relationships between the concepts further we require the following fundamental constructs. If A is an $n \times n$ matrix we associate with it a directed graph $D(A) = (V, \overline{E})$ where $V = \{1, 2, \ldots, n\}$ is the vertex set, $V(D)$, and the arc $(i,j) \in \overline{E}$ if and only if $a_{ij} \neq 0$, for $i \neq j$. Since this construct does not take account of whether or not a diagonal element of A is zero, we denote by V_0 the subset of vertices of V corresponding to $a_{ii} \neq 0$. Thus we have associated with A both $D(A)$ and a distinguished subset $V_0 \subseteq V$ of vertices corresponding to nonzero diagonal elements of A.

We will call a digraph D a *tree* if it has no cycles of length greater than two. Recall that D is called *symmetric* if whenever the arc $(i,j) \in \overline{E}$, the arc $(j,i) \in \overline{E}$ also. If D is symmetric we define $G(D) = (V, E)$ where $V(G) = V(D)$, i.e., G has the same vertex set as D, and where $[i,j] \in E$ if and only if both (i,j) and (j,i) belong to $\overline{E}$.

Clearly $D(A)$ is symmetric if and only if $a_{ij} \neq 0$ implies $a_{ji} \neq 0$ for $i \neq j$. In this case we call A *combinatorially symmetric*. Then we set $G(A) = G(D(A))$ and call $G(A)$ the *graph of* A. Recall that a graph G is a tree (forest) if it is connected (not connected) and has no cycles.

If D is not symmetric we define $G_0(D) = (V, E_0)$ where V is the Vertex set of D and E_0 consists of edges corresponding to symmetric pairs of arcs in D. Thus $G_0(D)$ represents the symmetric part of D. Again we can define $G_0(A) = G_0(D(A))$. We call $G_0(A)$ the *combinatorially symmetric part* of A.

We remind the reader that the matrix A is irreducible if and only if $D(A)$ is strongly connected. Let D be a tree. Then D is strongly connected if and only if it is symmetric and $G(D)$ is a tree.

Now we will call A a *tree* if $D(A)$ is a tree. Thus when A is a tree it is either reducible or combinatorially symmetric with $G(A)$ a tree.

In addition to the graphs and digraphs defined above we also require signed graphs and digraphs. Let us define $S(A) = (V, \overline{E}; \tau)$ where $V = \{1, 2, \ldots, n\}$, $(i, j) \in \overline{E}$ if and only if $a_{ij} \neq 0$ and $\tau : \overline{E} \to (+, -)$ is defined by $\tau(i, j) = +$ if $a_{ij} > 0$ and $\tau(i, j) = -$ if $a_{ij} < 0$. Thus $S(A)$ is derived from the *underlying digraph* $D(A)$ by introducing the mapping $\tau : \overline{E} \to (+, -)$. In the same way we can introduce $H(A)$ ($H_0(A)$) by adjoining a mapping $\tau : E \to (+, -)$ ($\tau : E_0 \to (+, -)$) to $G(A)$ ($G_0(A)$). This time the mapping is defined by $\tau[ij] = \tau(i,j)\tau(j,i)$ so that a positive two-cycle of $S(A)$ maps into a positive edge of $H(A)$ ($H_0(A)$) and a negative two-cycle into a negative edge of $H(A)$ ($H_0(A)$).

Here are two examples to illustrate all of our concepts.

Example 1. Let A have sign pattern

$$A = \begin{bmatrix} x & + & 0 & 0 & 0 & 0 & 0 \\ - & 0 & - & 0 & 0 & 0 & 0 \\ 0 & - & 0 & + & 0 & - & 0 \\ 0 & 0 & - & x & + & 0 & 0 \\ 0 & 0 & 0 & + & 0 & 0 & 0 \\ 0 & 0 & - & 0 & 0 & 0 & - \\ 0 & 0 & 0 & 0 & 0 & + & x \end{bmatrix},$$

where an x on the principal diagonal denotes a nonzero element which may be either positive or negative. We have $D(A)$, $G(A)$, $S(A)$, and $H(A)$ as shown in Figure 1(a), (b), (c), (d), respectively. Here $\longrightarrow$ and $\underline{\qquad}$ denote, respectively, either an unsigned arc or edge or a positive arc or edge, and $-\!-\!>$ and $-\,-\,-$ denote, respectively a negative arc or edge. The vertices in V_0 have the form ⊗ and those not in V_0 have the form ○. Observe that the matrix A is combinatorially symmetric.

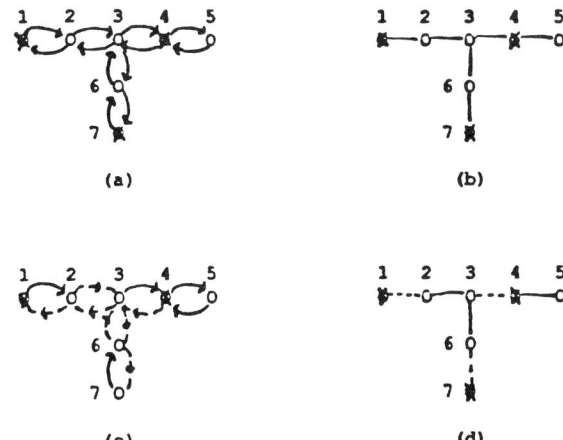

FIGURE 1. GRAPHIC ENTITIES ASSOCIATED WITH A IN EXAMPLE 1.

Example 2. Let A have sign pattern

$$A = \begin{bmatrix} x & + & 0 & 0 & 0 & 0 & 0 \\ + & 0 & - & 0 & 0 & 0 & 0 \\ 0 & + & 0 & - & + & + & - \\ 0 & 0 & 0 & 0 & + & 0 & + \\ 0 & 0 & 0 & + & 0 & 0 & 0 \\ 0 & 0 & + & 0 & 0 & 0 & - \\ 0 & 0 & 0 & 0 & 0 & 0 & 0 \end{bmatrix},$$

The various graphs $(D(A), G_0(A), S(A), H_0(A))$ are shown in Figure 2 (a), (b), (c), (d). This matrix is obviously not combinatorially symmetric. Observe here that $G_0(H_0)$ has three components (trees).

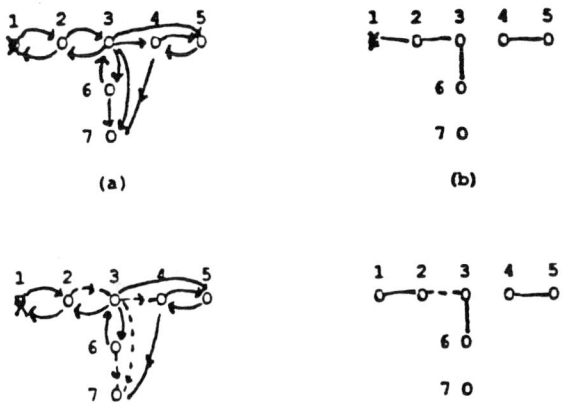

FIGURE 2. GRAPHIC ENTITIES ASSOCIATED WITH A IN EXAMPLE 2.

We have not made provisions for keeping track of the signs of the elements a_{ii}, $i = 1, \ldots, n$, of A, but this will usually be done separately.

Also, we will say that A has a *spanning tree* if $G_0(A)$ has a spanning tree. This will happen if and only if $G_0(A)$ is connected.

3. Equivalent conditions for qualitative stability. The problem of finding equivalent conditions for qualitative stability (sign stability) has been solved by Jeffries, Klee, and van den Driessche and an elegant version of their result has recently been published [JKV87]. It turns out the solution is purely qualitative and can be expressed by using the digraphs $D(A)$ and $S(A)$ and the graph $G_0(A)$. To do this we must introduce two coloring algorithms for $G_0(A)$.

A vertex coloring of $G_0(A)$ is called a δ-*coloring* with colors black and white if

(i) each distinguished vertex (i.e., each vertex in V_0) is colored black
(ii) no black vertex has exactly one white neighbor, and
(iiiδ) each white vertex has a white neighbor.

A vertex coloring of $G_0(A)$ is called an ϵ-*coloring* with colors black and white if (i) and (ii) are satisfied and

(iiiϵ) no white vertex has a white neighbor.

Note that the *trivial coloring* in which all vertices are black, is both a δ-coloring and ϵ-coloring.

We can now state the qualitative stability theorem.

THEOREM 1. [JKV87]. *The real matrix $A = [a_{ij}]$ is sign stable if and only if it satisfies the following five conditions:*

(α) Each vertex in V_0 corresponds to a negative diagonal element of A;

(β) $S(A)$ has only negative two-cycles;

(γ) $D(A)$ is a tree;

(δ) Each δ-coloring of $G_0(A)$ is trivial;

(ϵ) Each ϵ-coloring of $G_0(A)$ is trivial.

Observe that the matrix of Example 1 satisfies conditions (α), (δ), and (ϵ). On the other hand the matrix of Example 2 satisfies (γ) but has both a nontrivial δ-coloring and a nontrivial ϵ-coloring.

It is obvious that sign stable matrices are not Hicksian stable in general because they can have zero elements on the principal diagonal. For example, the matrix $A = [a_{ij}]$ with $a_{i,i+1} > 0$, $a_{i+1,i} < 0$, for $i = 1, 2, \ldots, n-1$, $a_{11} < 0$, $a_{ii} = 0$, $i = 2, \ldots, n$, and $a_{ij} = 0$ if $|i - j| > 1$ is sign stable and clearly not Hicksian stable.

4. Qualitative connections between sign and Hicksian stability. The starting point for determining qualitative conditions for Hicksian stability is to consider the set of all real $n \times n$ square matrices which are qualitatively invertible. These matrices have come to be called L-matrices. The following result of Bassett, Maybee, and Quirk characterizes them. Here a signed permutation matrix is any matrix obtained from the identity matrix by permuting rows and multiplying some of them by -1.

THEOREM 2. [BMQ68]. *The $n \times n$ real matrix A is an L-matrix if and only if there exists a signed permutation matrix P_+ such that the matrix $B = AP_+$ satisfies the following conditions:*

(a) $b_{ii} < 0$, $i = 1, 2, \ldots, n$;

(b) $S(B)$ has only negative cycles.

THEOREM 3. [BMQ68]. *If A is an L-matrix in normal form, then A is Hicksian stable.*

Now, if A is sign stable it is clearly an invertible matrix. On the other hand the conditions (α) through (ϵ) of Theorem 1 are all qualitative so every sign stable matrix must be an L-matrix. We can therefore state the following result.

COROLLARY OF THEOREM 2. *If the matrix A is sign stable there exists a signed permutation matrix P_+ such that AP_+ is Hicksian stable.*

Thus we can say that every sign stable matrix is qualitatively equivalent to a Hicksian stable matrix.

Example 3. Let A have the sign pattern shown.

$$\begin{bmatrix} - & - & 0 & 0 & 0 & 0 & 0 \\ + & 0 & - & 0 & 0 & 0 & 0 \\ 0 & + & 0 & + & 0 & - & 0 \\ 0 & 0 & - & - & + & 0 & 0 \\ 0 & 0 & 0 & - & 0 & 0 & 0 \\ 0 & 0 & + & 0 & 0 & 0 & + \\ 0 & 0 & 0 & 0 & 0 & - & - \end{bmatrix} ; \quad \begin{bmatrix} 1 & 0 & 0 & 0 & 0 & 0 & 0 \\ 0 & 0 & -1 & 0 & 0 & 0 & 0 \\ 0 & 1 & 0 & 1 & 0 & 0 & 0 \\ 0 & 0 & 0 & 0 & 1 & 0 & 0 \\ 0 & 0 & 0 & -1 & 0 & 0 & 0 \\ 0 & 0 & 0 & 0 & 0 & 0 & 1 \\ 0 & 0 & 0 & 0 & 0 & -1 & 0 \end{bmatrix}$$

It is easy to verify that A is sign stable and it is clearly not Hicksian stable. Applying the signed permutation matrix shown leads to the matrix below.

$$\begin{bmatrix} - & 0 & + & 0 & 0 & 0 & 0 \\ + & - & 0 & 0 & 0 & 0 & 0 \\ 0 & 0 & - & 0 & + & 0 & - \\ 0 & - & 0 & - & - & 0 & 0 \\ 0 & 0 & 0 & 0 & - & 0 & 0 \\ 0 & + & 0 & 0 & 0 & - & 0 \\ 0 & 0 & 0 & 0 & 0 & + & - \end{bmatrix}$$

It is easily verified that this matrix satisfies the conditions (a) and (b) of Theorem 2 so that it is an L-matrix in normal form and so Hicksian stable.

Since the sign stable matrices are a subclass of the class of L-matrices, the following interesting question can be asked. What special properties of a sign stable matrix in normal form exist which permit us to identify it among the L-matrices as being sign stable? This question is especially interesting in view of the fact that Jeffries, Klee and van den Driessche have developed fast algorithms for identifying sign stable matrices [JKV87], but it is not known whether or not the identification of L-matrices is NP-hard.

We note next the following result.

COROLLARY 4. *The sign stable matrix A is also Hicksian stable if and only if $V_0 = V$.*

Proof. Suppose V_0 is a proper subset of V. Then $a_{ii} = 0$ for at least one value of i and A is not Hicksian stable. Thus the only if part of the theorem is true. On the other hand, if $V_0 = V$, then conditions (β) and (γ) guarantee that A is an L-matrix in normal form. It follows by Theorem 3 that A is Hicksian stable. □

We will have more to say on the connection between sign stability and Hicksian stability in a different context below.

5. Connections between convergence and stability. In a 1945 paper which has now become a classic the mathematical economist Lloyd Metzler [M45] showed that stability, Hicksian stability, and convergence were all equivalent concepts for a class of matrices of special interest to economists. Not surprisingly, this class of matrices has come to be called Metzler matrices. We will not reproduce

Metzler's work here because if has been well reviewed elsewhere (see [Q81], for example). Metzler did note one simple fact of general interest, however. Namely, that if A is a convergent matrix the matrix $A - I$ is stable. For the class of Metzler matrices the converse is also true and both matrices are Hicksian stable.

The observation that A convergent implies that $A - I$ is stable is, of course, true in general and provides a method of passing from examples of convergent matrices to examples of stable matrices. Unfortunately the status of research on these matters is that much more work has been done on the characterization of stable matrices than on the characterization of convergent matrices and we would like to have methods of passing from stable to convergent rather than the converse.

To this end consider the linear fractional transformation

$$\lambda = \frac{z+1}{z-1}$$

of the complex z-plane to the complex λ-plane. Writing $\lambda = \mu + i\gamma$ and $z = x + iy$, we obtain

$$\mu + i\gamma = \frac{x^2 + y^2 - 1}{(x-1)^2 + y^2} - i\frac{2y}{(x-1)^2 + y^2}$$

This shows that z inside, on, or outside the unit circle in the complex z-plane maps into $\mu < 0$, $\mu = 0$, $\mu > 0$, respectively in the λ-plane, and conversely. The above form of the transformation is not particularly useful to us, so let us set

(3) $$\lambda = 1 + \frac{2}{z-1},$$

a form more convenient to our purpose. This suggests setting

$$A = I + 2(B - I)^{-1}$$

wherein A is a stable matrix if and only if B is a convergent matrix.

Now rewrite the above relation as

$$A - I = 2(B - I)^{-1}$$

whence

$$(A - I)(B - I) = (B - I)(A - I) = 2I.$$

Define

(4) $$A_1 = \frac{1}{\sqrt{2}}(A - I)$$
$$B_1 = \frac{1}{\sqrt{2}}(B - I).$$

Then the above relation becomes

$$A_1 B_1 = B_1 A_1 = I,$$

and it follows that

(5) $$B_1 = A_1^{-1}.$$

Thus we may pass from stable matrices to convergent matrices and vice-versa by the investigation of properties of inverses. We are now in a position to determine how stability translates to convergence.

The linear fractional transformation we have just used to derive the relation (5) has been used previously by other authors. A summary of its use can be found in the monograph [BS70]. However, these authors seem to have always written the relationship between a stable matrix A and a convergent matrix B in the form

$$A = (B + I)(B - I)^{-1}.$$

and the converse in the form

$$B = (A + I)(A - I)^{-1}.$$

Written this way, the relation seems more difficult to use. This is especially true in the qualitative case because the diagonal elements of $A + I$ will have ambiguous signs in general.

6. Properties of A^{-1} for A sign stable.
For A sign stable it is clear that $A_1 = \frac{1}{\sqrt{2}}(A - I)$ is an L-matrix in normal form. Therefore A_1 is Hicksian stable and so is A_1^{-1}. Let us set $A_1^{-1} = [\alpha_{ij}]$.

Before we investigate more carefully the properties of the matrix A_1^{-1}, it is important to note that in passing from the matrix A to the matrix A_1 we have left the area of purely qualitative analysis and moved into a partially quantitative area. This happens because the subtraction of the identity matrix from A is not a qualitative operation. Moreover, a little reflection shows that in moving from qualitative stability to convergence it is essential to introduce some quantitative restrictions. This is because convergence cannot be regarded as a qualitative problem. Indeed, if a matrix has all its eigenvalues inside the unit circle, then we can surely change the magnitudes of its elements in such a way as to move all the nonzero eigenvalues outside the unit circle.

It is of interest to compute the principal minors of A_1 in terms of those of A itself. To this end let J be an index set $J = (j_1, j_2, \ldots, j_q)$ with $1 \leq j_1 < j_2 < \cdots < j_q \leq n$. For an arbitrary matrix B let $E_k(B)$ denote the sum of the $k \times k$ principal minors of B, and let $p_B(\lambda)$ be the characteristic polynomial of B. Then we have

$$\det A_1(J) = p_{A(J)}(1)/2^{q/2}$$
$$= (-1)^q \frac{1 - E_1(A(J)) + E_2(A(J)) + \cdots \pm E_q(A(J))}{2^{q/2}}$$

Now each principal minor of A of order k has sign $(-1)^k$ or the value zero. Thus we can write

(6) $$\det A_1(J) = (-1)^q \frac{1 + |E_1(A(J))| + \cdots + |E_q(A(J))|}{2^{q/2}}$$

Next we observe that (see [LM66] or [M74]) there exists a positive diagonal matrix D such that $D^{-1}A_1D$ (and also $D^{-1}A_1D$) has symmetric part which is symmetric in absolute value. Since D has positive diagonal elements, the diagonal elements of A_1 are the same as those of $D^{-1}A_1D$ and the sign of every element of $D^{-1}A_1D$ is the same as the sign of the corresponding element of A_1. In the case where A is irreducible, A and A_1 are themselves combinatorially symmetric so that $D^{-1}A_1D$ is the sum of a diagonal matrix and a skew-symmetric matrix. In any case we can write

(7) $$D^{-1}A_1D = \widehat{K} + A_2$$

where $\widehat{K}$ is the sum of a diagonal matrix and a skew-symmatric matrix and A_2 is combinatorially asymmetric, i.e., symmetrically placed elements are either both zero or both nonzero. Now we can find a permutation matrix P such that

(8) $$P^T D^{-1} A_1 D P = \begin{bmatrix} A_{11} & 0 & & 0 \\ A_{21} & A_{22} & & 0 \\ & & & 0 \\ A_{q1} & A_{q2} & \cdots & A_{qq} \end{bmatrix}.$$

Here each diagonal block is irreducible so A_{ii} is the sum of a diagonal matrix and a skew-symmetric matrix, $A_{ii} = D_i + K_i$. Moreover, $G(A_{ii})$ is a tree. (Please remember that one or more, even all, of the diagonal blocks can be of size one in which case $A_{ii} = d_i$ for some negative number d and $G(A_{ii})$ is trivial.)

Let us turn our attention first to finding the inverse of A_1 in the case where $A_1 = D_1 + K_1$ and $G(A_1)$ is a tree. In this case given any pair of vertices i and j of $G(A_1)$ there exists exactly one path in G joining them. Call this path $p(i,j)$, denote its length by l_{ij}, and denote the cominor of $p(i,j)$, i.e., the principal minor of A_1 in the rows and columns corresponding to indices not in the path p_{ij}, by $\det A_1(p(ij))$. Note that $A_1(p(ij))$ is an $((n-1)(l_{ij}+1)) \times ((n-1)(l_{ij}+1))$ order submatrix of A_1. Finally let $A_1[p(ij)]$ denote the product of the elements of A_1 corresponding to the arcs of the path $p(i,j)$ (see [M0VW88] for discussion of all of these ideas. It follows that for $i \neq j$ the element α_{ij} of $D^{-1}A_1D$ is given by

(9) $$\alpha_{ij} = (-1)^{l_{ij}} A_1[p(ij)] \det A_1(p(ij))/\det A_1.$$

Of course, for $i = j$ we have

(10) $$\alpha_{ii} = \det A_1(i)/\det A_1$$

where $A_1(i)$ is the submatrix of A_1 obtained by elimination of row i and column i.

Now observe that, when l_{ij} is even, $p(i,j)$ and $p(j,i)$ have the same sign and, when l_{ij} is odd, $p(i,j)$ and $p(j,i)$ have opposite signs. We can therefore state the following result.

THEOREM 5. *Let the sign stable matrix A be irreducible. Then there exists a positive diagonal matrix D and a permutation matrix P such that*

(11) $$P^T D^{-1} A_1^{-1} DP = \begin{bmatrix} A_{11}^1 & A_{12}^1 \\ A_{21}^1 & A_{22}^1 \end{bmatrix}$$

where $A_{21}^1 = -(A_{12}^1)^T$, A_{11}^1 *and* A_{22}^1 *are symmetric matrices with negative elements on the principal diagonal, and every element* $\alpha_{ij} \neq 0$ *for* $1 \leq i \leq n, 1 \leq j \leq n$.

For convenience let us set $\tilde{A}_1 = P^T D^{-1} A_1^{-1} DP$. Obviously $\tilde{A}_1$ is combinatorially symmetric. From the facts that $A_{21}^1 = -(A_{12}^1)^T$ and A_{11}^1 and A_{22}^1 are symmetric we deduce that the vertices of $H(\tilde{A}_1)$ inverse can be partitioned into two subsets V_1 and V_2 so that edges joining two vertices in V_1 or two vertices in V_2 are positive and edges joining a vertex in V_1 and a vertex in V_2 are negative. It follows that $H(\tilde{A}_1)$ has all cycles positive. Note also that $V_1 \neq \emptyset \neq V_2$.

THEOREM 6. *Let A be an irreducible sign stable matrix. Then* A_1^{-1} *is a full matrix with all cycles positive and with negative diagonal elements.*

Proof. The remarks above show that $\tilde{A}_1^{-1}$ satisfies the conditions, hence A_1^{-1} does also. □

Let us continue to suppose that A is irreducible. To simplify the notation we set $A_1^{-1} = B_1$ as in Section 5 and observe that B_1^{-1} is a tree. Therefore, following Klein [K82] and Barrett and Johnson [BJ84], we set $B_{1\langle ij \rangle} = \det B_1[i,j]$, i.e., the second order principal minor of B_1 in rows and columns i and j, if $[i,j]$ is an edge of $G(B_1^{-1}) = G(A)$. For each k let d_k be the degree of the vertex k in $G(A)$ and let $B_{1\langle k \rangle}$ be the value of the k-th diagonal element of the matrix B_1. For $G(A) = (V, E)$ let us set $E(A)$ equal to E.

THEOREM 7. *Let A be irreducible. Then*

(12) $$\det B_1 = \frac{\displaystyle\prod_{[i,j] \in E(A)} B_{1\langle i,j \rangle}}{\displaystyle\prod_{k=1}^{n} B_{1\langle k \rangle}^{d_k - 1}}$$

Note that in the formula (12) the numerator contains exactly $n - 1$ factors.

7. A uniqueness question. The results of the last section provide considerable detail about the matrix B_1. It is natural to ask if they, in fact, characterize this matrix. We will show that they do so in the class of L-matrices.

THEOREM 8. *Let* A_1 *be an L-matrix in normal form having a cycle of length greater than two. Then* $A_1^{-1} = B_1$ *either has an element of undetermined sign or else* $G_0(B_1)$ *has a negative cycle.*

Proof. Suppose first that $D(A_1)$ has a cycle c of length $k \geq 3$ and that the transposed cycle c^T does not belong to $D(A_1)$. Without loss of generality we may suppose that vertices 1, 2, and 3 of $D(A_1)$ belong to c and that (1,2) and (2.3) belong to c and (2,1) is not an arc of $D(A_1)$. Now there is at least one path in $D(A_1)$ from each of these vertices to each other one, since they all belong to the same cycle. the path from vertex two to vertex one in c must have the opposite sign to the sign of the arc (1,2) because c is a negative cycle. If there is a path from vertex two to vertex one having the same sign as the arc (1,2) then the element b_{21}^1 is of undetermined sign and we are finished. Suppose therefore that no such alternative path exists. Then the elements b_{12}^1 and b_{21}^1 have opposite signs and the edge [1,2] of $G_0(B_1)$ is negative. By a similar argument either b_{23}^1 or b_{32}^1 has an undetermined sign or else the edge [2,3] of $G_0(B_1)$ is negative. finally the path (1,2,3) and the path from vertex three to vertex one in c must have opposite signs. It follows that either b_{12}^1 or b_{21}^1 has an undetermined sign or else the edge [1,3] of $G_0(B_1)$ is negative. Thus we either have an element of undetermined sign in B_1 or else the 3-cycle [1,2,3,1] of $G_0(B_1)$ is negative. It remains to consider the case where $D(A_1)$ has a cycle c of length $k \geq 3$ and is a symmetric digraph. In this event k must be an even integer for any such cycle c. But in this case, if i and j are two vertices belonging to c (hence also to c^T) an even distance apart in c then the paths from i to j in c and in c^T have opposite signs. It follows that the element b_{ij}^1 is of undetermined sign (so also is b_{ji}^1). □

COROLLARY OF THEOREM 8. *Let A_1 be an irreducible L-matrix in normal form. Then A_1^{-1} exists and is full with all cycles positive if and only if A_1 has no cycle of length greater than two.*

Let us continue to assume A_1 is irreducible and investigate more closely its relation to B_1. First we recall from the theory of L-matrices [LM83] that if A_1 is any L-matrix and $a_{ij}^1 \neq 0$, then b_{ij}^1 satisfies $a_{ij}^1 b_{ij}^1 < 0$.

Next we partition A_1 as a bipartite matrix [M88] so that it takes the form

$$\widetilde{A}_1 = \begin{bmatrix} D_1 & A_{12} \\ A_{21} & D_2 \end{bmatrix}$$

where D_1 is an $r \times r$ diagonal matrix with negative diagonal entries, D_2 is an $(n-r) \times (n-r)$ diagonal matrix with negative diagonal entries, A_{12} is an $r \times (n-r)$ matrix with $n-1$ nonzero entries, and $A_{21} = (-A_{12})^T$. We call this the bipartite form of A_1.

By Theorem 5 the same partition puts B_1 into the form specified in (11) where, in addition to the facts stated in the Theorem, whenever A_{12}^1 and A_{12} (A_{21}^1 and A_{21}) both have nonzero entries their product is negative.

Using all of these facts let us compute the product $A_1 B_1$. This is, of course, just the identity matrix. But it is of interest to observe that many of the individual elements are unambiguously signed and that the location of such elements and their signs can be a priori determined. We have

$$\widetilde{A}_1 B_1 = \begin{bmatrix} D_1 A_{11}^1 + A_{12}(-A_{12}^1)^T & D_1 A_{12}^1 + A_{12} A_{22}^1 \\ (-A_{12})^T A_{11}^1 + D_2(-A_{12}^1)^T & (-A_{12})^T A_{12}^1 + D_2 A_{22}^1 \end{bmatrix}$$

Now observe that each entry of the products $D_1 A_{11}^1$, $D_1 A_{12}^1$, $D_2(-A_{12}^1)^T$ and $D_2 A_{22}^1$ is uniquely determined as to signs. In fact, we have

$$\text{sign } (D_1 A_{11}^1)_{ij} = -\text{sign } (A_{11}^2)_{ij'}$$
$$\text{sign } (D_1 A_{12}^1)_{ij} = -\text{sign } (A_{12}^2)_{ij'}$$
$$\text{sign } (D_2(A_{12}^1)^T)ij = -\text{sign } (A_{12}^1)_{ij'}^T$$
$$\text{sign } (D_2 A_{22}^1)_{ij} = -\text{sign } (A_{22}^1)_{ij}.$$

The remaining four factors will in general have some terms with a priori determined signs and some terms whose signs must be determined from the fact that $\widetilde{A}_1 B_1$ is the identity.

To see which terms have a priori determined signs and which do not we remember that we are dealing with the irreducible case here and we represent $G(A)$ in bipartite form. The first r integers $V_1 = \{1, 2, \ldots, r\}$ are in one set and the remaining $n - r$ integers $V_2 = \{r + 1, \ldots, n\}$ are in the other set of the partition of the vertices of G.

Consider an element of $A_{12}(-A_{12}^1)^T$ or of $A_{12} A_{22}^1$. Let the element be denoted by b_{jk}. If $j \neq k$ and the degree of the vertex j is greater than one then b_{jk} will not be a priori determined. It will also not be a priori determined in the case of the product b_{jj} when it is the diagonal element of $A_{12} A_{22}^1$. For elements of $(-a_{12})^T A_{11}^i$ and $(-A_{12})^T A_{12}^1$, the same result holds if the degree of the vertex k is greater than one.

We can summarize all of the results on the product $\widetilde{A}_1 \widetilde{B}_1$ as follows.

THEOREM 9. *The product $\widetilde{A}_1 \widetilde{B}_1$ can be written as a sum $\widetilde{A}_1 \widetilde{B}_1 = C_1 + C_2$ where*

$$\text{sign } C_1 = \begin{bmatrix} -\text{sign } (A_{11}^1)_{ij} & -\text{sign } (A_{12}^1)_{ij} \\ \text{sign } (A_{12}^1)_{ij}^T & -\text{sign } (A_{22}^1)_{ij} \end{bmatrix}$$

and $\text{sign } C_2 = [c_{ij}^2]$ is uniquely determined a priori if and only if $i = j$ or if either degree of i as a vertex in V_1 is one or the degree of k as a vertex in V_2 is one. The signs of all diagonal elements of both C_1 and C_2 are positive and the a priori determined signs of all other elements of C_2 are the negative of the signs of the corresponding elements of C_1. Here $V_1 = \{1, 2, \ldots, r\}$, $V_2 = \{r + 1, \ldots, n\}$.

Notice that what is really happening here can be explained by saying that the elements of the product $\widetilde{A}_1 \widetilde{B}_1$ can be classified on the basis of combinatorial information about A into those that are the sum of two terms and those that are not. The details about the signs are not particularly surprising. What is impressive is the amount of insight derivable upon combinatorial and purely qualitative grounds.

8. Properties of B for A sign stable. We have now obtained enough information about the matrix B_1 so that we can make some nontrivial statements about the matrix B itself. Our previous results together with the relation $B = 2 B_1 + I$ enable us to formulate the following.

THEOREM 10. *Let A be a sign stable matrix. Then there exists a positive diagonal matrix D and a permutation matrix P such that*

$$\widetilde{B} \equiv P^T D^{-1} B_1 D P = \begin{bmatrix} \widetilde{B}_{11} & 0 & \cdots & 0 \\ \widetilde{B}_{21} & \widetilde{B}_{22} & \cdots & 0 \\ \cdots\cdots\cdots\cdots\cdots\cdots\cdots \\ \widetilde{B}_{q1} & \widetilde{B}_{q2} & \cdots \widetilde{B}_{qq} \end{bmatrix}$$

where each diagonal block is an $r_k \times r_k$ matrix with $\widetilde{B}_{kk}$) a full matrix with positive cycles and which may also be written in the form

$$\widetilde{B}_{kk} = \begin{bmatrix} B_{11}^k & B_{12}^k \\ -(B_{12}^k)^T & B_{22}^k \end{bmatrix}$$

with B_{11}^k and B_{22}^k symmetric (square) matrices.

Notice that Theorem 10 does not make any statement about the principal minors of B (or $\widetilde{B}$) or even about the diagonal elements.

Unfortunately, the results we have derived are not sufficient to characterize B as being uniquely derived from a purely qualitative matrix. the results that are missing involve a characterization of which non-principal minors of B must be zero in order to guarantee that the inverse has the desired tree structure. This is a topic of current research by the author and others and we hope to be able to report on the results soon.

BIBLIOGRAPHY

[BJ84] WAYNE BARRET AND CHARLES JOHNSON, *Determinantal formulae for matrices with sparse inverses*, Linear Algebra and Its Applications 56, 73–88.

[BMQ68] LOWELL BASSETT, JOHN MAYBEE AND JAMES QUIRK, *Qualitative economics and the scope of the correspondence principle*, Econometrika 26, 544–563.

[BS70] S. BARNETT AND C. STOREY, *Matrix Methods in Stability Theory*, Nelson and Sons, London.

[H39] SIR JOHN HICKS, *Value and Capital*, London.

[JKV87] CLARK JEFFRIES, VICTOR KLEE, AND PAULINE VAN DEN DRIESSCHE, *Qualitative stability of linear systems*, Linear Algebra and Its Applications 77, 1–48.

[K82] D. KLEIN, *Tree diagonal matrices and their inverses*, Linear Algebra and Its Applications 42, 109–117.

[M45] LLOYD METZLER, *Stability of multiple markets: the Hicks conditions*, Econometrika 13, 277–292.

[M66] J. MAYBEE, *New generalizations of Jacobi matrices*, SIAM J. Applied Math 14, 1032–1037.

[M74] J. MAYBEE, *Combinatorially symmetric matrices*, Linear Algebra and Its Applications 8, 529–537.

[M0VW88] J. MAYBEE, D. OLESKY, P. VAN DEN DRIESSCHE, AND G. WIENER, *Matrices, digraphs, and determinants*, Forthcoming.

[Q81] JAMES QUIRK, *Qualitative stability of matrices and economic theory*, in *Computer Assisted Analysis and Model Simplification*, H.J. Greenberg and J.S. Maybee, editor, Academic Press, New York.

[S47] PAUL A. SAMUELSON, *Foundations of Economic Analysis*, Cambridge, Massachusetts.

TREE STRUCTURES IN IMMUNOLOGY*

J. K. Percus†

Abstract. An important component of the behavior of cell populations is often associated with branching processes in time, physical space, or more abstract configuration spaces. Several examples are presented, arising from studies in mathematical immunology. First is simple stem cell proliferation, with the hemopoietic stem cell as prototype, to introduce concepts, theorems, and approximations. Extension to multitype branching is then considered, as in macrophage production, with the analysis restricted to that of population moments. A further area of application is that of somatic mutations contributing to the fine scale diversity of antibody production, which is modeled to accommodate the machinery developed above. Finally, the immune network is introduced as an example in which branching in species space may capture a fair amount of the phenomenology.

1. Introduction. *Combinatorics* [1] deals with the enumeration of the discrete states Φ that can be attained by a defined system, generally in the form of weighted sums of state functions

$$(1.1) \qquad F_x = \sum_\Phi W_x(\Phi) S(\Phi).$$

It is thereby coextensive with probability theory. Various aspects of biology which focus on cells or even macromolecules as units have combinatorics as their natural framework; here we will be primarily interested in situations which arise in *immunology* [2]. There are fascinating topics, such as that of antigen (Ag) binding by antibody (Ab), which are blatantly combinatorial in nature, but which we will not consider because of their innate complexity. Instead, we will focus on those which involve temporal or spatial *branching* [3], and hence are associated with the general area of tree structures.

2. Simple Stem Cell Proliferation. The immune system of a functioning mammal can be traced back to a pluripotent hemopoietic *stem cell* (HSC), some of whose repertoire is presented in Fig. 1. Stem cells receive external instructions which tell them how often to reproduce themselves on division, and how often to differentiate to one or more new cell types which may be, but are not necessarily, non-proliferating cells, or *end cells*. This ratio, a principal determinant of population growth, may be controlled by cell density or by longer range biochemical interactions. The most primitive model, and the one we will in principle be able to assess, is that of a single differentiation channel, describable as a pure homogeneous

---

*Supported in part by DOE contract DE-AC02-7603077.
†Courant Institute of Math. Sci. and Physics Dept., New York University 251 Mercer St., New York, NY 10012.

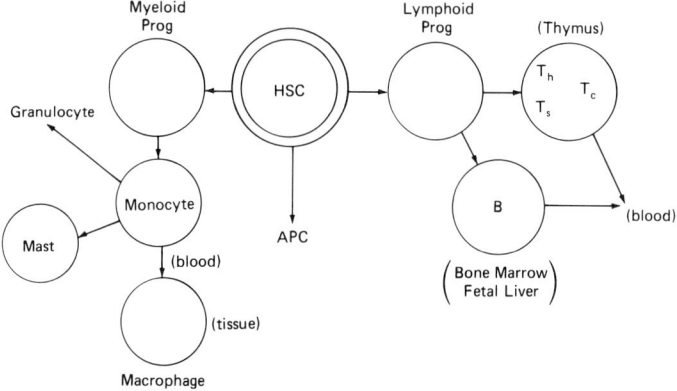

Fig. 1. Repertoire of hemopoietic stem cell, via myeloid and lymphoid progenitors (Prog), antigen-presenting cells (APC); arena of action in parentheses.

Fig. 2. Simple Stem Cell Proliferation; $q = 1 - p$

(no age or cycle time dependence) Markov chain, controlled by a single probability p, and with the idealized assumption of constant generation time.

Interestingly, some of the cleanest work has been done with hydra [4], most of whose cell types are self proliferating, but one pair – the neurons and the nematocytes, or poison dart vehicles – is produced from the interstitial stem cells, a close relation between nerve and primitive immune system which is maintained as one goes up the phyla.

Let us only count stem cells. Then at division, a cell either splits into two more or "dies". This is an example of a single type branching process, intensively studied since Galtou and Watson [5]. If a stem cell is given a multiplicative weight of x, the next generation has x^2 with probability p, or x^0 with probability q, and is then represented by the transformation

(2.1) $$x \to f(x) = q + px^2.$$

At the next division, then, one would have

(2.2) $$f(x) \to f_2(x) = q + p(q + px^2)^2$$
$$= q(1 + pq) + 2p^2qx^2 + p^3x^4, \text{ etc.}$$

(for an exponential distribution of division times, one would use instead $x \to f(x) = x + [q + (1 - p - q)x + px^2]dt$ in each interval dt). In general, for k cells produced with probability p_k, (2.1) would be replaced by

(2.3) $$x \to f(x) = \sum p_k x^k,$$

with the normalization restriction

(2.4) $$f(1) = 1.$$

If Z_N is the number of cells at generation N, starting with one at generation 0, (2.3) is equivalent to

(2.5) $$p_k = Pr(Z_1 = k | Z_0 = 1)$$
$$\text{and } f(x) = E(x^{Z_1} | Z_0 = 1).$$

Correspondingly, for the Nth generation,

(2.6) $$f_N(x) = \sum p_{N,k} x^k$$
$$= E(x^{Z_N} | Z_0 = 1).$$

It is clear that

(2.7) $$f_N(x) = f_{N-1}(f(x)),$$

and so, together with $f_0(x) = x$, that

(2.8) $$f_N(x) = \overbrace{f \cdot f \cdot f \cdots f(x)}^{N},$$

leading to the general

(2.9) $$f_{A+B}(x) = f_A(f_B(x)).$$

The explicit structure of the chain is not easy to find in detail, but is certainly limited by the behavior of its moments. We define

(2.10) $$m = E(Z_1 | Z_0 = 1)$$
$$= \sum k p_k = f'(1),$$

the *multiplication factor*, as well as the second factorial moment

(2.11)
$$Q = \sum k(k-1)p_k = f''(1)$$
$$= \sigma^2(Z_1|Z_0 = 1) + m^2 - m.$$

Not surprisingly, at the Nth generation, $m_N = \frac{d}{dx}f_N(x)|_{x=1} = \frac{d}{dx}f(f_{N-1}(x))|_{x=1} = f'_{N-1}(1)f'(1) = m_{N-1}m$, so that

(2.12)
$$m_N = m^N.$$

Similarly, $Q_N = \frac{d^2}{dx^2}f_N(x)|_{x=1} = \frac{d}{dx}[f'_{N-1}(x)f'(f_{N-1}(x))]_{x=1} = f''_{N-1}(1)f'(1) + (f'_{N-1}(1))^2 f''(1)$, from which

(2.13)
$$Q_N = m\frac{m^N - 1}{m - 1}Q.$$

There are general properties one can prove [5], suggested by the iterated multiplication; they pertain to the limit $N \to \infty$, and have only weak subsidiary restrictions:

a) $\lim W_N = Z_N/m^N$ converges to a random variable.
b) if $p_j \neq 0$, $\lim p_{N,k}/p_{N,j} = \pi_k$ exists for $k \geq 1$.
c) $m < 1$ (subcritical): $\lim Pr(Z_N = k|Z_N > 0)$ exists.
d) $m = 1$ (critical): $Pr(Z_N/N > z|Z_N > 0) \to e^{-2z/\sigma^2}$

$$\frac{1}{N}(\frac{1}{1 - f_N(x)} - \frac{1}{1 - f(x)}) \to \frac{1}{2}\sigma^2$$

e) $m > 1$ (supercritical): if $f(x_0) = x_0 \neq 1$, and $\gamma = f'(x_0)$, then $\gamma^{-N}(f_N(x) - x_0)$ converges.

But these are not of much help in specific cases.

Are there then any solvable examples to serve as guides? One function that iterates easily is

(2.14)
$$f(x) = -1 + 2x^2 = \cos(2\cos^{-1}x)$$
$$\text{so that } f_N(x) = \cos(2^N\cos^{-1}x),$$

but this is not a probability generating function. For an exponential probability distribution, one does have the rational function

(2.15)
$$f(x) = 1 + \frac{m(x-1)}{1 - \frac{Q}{2m}(x-1)},$$

here parametrized so that

(2.16)
$$f(1) = 1, \ f'(1) = m, \ f''(1) = Q,$$

and iterating at once to

(2.17) $$f_N(x) = 1 + \frac{m^N(x-1)}{1 - \frac{Q}{2m}\frac{m^N-1}{m-1}(x-1)}.$$

We will take advantage of this.

Can we at least carry out well defined approximations, for example expand about one of the fixed points $f(x_0) = x_0$ (i.e. $x_0 = 1$ or q/p for our basic case), or the critical point $m = 1$? Several suggestions have been made. Feller, for $m > 1$ and $N \gg 1$, observed that m_N and Q_N – suitably scaled – are moments of a simple diffusion process, and obtained an asymptotically valid result. In the same region, and in the same vein, it is easy to see that if $|f'(x_0)| < 1$ for a fixed point, then x_0 is a stable fixed point, to which $f_N(x)$ converges. Thus, the difference equation in N can be converted to a differential equation. Or, one can use the rational fraction expansion

(2.18) $$f(x) = x_0 + \gamma(x - x_0) + \frac{1}{2}Q_0(x - x_0)^2 + \ldots$$
$$= x_0 + \frac{\gamma(x - x_0)}{1 - \frac{Q_0}{2\gamma}(x - x_0)} + \ldots,$$

which can be iterated. But then not even m_N need be reproduced. To get m_N correctly, and higher moments with good accuracy, one can instead expand about the fixed point $x = 1$, even when unstable, obtaining $f_N(x)$ of (2.17). And this may be the best one can do with a fixed point expansion.

A critical point, $m = 1$, expansion can be systematized much more easily. In fact, $m = 1$, $p = \frac{1}{2}$ in our prototype [7,8], is often closely approximated in real systems, since it is the threshhold between proliferation and non-proliferation of the differentiated type. For this purpose, we need another concept. Suppose $m > 1$. Then m^N establishes the scale for Z_N, suggesting that we consider

(2.19) $$\phi_N(s) = E(e^{-sZ_N/m^N})$$
$$= f_N(e^{-s/m^N}).$$

Clearly then $\phi_{N+1}(ms) = f(\phi_N(s))$. Furthermore, ϕ_N converges, as can be shown, to some asymptotic ϕ, which thereby satisfies

(2.20) $$\phi(ms) = f(\phi(s))$$
$$\phi(0) = 1, \ \phi'(0) = -1,$$

uniquely determining ϕ. If ϕ is known – it is necessarily monotonically decreasing – all is known, for ϕ^{-1} exists and we have

(2.21) $$f(x) = \phi(m\phi^{-1}(x)),$$

a nonlinear similarity transformation, with the consequence that

(2.22) $$f_N(x) = \phi(m^N \phi^{-1}(x)).$$

How does this help? Suppose that $r = m - 1$ is small, and that we imagine ourselves in the vicinity of $x = 1$. Then set

(2.23)
$$f(x) = x + rU(x)$$
so that $U(1) = 0$, $U'(1) = 1$.

Hence $(1+r)s(\phi) = s(\phi+rU(\phi)) = s(\phi)+rU(\phi)s'(\phi)+\frac{1}{2}r^2U(\phi)^2s''(\phi)+\ldots$, or $s(\phi) = U(\phi) \, s'(\phi) + \frac{1}{2}rU(\phi)^2 s''(\phi) + \ldots = U(\phi)(1+\frac{1}{2}rU(\phi)\frac{d}{d\phi}\ldots)s'(\phi)$. We write this as $s'(\phi) = (1+\frac{1}{2}rU(\phi)\frac{d}{d\phi}+\ldots)^{-1}(s(\phi)/U(\phi)) = (1-\frac{1}{2}rU(\phi)\frac{d}{d\phi}+\ldots)(s(\phi)/U(\phi))$. Since $s/U \to -1$ as $\phi \to 1$, we can integrate this first order equation at once to obtain the leading approximation

(2.24)
$$s(\phi) = -U(\phi)\exp(\frac{1}{1+\frac{1}{2}r}\int_1^\phi \frac{1-U'(\phi)}{U'(\phi)}d\phi).$$

For example, for $f(x) = q + px^2$, we get

(2.25)
$$\phi(s) = \frac{q}{p} + \frac{r}{pr+ps} + \frac{r^3}{(r+ps)^2}s\ell n(\frac{r}{r+ps}) + \ldots,$$

which is quite decent even in the extreme case $p = 1$, $q = 0$, $r = 1$, for which the exact solution is trivially found.

3. Macrophage Production. *Macrophages* have a vital role both in antigen presentation to start the immune system going in the dominant T-cell dependent mode, and as "effectors", i.e. killers, to do the final mopping-up job. They emerge from the blood forming cells in the bone marrow via a stem cell S, differentiate to a rapidly proliferating precursor M, which then produces the non-proliferating macrophage end cells E. There is said to be evidence [9] that each daughter can choose to differentiate independently of the other. Uncritically assuming this – it does not affect the analysis – the resulting multi-stage branching process per generation is shown in Fig. 3 [10]. Weighting S, M, and E by x, y, and z, the weighted transitions per generation then take the form

(3.1)
$$\begin{aligned} x &\to (px+qy)^2 \\ y &\to (p'y+q'z)^2 \\ z &\to z \end{aligned}$$

If this were an *irreducible* process, in which an initial cell of any type would eventually produce cells of all types, much of the exact classical theory of one type would carry over without difficulty [11] – not the approximations, to be sure. But this system is hierarchical, making things more difficult in theory, but perhaps easier in practice.

Let us now define $\mathbf{Z}_N$ as the cell number vector (S, M, E) at stage N, and look in sequence at the three basic initial conditions.

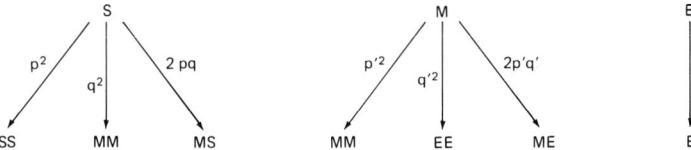

Fig. 3. Macrophage Proliferation Hierarchy.

A. If $\mathbf{Z}_0 = (0,0,1)$, then $\mathbf{Z}_N = (0,0,1)$ forever.

B. $\mathbf{Z}_0 = (0,1,0)$. This of course functions as a monotype if only M is counted. If E is also counted, then its weight z comes along pretty much for the ride:

$$(3.2) \qquad f_N(y,z) = f_{N-1}(f(y,z),z)$$
$$f(y,z) = (p'y + q'z)^2, \quad f_0(y,z) = y.$$

To be sure, $f(1,z) \neq 1$, but it is easy enough to normalize. Define

$$(3.3) \qquad f_N(\lambda y, z) = \lambda g_N(y,z).$$

Then it is still true that $g_N(y,z) = g_{N-1}(g(y,z),z)$ as well as $g_0(y,z) = y$, and the condition that $g(1,z) = 1$ becomes simply that of $(p'\lambda + q')^2 = \lambda$ or

$$(3.4) \qquad \lambda = \frac{1}{2p'^2}[1 - 2p'q' + \sqrt{1 - 4p'q'}]$$

We thus return to a monotype situation.

The monotype format, with its explicit z-dependence, is more sensitive to approximation than our z-independent prototype of Sec. 2 would be. What can we find reliably? The moments of Z_N^E are an example, but these are not separable from those of Z_N^M (vector index indicated by superscript). We expand the formalism appropriately by first writing

$$(3.5) \qquad m_{ij} = E(Z_1^j | \{Z_0^k = \delta_{ik}\})$$
$$= \frac{\partial f^i}{\partial x_j}(1),$$

where **x** denotes (x, y, z), $x_i \to f^i(\mathbf{x})$ in one generation, and **1** is the vector of all ones. Since $f_N^i(\mathbf{x}) = f_{N-1}^i(\mathbf{f}(\mathbf{x}))$, we can transform the complete process to vectorial notation

(3.6)
$$\mathbf{f}_N(\mathbf{x}) = \overbrace{\mathbf{f} \cdot \mathbf{f} \cdots \mathbf{f}(\mathbf{x})}^{N},$$
$$\mathbf{f}_0(\mathbf{x}) = \mathbf{x},$$

and it follows just as in the monotype case that after N generations

(3.7)
$$m_{N,ij} = (m^N)_{ij}.$$

In the present two-dimensional M-E case, for example, it is readily seen that

(3.8)
$$\mathbf{m} = \begin{pmatrix} 2p' & 2q' \\ 0 & 1 \end{pmatrix}.$$

If $p' \neq \frac{1}{2}$, **m** has two eigenvalues, $2p'$ and 1, so that

(3.9)
$$\mathbf{m}^N = (2p')^N (I - \mathbf{m})/(1 - 2p') + (2p'I - \mathbf{m})/(2p' - 1)$$

so that the mean E production from an initial M cell is given by

(3.10)
$$m_{N,ME} = 2q' \frac{(2p')^N - 1}{2p' - 1},$$

the expected $2p'$ multiplication. But if $p' = \frac{1}{2}$, $\mathbf{m} = \begin{pmatrix} 1 & 1 \\ 0 & 1 \end{pmatrix}$, so $\mathbf{m}^N = \begin{pmatrix} 1 & N \\ 0 & 1 \end{pmatrix}$, and

(3.11)
$$m_{N,ME} = N,$$

a steady non-exponential growth at threshhold.

C. For the full macrophage problem, $\mathbf{Z}_0 = (1, 0, 0)$, there are three components, and the moment problem is scarcely harder, although the corresponding $Q_{N,ij}^\alpha = E(Z_N^i Z_N^j | Z_0^k = \delta_{\alpha k}) - \delta_{ij} E(Z_N^i | Z_0^k = \delta_{\alpha k})$ is a bit of a mess [12]. In fact, the hierarchical structure simplifies matters in both cases. We choose to use the recursion

(3.12)
$$\mathbf{m}_N = \mathbf{m}_{N-1} \mathbf{m}$$

via the generating function

(3.13)
$$\mathbf{m}(\lambda) = \sum_0^\infty \mathbf{m}_N \lambda^N,$$

so that

(3.14)
$$\mathbf{m}(\lambda) = I + \lambda \mathbf{m}(\lambda) \mathbf{m}.$$

Now

(3.15) $$\mathbf{m} = \begin{pmatrix} 2p & 2q & 0 \\ 0 & 2p' & 2q' \\ 0 & 0 & 1 \end{pmatrix},$$

and we obtain in sequence: $m_{ss}(\lambda) = 1 + 2p\lambda m_{ss}(\lambda)$, so that

(3.16) $$m_{ss}(\lambda) = \frac{1}{1 - 2p\lambda}, \quad m_{N,ss} = (2p)^N;$$

$m_{sM}(\lambda) = 2\lambda m_{ss}(\lambda) + 2p'\lambda m_{sM}(\lambda)$, so that

(3.17) $$m_{N,sM} = \frac{q}{p - p'}((2p)^N - (2p')^N);$$

$m_{sE}(\lambda) = 2q'\lambda m_{sM}(\lambda) + \lambda m_{sE}(\lambda)$, so that

(3.18) $$m_{N,sE} = \frac{2qq'}{p - p'}\left(\frac{(2p)^N}{p - q} - \frac{(2p')^N}{p' - q'}\right) + \frac{4qq'}{(p - q)(p' - q')},$$

which can certainly be used in principle to check the homogeneous Markov chain assumption.

The situation with respect to approximations is less clear than in the monotype case. With some effort, moments higher than $\mathbf{m}$ can be found, and one can try to incorporate them into rational fractions, e.g.

(3.19) $$\mathbf{f(x)} = 1 + \mathbf{m(x - 1)} + \frac{1}{2}(\mathbf{x - 1})\mathbf{Q(x - 1, x - 1)} + \cdots$$
$$= 1 + \frac{\mathbf{m(x - 1)}}{1 - (1/2)\mathbf{K} \cdot (\mathbf{x - 1})} + \cdots,$$

the final form being one that can be iterated. But reduction to this form is achieved only in special cases ($p = p' = 1$ in the above) and so the degree of approximation suffers. However, it should be pointed out that the $\phi(\mathbf{s})$ formulation patterned after (2.20) et seq. goes through in total analogy as do its associated approximations.

4. Antibody Diversity. The immune system has phenomenal specificity, mounting attacks against perhaps 10^9 different antigens (Ag), thereby demanding many more specific immunoglobin (Ig) molecules than could be coded in our DNA. Let us concentrate upon the B-lymphocytes, which are required for nonviral attacks (the T-cell story goes similarly). The diversity is now known to arise in two ways, a major component being present when the B-cell first meets its antigen prey, followed by fine tuning during proliferative formation of memory and plasma cells. The Ig molecule has two heavy (H) and two light (L) chains, each composed of a constant region (C) which determines the Ig class (Ig M, Ig G, Ig E, ...) characterizing

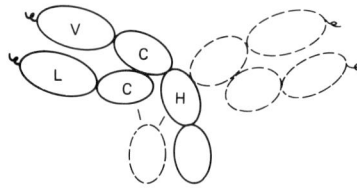

Fig. 4. Immunologlobin Molecule.

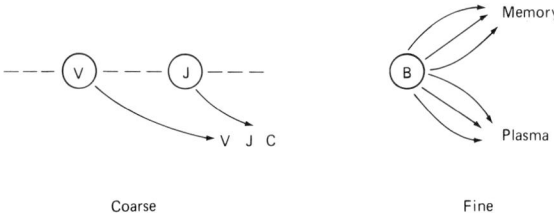

Fig. 5. Sources of Diversity.

the general mode of action, joined to a variable region (V) containing the receptor proper, or antibody (Ab).

The variable, joining (J), and constant regions are housed far apart on the DNA, with a few different J-fragments and many different V-fragments. These join 1:1:1, with many combinations available, as well as some leeway in joining location, to induce the surface Ig of the virgin (before meeting antigen) B lymphocyte.

Next is the fine-tuning. The B-cell that traps suitable Ag clonally explodes, perhaps with T-cell help, producing populations of memory cells for a next encounter (with altered C) and plasma cells for immediate attack.

But the V-regions have mutational "hot spots", producing via *somatic mutation* (not via gametes) a further selection of "nearby" Ig's to react more strongly to Ag that might have been reacted to only weakly to start.

We now ask for the fine-tuning "consanguinity" distribution after the elapse of time. One can imagine a mutation probability p at each division, but divisions are not synchronous, really have a spread, and of course don't produce precisely one mutation every 1/p divisions. We will therefore represent the dynamics by a model in which proliferation continues for a mutation time unit, producing t cells with probability p_t – all of which then mutate – with appropriate monotype generating function

(4.1) $$f(x) = \sum p_t x^t.$$

Question: if the cell population is M after n time units, what is the expected number of all pairs, $E_{nk}(M)$, whose most recent common ancestor was k generations prior? The resulting progeny are of course separated by 2k mutational events. In particular, we seek the generating function

(4.2) $$F_{nk}(x) = \sum_M x^M E_{nk}(M),$$

where weighting by the probability of M is implied.

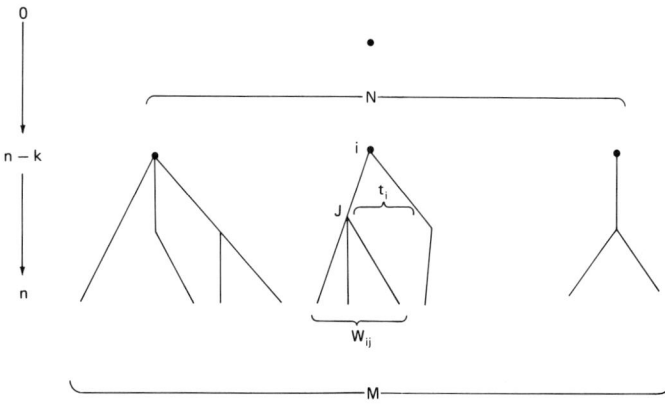

Fig. 6. Mutational Branching Pattern.

We of course have to start at generation $n-k$, where there are N cells, distributed according to

(4.3) $$f_{n-k}(x) = \sum p_{n-k,N} x^N.$$

If these have $t_1, \ldots, t_N$ descendents in generation $n - k + 1$, let W_{iJ} denote the number of descendents in generation n of the Jth "daughter" of cell number i, $J \leq t_i$. Since

(4.4) $$M = \sum_{j=1}^{N} \sum_{K=1}^{t_j} W_{jK},$$

$$\#_k = \sum_{i=1}^{N} \sum_{I \neq J}^{t_i} W_{iI} W_{iJ},$$

where $\#_k$ is the number of k-separated pairs, we must find

$$(4.5) \quad F_{nk}(x) = \sum_{N,\mathbf{t},\mathbf{W}} P(N,\mathbf{t},\mathbf{W}) \sum_{i=1}^{N} \sum_{I \neq J}^{t_i} W_{iI} W_{iJ}\, x^{\sum_{j=1}^{N} \sum_{K=1}^{t_j} W_{jK}}$$

Simplest is to decompose this into its contributions from each of the N channels:

$$(4.6) \quad F_{nk}(x) = \sum_{N} \sum_{M_1,\ldots,M_N} x^{\sum_{i=1}^{N} M_i} \sum_{i=1}^{N} E_{kk}(M_i) p_{n-k,N} \prod_{j \neq i} p_{k,M_j}$$

$$= F_{kk}(x) \sum_{N} N\, p_{n-k,N} (\sum p_{k,M} x^M)^{N-1}$$

$$= F_{kk}(x) f'_{n-k}(f_k(x)).$$

There remains only the computation of

$$F_{kk}(x) = \sum_{t,\mathbf{W}} P(t,\mathbf{W}) \sum_{I \neq J}^{t} W_I W_J\, x^{\sum_{K=1}^{t} W_K}$$

$$= \sum_{t} p_t \sum_{W_1,\ldots,W_t} \prod_{i=1}^{t} p_{k-1,W_i} \sum_{I \neq J}^{t} W_I W_J\, x^{\sum_{1}^{t} W_k}$$

$$= \sum_{t} t(t-1) p_t \prod_{i=3}^{t} (\sum_{W} p_{k-1,W} x^W)(\sum_{W} W p_{k-1,W} x^W)^2$$

$$= f''(f_{k-1}(x))(x f'_{k-1}(x))^2, \text{ and we conclude that}$$

$$(4.7) \quad F_{nk}(x) = (x f'_{k-1}(x))^2 f'_{n-k}(f_k(x)) f''(f_{k-1}(x))$$

Note that, as a special case, if we don't condition on M, then we can let $x = 1$; since $f_k(1) = 1$, $f'_k(1) = m^k$, $f''(1) = Q$, we then have

$$(4.8) \quad F_{nk} = m^{n+k-2} Q.$$

What we really want is the conditional probability, without the M-weight, i.e.

$$(4.9) \quad \begin{aligned} E_n(k|M) &= E_{nk}(M)/p_{n,M} \\ &= \frac{(x f'_{k-1}(x))^2 f'_{n-k}(f_k(x)) f''(f_{k-1}(x)) \|_M}{f_{n-k}(f_k(x)) \|_M}, \end{aligned}$$

with the notation $g(x)\|_M = \text{coef } x^M$ in $g(x)$. This simplifies materially in the "steady state" asymptotic limit $n \to \infty$, for we recall that

$$(4.10) \quad p_{n,s}/p_{n,1} \to \pi_s$$

converges, so that if

$$(4.11) \quad P(x) = \sum \pi_s x^s,$$

then (4.9) reduces to

$$(4.12) \qquad E(k|M) = \frac{(xf'_{k-1}(x))^2 f''(f_{k-1}(x)) P'(f_k(x)) \|_M}{P(f_k(x)) \|_M}.$$

Since $P(f_k(x)) = P(x) \lim p_{n+k,1}/p_{n,1}$, (4.12) transforms without difficulty to the neater form

$$(4.13) \qquad E(k|M) = \frac{x^2 P'(x) \frac{d}{dx} \ln f'(f_{k-1}(x)) \|_M}{P(x) \|_M}.$$

Hence, any decent information on $f_k(x)$, via fixed point or critical point expansions for example, yields an explicit estimate. Extension to true asynchronous division, it should be pointed out, is reasonably routine.

5. The Immune Network. Having seen a few trees, let us stand back and look at the forest. A basic picture of the immune response to bacterial antigen is shown [13] in Fig. 7.

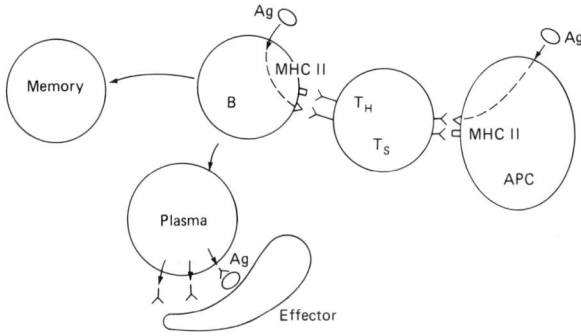

Fig. 7. Immune System Processing of Bacterial Antigen.

It starts with the antigen presenting cell, which must also present the host's signature in the form of a class II protein of the major histocompatibility complex (MHC). The activated T-helper (or T-suppressor for reverse action), on seeing the double antigen on the B-cell surface, incites B-cell division, and the process ends with the eradication of the Ag by an effector cell. A simpler response (Fig. 8) is also available for virally infected cells, which are recognized and killed by cytotoxic T-cells.

But not only foreign bodies act as antigen. So do antibodies and normal cell components. It is thought that reactability to the latter is eliminated during T-cell

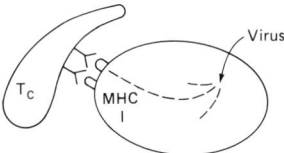

Fig. 8. Immune System Processing of Viral Antigen.

maturation (if they react, they don't proliferate), but the former creates a whole network of interactions [14]: The *paratope* part of antibody triggers a clonal explosion of its own B-cell when bound to antigen; other parts are *epitopes*, serving as antigens for other lymphocytes, and whose excision instead decreases proliferation of the host B-cell, perhaps to zero. Some epitopes are widespread, others specific to the antibody in question, and called *idiotopes*. This can apply to both B and T populations. An ultrasimplified model identifies a *functional unit* – the whole immune response – by its principal recognition site x (standing for molecular species), e.g. paratope for B-cell recognition, and asks how the concentration $\rho(x)$ of x-units grows or decays in response to all other environmental concentrations $\rho(y)$, assuming complete mixing to avoid any spatial dependence. Embedded within this picture is the old idiotype, anti-idiotype, anti-anti-idiotype, etc. picture.

Assuming neither direct competition nor synergism between sites on the functional unit, and no time delay across it (!), the kinetics of concentration growth might be expected to take the form [16]

(5.1) $$\dot\rho(x,t) = \alpha(x) - \mu(x)\rho(x,t) + \rho(x,t)[\int K(x,y)\rho(y,t)dy + a(x,t)]$$

where $\alpha(x)$ is the birth rate, $\mu(x)$ the death rate, and $\phi(x,t)$ the external antigen concentration. There is some evidence for symmetry [17]

(5.2) $$K(x,y) = K(y,x)$$

in which case we also have the variational form

(5.3) $$\dot\rho(x,t)/\rho(x,t) = \frac{\delta}{\delta\rho(x,t)}\{\int[\int_{\alpha_x(\rho)}^{\rho(x,t)}\frac{d\rho}{\rho}]dx \\ + \frac{1}{2}\iint K(x,y)\rho(x,t)\rho(y,t)dxdy + \int a(x,t)\rho(x,t)dx\},$$

where

(5.4) $$\alpha_x(\rho) = \alpha(x) - \mu(x)\rho$$

is the net birth rate. Steady state analysis has been made [18] for a number of special forms for $K(x, y)$, resulting in possible expansations of the transition from the virgin system to one with a ramified distribution of activated and low or high dose repressed (tolerant) concentrations after antigenic exposure. These models were typically in one-dimensional x-space, which seems a drastic oversimplification (although we have heard about quasi one-dimensionality of nets before in these proceedings). The connections are presumably on a space of very high dimensionality, so that a much better model might be a Cayley tree, or perhaps a Bethe lattice, to avoid imposed inhomogeneity and introduce presumed discreteness.

We can also recognize (5.3) as a mean field approximation to a simple model which can be analyzed in fair detail. The point is this. Dissipative bulk motion of a fluid follows a law of motion typically of the form

(5.5) $$\dot{\rho}(x,t) = -\gamma(x,t)\frac{\delta F[\rho(\cdot,t)]}{\delta \rho(x,t)},$$

where F is the Helmholtz free energy of the fluid, which can equally well be a lattice gas. If the lattice has a site-site interaction

(5.6) $$\phi(x,y) = \phi_0(x,y) - \phi_1(x,y),$$

where ϕ_0 is short and ϕ_1 long range, then in the presence of an external potential u, F appears in mean field approximation [19] as

(5.7) $$F[\rho] = \int f_0(\rho(x))dx - \frac{1}{2}\iint \phi_1(x,y)\rho(x)\rho(y)dxdy + \int u(x)\rho(x)dx,$$

where f_0 is the "core" specific free energy. Comparing with (5.3), we are thus led to the association

(5.8) $$\alpha(\rho) \sim -\rho f_0'(\rho)$$
$$K \sim \phi_1$$
$$a \sim -u.$$

Even the ideal gas $f_0 = \rho \ln \rho - \rho$ would give rise to $\alpha = -\rho \ln \rho$, tolerably similar to $\alpha - \mu\rho$, and an additional core term $\Delta f_0 = -\rho \ln(1 - b\rho)$ allows for fine tuning.

The major determinant in such models is the kernel K. In a very general empirical fashion, one would imagine that K would correspond to short-range repulsion, long-range attraction, precisely the form to give a phase transition in the steady state of (5.5). Thus, one can readily produce [16] a situation in which an initial

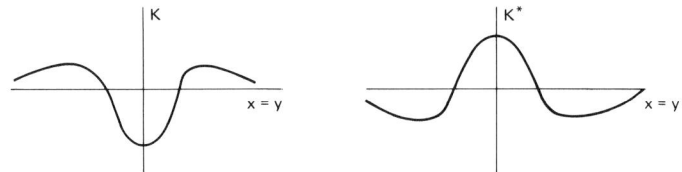

Fig. 9. Site-site Interactions in Single and Dual Fluid Models.

weakly unstable concentration density flips locally to a high or low holding pattern under antigenic stimulation.

But in fact, we know a little more biochemistry: molecules create a fit when they have complementary structure, i.e. are in part of dual structures. Hence the form [18]

(5.9) $$\iint K^*(x,y)\, \rho(x,t)\, \rho^*(y,t) dx dy$$

in (5.3) makes more sense, with K^* as in Fig. 9, and $\rho^*(y,t) \equiv \rho(y^*,t)$, where y^* is dual to y.

What all of this has to do with combinatorics is that, as suggested above, since (5.3) is the mean field approximation to a more simply defined lattice model, we may choose to investigate the latter instead. To indicate the possibilities, let us look at the most basic high-dimensional example, that of identical cores on a Bethe lattice with coordination number c ($c = 3$ in Fig. 10) and next nearest neighbor exclusion.

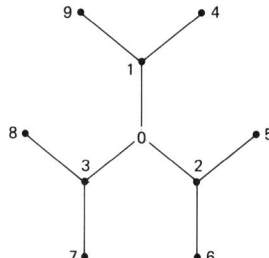

Fig. 10. Exclusion Zone for Site 0 on $c = 3$ Bethe Lattice.

The statistical state of this tree structure is most easily analyzed by quoting an appropriate phrasing of the potential distribution theorem [20,21]. Let $\nu(i) = (0,1)$

denote the stochastic occupation variable of site i, with each configuration given an intrinsic weight of $\exp \sum_i (\mu - u(i))\nu(i)$ (technically, $u(i)$ is external potential in temperature units and μ, the chemical potential, controls the overall level of occupancy), and define $\rho(i) = E(\nu(i))$. Then choosing a reference site 0, one has

$$(5.10) \qquad \rho(0)e^{(u(0)-\mu)} = E((1-\nu(0))\prod_{<0,j>}(1-\nu(j)))$$

$$\text{or } \mu - u(0) = \ell n\ \rho(0) - \ell n\ E((1-\nu(0))\prod_{<0,j>}(1-\nu(j)))$$

where j in $<0,j>$ runs over the exclusion neighborhood of 0.

The key to the evaluation of (5.10) is the observation that, fixing the occupation of any vector set whose surface with respect to its external complement is of minimum diameter two (surface vertices of the complement are at least third neighbors via paths in the fixed set) will decompose the complement into independent subsets. Let $|j|$ denote the distance from j to 0, and abbreviate $E(\prod_j(1-\nu_j))$ as $E(\prod_j j)$. Then since $1 - \nu(i)$ fixes site i at $\nu = 0$, and all kernels have value 0 or 1, we have

$$(5.11) \qquad E(\prod_{|j|\leq 2} j) = E(\prod_{|j|=2} j | \prod_{|j|\leq 1} j)\ E(\prod_{|j|\leq 1} j)$$

$$= \prod_{k=1}^{c} E(\prod_{j\in B_k}^{|j|=2} j | \prod_{|j|\leq 1} j) E(\prod_{|j|\leq 1} j),$$

where B_k is the kth arm of site 0. Continuing, this becomes

$$\prod_{k=1}^{c} E(\prod_{j\in B_k}^{|j|=2} j \prod_{|j|\leq 1} j)/E(\prod_{|j|\leq 1} j)^{c-1}$$

$$= \prod_{k=1}^{c} [E(\prod_{j=B_k}^{|j|=2} j \prod_{|j|=1}^{j\neq k} j|k_0)\ E(k0)]/E(\prod_{|j|\leq 1} j)^{c-1}$$

$$(5.12) \qquad = \prod_{k=1}^{c} [E(\prod_{j\in B_k}^{|j|=2} j|k0)\ E(\prod_{|j|=1}^{j\neq k} j|k0)\ E(k0)]/E(\prod_{|j|\leq 1} j)^{c-1}$$

$$= \prod_{k=1}^{c} [E(\prod_{j\in B_k}^{|j|\leq 2} j)\ E(\prod_{|j|\leq 1} j)/E(k0)]/E(\prod_{|j|\leq 1} j)^{c-1}$$

$$= \prod_{k=1}^{c} [E(\prod_{j\in B_k}^{|j|\leq 2} j)/E(k0)]\ E(\prod_{|j|\leq 1} j).$$

But each site cluster in (5.12) is of diameter ≤ 3, so that e.g. $E(\prod_{|j|\leq 1}(1-\nu(j))) = E(1-\sum_{|j|\leq 1}\nu(j))$, all $E(\nu(j)\nu(k))$, $k\neq j$, vanishing. We conclude that

$$(5.13) \qquad \mu - u(0) = \ell n\ \rho(0) - \sum_{k=1}^{c} \ell n(1 - \sum_{j\in B_k}^{|j|\leq 1}\rho(j)) - \ell n(1 - \sum_{|j|\leq 1}\rho(j))$$

$$+ \sum_{k=1}^{c} \ell n(1 - \rho(0) - \rho(k)).$$

Imagine the system as statistically uniform: $u(j) = 0$, $\rho(j) = \rho$. Then (5.13) reduces to

(5.14) $$\mu = \ln \rho - (c+1)\ln(1 - (c+1)\rho) + c \ln(1 - 2\rho)$$

Classical thermodynamics follows in the usual fashion, via the inverse compressibility relation

(5.15) $$\frac{dp}{d\rho} = \rho \frac{d\mu}{d\rho} = 1 + (c+1)^2 \frac{\rho}{1-(c+1)\rho} - 2c\frac{\rho}{1-2\rho}.$$

The significance of (5.15) is that, since $dP/d\rho > 0$ always, it shows that there is no two-phase instability built in. In other words, longer range attractive interaction is not mere Ising on the cake, which is no surprise. However, such attractions can be modeled in standard ways: making c formally less than one, applying a mean field tail, or literally appending negative site-site interactions. And of course, the dual fluid representation mentioned above can be incorporated without much fuss. It seems clear that the door is open to a wealth of simple models, equilibrium and dynamical, with more than marginal relevance to immune networks.

Acknowledgments

The work reported stems in considerably degree from joint investigations with M. Altman, S. Childress, and A. Robledo.

REFERENCES

(1) This is the framework adopted e.g. in J. K. Percus, "Combinatorial Methods," Springer, New York, 1971.
(2) A quite up-to-date and accessible survey is to be found in I. Roitt, J. Brostoff and D. Male, "Immunology," Mosby, 1985.
(3) See e.g., P. Jagers, "Branching Processes with Biological Applications," Wiley, 1975.
(4) See e.g., Th. L. Lentz, "The Cell Biology of Hydra," North-Holland, 1966.
(5) For an extended discussion, see K. B. Athreya and P. E. Ney, "Branching Processes," Springer-Verlag, 1972.
(6) See e.g., J. K. Percus, "Combinatorial Methods in Developmental Biology," Courant Inst. Math. Sci, 1977.
(7) L. M. Milne-Thomson, "The Calculus of Finite Differences," MacMillan, 1951, p. 341.
(8) J. K. Percus, "The Mathematics of Immunology," Courant Inst. Math. Sci., 1988.
(9) T. Suda, J. Suda, and M. Ogawa, Proc. Natl. Acad. Sci. USA **81** (1984), p. 2520.
(10) An extensive analysis is presented in C. A. Macken and A. S. Perelson, "A Branching Process Model of Cell Proliferation and Differentiation," Springer-Verlag, 1988.
(11) C. J. Mode, "Multiple Branching Processes," Elsevier, 1971.
(12) See reference 10 as well as C. A. Macken, A. S. Perelson, and C. C. Stewart, in "Modeling of Biomedical Systems," (editor, J. Eisenfeld and M. Witten), IMACS Transactions, 1985.
(13) See e.g., S. Tonegawa, Sci. American **253** (1985), p. 122.
(14) N. K. Jerne, Ann. Immuno. (Inst. Pasteur) **125C** (1974), p. 373.
(15) P.H. Richter, in "Theoretical Immmunology" (editors, Bell, Perelson and Pinkley), Dekker, 1978.
(16) J. K. Percus, in "Sante Fe Conference on Theoretical Immunology," (editor, A. S. Perelson), Addison-Wesley, 1988.
(17) G. W. Hoffmann and A. Cooper-Willis, in "Mathematical Modeling in Immunity and Medicine," North-Holland (editor, Marchuk and Belykh), 1983.
(18) L. Segel and A. S. Perelson, in reference 16.
(19) See e.g. J. K. Pecus and G. O. Williams, in "Fluid Interfacial Phenomena," (editor, C. A. Croxton) Wiley, 1986.
(20) J. L. Jackson and L. S. Klein, Phys. Fluids **7** (1964), p. 228.
(21) B. Widom, J. Chem. Phys. **39** (1963), p. 2808.

MEANINGLESS STATEMENTS, MATCHING EXPERIMENTS, AND COLORED DIGRAPHS

(APPLICATIONS OF GRAPH THEORY AND COMBINATORICS TO THE THEORY OF MEASUREMENT)

FRED S. ROBERTS†

Abstract. This paper considers applications of graph theory and combinatorics to the theory of measurement, an interdisciplinary subject developed with the goal of putting the process of measurement on a firm mathematical foundation. Many of the mathematical problems arising from measurement theory are interesting problems in graph theory and combinatorics. After presenting a brief introduction to measurement theory, this paper discusses three questions in measurement theory and the resulting mathematical problems. These problems deal with classifying automorphisms of colored digraphs, specifying certain invariant semiorders and indifference graphs, and identifying certain homogeneous order relations.

1. Introduction. Attempts to understand the nature of measurement have a long history. Some of the early influential works on the subject are those by Helmholtz [1887] and by Campbell [1920, 1928, 1938]. Modern attempts to put measurement on a firm mathematical foundation have their stimulus in the work of Stevens [1946, 1951, 1959] and in the foundational papers by Scott and Suppes [1958] and Suppes and Zinnes [1963]. The social and biological sciences have played a major role in stimulating the development of the theory of measurement, as it is these sciences which have given rise to totally new types of measurement scales. (Two examples of these are the semiorders introduced by Luce [1956], about which we shall have more to say below, and the scales arising in conjoint measurement that were introduced by Debreu [1960] and Luce and Tukey [1964].) As the modern theory of measurement has evolved, it has become a mathematically sophisticated subject. The mathematical techniques used by measurement theorists have included methods of algebra, analysis, functional equations, and probability theory. Methods of discrete mathematics have played an important role in the study of semiorders and the related interval orders (see below); in some problems involving the measurement of subjective probability (see for example Kraft, Pratt, and Seidenberg [1959], Luce [1967], and Scott [1964]); and in some work on algebraic difference measurement and conjoint measurement (see for example Scott [1964]). In recent years, an increasing number of mathematical problems arising from questions in measurement theory have been interesting problems in graph theory and combinatorics. It is the purpose of this paper to present a selection of questions in measurement theory and the resulting problems in graph theory and combinatorics.

The mathematical problems we present are concerned with identifying certain homogeneous order relations, specifying certain invariant semiorders and indifference graphs, and classifying automorphisms of colored digraphs. The paper by

---

†Department of Mathematics, Rutgers University, New Brunswick, N.J. 08903

Fishburn and Roberts [1989] surveys other such problems which arise from measurement theory and involve combinatorial and number-theoretic properties of certain sequences of integers.

The paper is organized as follows. In the next section, we present some general background in the theory of measurement. Section 3 studies the concept of a meaningful statement. Informally, a statement involving scales of measurement is meaningful if its truth value is unchanged whenever all the scales are replaced by other acceptable scales. The meaningfulness of scales which map into certain homogeneous relational systems is considered. Section 4 is concerned with conclusions from order and matching experiments, and the meaningfulness of such conclusions. Here, we consider statements involving the functions which provide real-valued representations of the order relations called semiorders and the graphs called indifference graphs. Section 5 introduces a systematic classification of scales of measurement through the use of two parameters describing a certain group of automorphisms, and studies this classification for colored digraphs. Finally, Section 6 provides some closing remarks.

2. The Theory of Measurement. The approach we will take to the theory of measurement is sometimes called the representational approach. It is summarized in the books by Pfanzagl [1968], Krantz, et al. [1971], Roberts [1979], Narens [1985], Suppes, et al. [1988], and Luce, et al. [1988].

It is clear that measurement has something to do with numbers (though a modern point of view is that this is not necessary). Examination of the paradigm examples of physics suggests that in measurement, we are assigning numbers to objects being measured in a way which "preserves" certain empirical relations. For instance, in measurement of temperature, we seek to preserve the relation "warmer than" and in measurement of mass the relation "heavier than." Thus, suppose A is a set of objects and R is a binary relation on A. The relation aRb is interpreted to mean that a is warmer than b, a is heavier than b, etc. Then we seek a real-valued function f on A so that for all a, b in A,

$$(1) \qquad aRb \leftrightarrow f(a) > f(b).$$

Equation (1) makes precise what we mean in these instances to preserve the relation R. In the case where R is interpreted as preference, f is called a *utility function* (or an *ordinal utility function*).

In measurement of mass, there is more being preserved than a relation R. Namely, there is an operation of combination of objects and this operation is being preserved in the sense that mass is additive over this operation. More precisely, suppose o is a binary operation, with aob interpreted to mean a combined with b. Then we ask for a function f which satisfies not only (1) but also

$$(2) \qquad f(aob) = f(a) + f(b).$$

There does not seem to be a similar requirement in measurement of temperature.[1] If we place such a requirement on the measurement of preference, we call the corresponding utility function a *cardinal utility function*.

Abstracting from these examples, we talk about a *relational system* $\mathcal{A} = (A, R_1, R_2, \ldots, R_p, o_1, o_2, \ldots, o_q)$, where A is a set, each R_i is a relation on A (not necessarily binary) and each o_i is a binary operation on A. We then introduce the concept of a homomorphism from $\mathcal{A}$ into another relational system $\mathcal{B} = (B, R'_1, R'_2, \ldots, R'_p, o'_1, o'_2, \ldots, o'_q)$, where R_i and R'_i are both n_i-ary relations (same n_i) and o_i and o'_i are both binary operations. A *homomorphism* is a function $f : A \to B$ which preserves all the relations and operations in the sense that for $i = 1, 2, \ldots, p$ and for all $a_1, a_2, \ldots, a_{n_i} \in A$,

$$R_i(a_1, a_2, \ldots, a_{n_i}) \leftrightarrow R'_i(f(a_1), f(a_2), \ldots, f(a_{n_i}))$$

and for $i = 1, 2, \ldots, q$ and for all $a, b \in A$,

$$f(a o_i b) = f(a) o'_i f(b).$$

For instance, in the way we have formulated the measurement of temperature, we are asking for a homomorphism from the relational system (A, R) into the relational system $(\mathbf{R}, >)$, and in the way we have formulated the measurement of mass, we are asking for a homomorphism from the relational system (A, R, o) into the relational system $(\mathbf{R}, >, +)$. In the case of mass, negative numbers do not arise, so we can also think of the homomorphism as being into $(\mathbf{R}^+, >, +)$, where $\mathbf{R}^+$ denotes the positive reals.[2]

Given a relational system $\mathcal{B}$ (usually with B a subset of $\mathbf{R}$), there are two basic problems of measurement theory. The first is the *representation problem*, which asks for conditions on $\mathcal{A}$ (necessary and) sufficient for the existence of a homomorphism from $\mathcal{A}$ into $\mathcal{B}$. A subproblem of this is to find an algorithm for obtaining the homomorphism. If f is a homomorphism from $\mathcal{A}$ into $\mathcal{B}$, $(\mathcal{A}, \mathcal{B}, f)$ is sometimes called a *scale*; sometimes just f is called a scale. This terminology is not consistent

---

[1] R.D. Luce (personal communication) has pointed out that in principle, one could think of combining two objects in certain situations. Suppose the identical objects are cubes of homogeneous material heated to varying degrees. We could place them in pairwise contact in an otherwise insulated environment, allow them to reach equilibrium, and assign a temperature half-way between the other two as the equilibrium temperature. However, for the combination operation to define a genuine operation, we would have to think of combining objects built up by a sequence of such combinations, and things could get messy. Besides, as Luce also points out, it would be difficult if not impossible to implement this scheme in practice.

[2] Note that although binary operations are ternary relations, they are sometimes singled out in the measurement theory literature because the notion of homomorphism as we have defined it does not treat a binary operation in the same way it would treat the corresponding ternary relation. The latter would require that

$$c = a o_i b \longleftrightarrow f(c) = f(a) o'_i f(b),$$

while we require only $\longrightarrow$. It might sometimes be necessary to treat functions of more than two arguments in this special way as well, though we have chosen not to do so. The whole issue disappears if we limit consideration to *isomorphisms*, i.e., one-to-one homomorphisms, as a number of authors do.

in the literature. For instance, following a convention initiated by Narens, Luce, et al. [1988] use the term *scale* for the collection of all homomorphisms from $\mathcal{A}$ into $\mathcal{B}$. A solution to the representation problem is called a *representation theorem*. The second basic problem of measurement theory is the *uniqueness problem*: Given a homomorphism from $\mathcal{A}$ into $\mathcal{B}$, how unique is it? A solution to the uniqueness problem is called a *uniqueness theorem*. A representation theorem gives conditions under which measurement can take place. A uniqueness theorem is used to classify scales and to determine what assertions arising from scales are meaningful, in a sense to be made precise shortly. We shall concentrate here on the theory of uniqueness.

Suppose f is a homomorphism from $\mathcal{A}$ into $\mathcal{B}$. An *admissible transformation* of f is a function $\phi : f(A) \to B$ so that $\phi \circ f$ is again a homomorphism from $\mathcal{A}$ into $\mathcal{B}$. Examples of admissible transformations are the transformations which change pounds into grams, inches into feet, and degrees Fahrenheit into degrees Centigrade. To give another example, note that the function $f(x) = 3x$ is a homomorphism from $(\mathsf{R}, >)$ into $(\mathsf{R}, >)$. The function $\phi : \mathsf{R} \to \mathsf{R}$ defined by $\phi(x) = x + 10$ is an admissible transformation of f, since $\phi \circ f(x) = 3x + 10$ is again a homomorphism.

Stevens [1946, 1951, 1959] had the idea that the class of admissible transformations could be used to classify the scale types of homomorphisms. The types of scales introduced by Stevens have found widespread use in the social and biological sciences. We briefly introduce them here. It should be mentioned that the Stevens notion of scale type only applies when all homomorphisms are related by admissible transformations. In this case, we say that the homomorphisms or scales are *regular*. Since most scales of measurement encountered in practice are regular (semiorders are an exception), the Stevens theory is quite widely applicable.

We say that a homomorphism f is a *ratio scale* if the class of admissible transformations of f consists exactly of the similarities $\phi(x) = \alpha x$ for some real number $\alpha > 0$. In a ratio scale, we may change only the unit. Examples of ratio scales are mass, temperature (on the absolute or Kelvin scale), and time intervals (such as years, seconds, etc.). We say f is an *interval scale* if the class of admissible transformations of f consists exactly of the positive linear transformations $\phi(x) = \alpha x + \beta$ for some $\alpha > 0$ and β real numbers. In an interval scale, we may change both the unit and the zero point. Examples of interval scales are temperature and time in a calendar sense (this is the year 1988). We say that f is an *ordinal scale* if the class of admissible transformations consists exactly of the strictly monotone increasing functions ϕ on $f(A)$. The Mohs scale of hardness is an example of an ordinal scale. Here, minerals receive an integer between 1 and 10, with the only significance of these integers being that if a receives a higher number than b, then a scratches (is harder than) b. Clearly any 10 numbers ordered in the same way could replace these 10 integers. (If a scale is thought of as a family of homomorphisms, then scale type can be defined in a similar way, as a characteristic of the family under function composition. See Luce, et al. [1988].)

Many uniqueness theorems in measurement theory are formulated to give the scale type of the resulting homomorphism. Examples of such uniqueness theorems are the following. (For proofs, see Roberts [1979].) If f is a homomorphism from

(A, R) into $(\mathbf{R}, >)$, then f defines an ordinal scale. If f is a homomorphism from (A, R, o) into $(\mathbf{R}, >, +)$, then f defines a ratio scale. The latter gives mass measurement as a ratio scale, which is what we expect. However, the former gives temperature measurement as an ordinal scale, which is not what we expect. In examining the former result, one observes that in measurement of temperature, we are capable not only of comparing temperatures in a relative sense, but of comparing temperature differences in a relative sense. Then one talks about a quaternary relation D on a set A, with $D(a, b, c, d)$ interpreted to mean that the difference in warmth between a and b seems greater than that between c and d. We then seek to preserve the relation (A, D) by finding a real-valued function f on A so that for all a, b, c, d in A,

$$D(a, b, c, d) \leftrightarrow f(a) - f(b) > f(c) - f(d).$$

This function f is a homomorphism from (A, D) into $(\mathbf{R}, \Delta)$, where

$$\Delta(x, y, u, v) \leftrightarrow x - y > u - v.$$

If f is such a homomorphism, one can prove, under some additional hypothesis such as a solvability hypothesis, that f defines an interval scale. This is what we expect about temperature.

Another way to look at the uniqueness of a measurement scale or homomorphism is to think of an admissible transformation as an automorphism of the image relational system $\mathcal{B}$ and to classify scales by certain properties of the group of these automorphisms. An *automorphism* of $\mathcal{B}$ is a one-to-one onto homomorphism from $\mathcal{B}$ to $\mathcal{B}$. For instance, an automorphism of the system $(\mathbf{R}, >)$ is a one-to-one onto function from $\mathbf{R}$ to $\mathbf{R}$ so that

$$a > b \leftrightarrow f(a) > f(b),$$

i.e., an onto monotone increasing function. Similarly, the automorphisms of the system $(\mathbf{R}^+, >, +)$ are exactly the similarities and the automorphisms of the system $(\mathbf{R}, \Delta)$ are exactly the positive linear transformations.

One of the most important applications of the theory of uniqueness in measurement is concerned with making sense of what assertions using scales of measurement are meaningful. A statement involving scales of measurement is meaningful if its truth or falsity is not an artifact of the particular versions of scales which are used. Formally, assuming all scales in question are regular, we say that a statement involving scales of measurement is *meaningful* if its truth value is unchanged whenever every scale is modified by an admissible transformation. This definition goes back to Suppes [1959] and Suppes and Zinnes [1963]. If scales can be irregular, it needs to be modified to the following more general form: A statement involving scales of measurement is *meaningful* if its truth value is unchanged whenever every scale is replaced by another acceptable scale. These definitions are reasonably well accepted. However, in certain situations they do not give a reasonable concept. For a discussion, see Roberts [1985], Falmagne and Narens [1983], and Luce, et al. [1988].

To illustrate the concept of meaningfulness, consider the statement S: "a weighs twice as much as b." S can be written as the statement

(3) $$f(a) = 2f(b).$$

Now (3) is meaningful if for every admissible transformation ϕ of f, (3) holds if and only if

(4) $$\phi \circ f(a) = 2\phi \circ f(b).$$

Since weight is a ratio scale, $\phi(x) = \alpha x$, for some real number $\alpha > 0$. Hence, (4) becomes

(5) $$\alpha f(a) = 2\alpha f(b),$$

which of course is equivalent to (3). Hence, the statement (3) is meaningful. Thus, for instance, it is meaningful to say that I weigh twice what the elephant in the circus weighs. Presumably this statement is false in pounds, and also in grams, kilograms, and any other measure of weight. Meaningfulness is not the same as truth. Meaningfulness is concerned with whether or not an assertion based on scales of measurement is an accident of the particular scales used.

Suppose next that S is the statement: "a is twice as warm as b." This statement can again be written as (3), but now (4) becomes

(6) $$\alpha f(a) + \beta = 2[\alpha f(b) + \beta],$$

and it is quite possible for (3) to hold and (6) to fail. For instance, a can be twice as warm as b in Fahrenheit, but not in Centigrade.

In general, for interval scales, such as temperature, it is meaningful to assert that $f(a) - f(b) > f(c) - f(d)$; that is, we can compare intervals. For ratio scales, we can compare ratios: $f(a)/f(b) > f(c)/f(d)$ is meaningful. For ordinal scales, we can compare size: $f(a) > f(b)$ is meaningful.

There are many interesting and subtle applications of the theory of meaningfulness in the literature. Some important applications involve dimensional invariance in physics, psychophysical scaling, statistical tests, index numbers, performance analysis of students, performance testing of alternative computer systems, and the understanding of the basic forms of scientific laws. For a survey of such applications, see Roberts [1985].

3. Systematic Study of the Meaningfulness of the Assertion f(a) > f(b). As the first example of a problem in measurement theory which has given rise to interesting graph–theoretical and combinatorial problems, we mention the problem of systematically studying the meaningfulness of different statements involving scales of measurement. Such a study would take a particular statement, and try to understand what properties of homomorphisms or their domain and range relational systems lead to meaningfulness of the statement. Not much is known about this problem, except for the statement $f(a) > f(b)$. (For general results about other statements, which show that the meaningfulness of a variety of generic statements involving scales is equivalent to the corresponding scales having a certain scale type, see Roberts and Rosenbaum [1986].)

In this section and the next, we consider the meaningfulness of the statement $f(a) > f(b)$. Here, we consider the case where f is a homomorphism from (A, R) into (R, S), where R and S are M-ary relations. To say that f is a homomorphism from (A, R) into (R, S) means that f is a real-valued function on A so that for all $a_1, a_2, \ldots, a_M \in A$,

$$(a_1, a_2, \ldots, a_M) \in R \leftrightarrow (f(a_1), f(a_2), \ldots, f(a_M)) \in S.$$

If f is a homomorphism from (A, R) into (R, S), the statement $f(a) > f(b)$ is meaningful if for all other homomorphisms g from (A, R) into (R, S),

$$f(a) > f(b) \leftrightarrow g(a) > g(b).$$

Even the meaningfulness of the comparison $f(a) > f(b)$ in the situation in question is not well understood without an additional assumption about the relation S. We shall make a special assumption about S which is of current interest in the measurement theory literature and to which we return in Section 5. This assumption, originally due to Narens [1981B], is that (R, S) is *M-point homogeneous* in the sense that whenever

(7) $\qquad x_1 > x_2 > \cdots > x_M \quad \text{and} \quad y_1 > y_2 > \cdots > y_M$

for real numbers x_i and y_i, there is an automorphism ϕ of (R, S) so that $\phi(x_i) = y_i$, $i = 1, 2, \ldots, M$. In this sense, $(\mathsf{R}, >)$ is M-point homogeneous for all M, since the automorphisms of $(\mathsf{R}, >)$ are the monotone increasing functions from R onto R and given any real numbers as in (7), there is always such a monotone increasing function which takes each x_i into the corresponding y_i. To give another example, (R, Δ) is 2-point homogeneous. To see why, recall that the automorphisms are all positive linear transformations $\phi(x) = \alpha x + \beta$, $\alpha > 0$, and given any $x_1 > x_2$ and $y_1 > y_2$, there always is such a positive linear transformation which maps x_1 into y_1 and x_2 into y_2. It follows that (R, Δ) is 1-point homogeneous. However, (R, Δ) is not 3-point homogeneous. To give a third example, $(\mathsf{R}^+, >, +)$ is 1-point homogeneous but not 2-point homogeneous, since the automorphisms are the similarities $\phi(x) = \alpha x, \alpha > 0$. (Note that M-point homogeneity is like *M-transitivity*, which requires that given any unordered, distinct $x_1, x_2, \ldots, x_M$ and unordered, distinct $y_1, y_2, \ldots, y_M$, there be an automorphism ϕ such that $\phi(x_i) = y_i$, $i = 1, 2, \ldots, M$.)

Before understanding what homomorphisms into M-point homogeneous systems (R, S) lead to meaningful comparisons $f(a) > f(b)$, we need to understand what M-ary relations (R, S) are M-point homogeneous. Suppose $\underline{M} = \{1, 2, \ldots, M\}$. A *ranking* of $\underline{M}$ is a strict weak order, that is, a linear order with ties, for instance

1,4,5,2-3,6, where the hyphen indicates 2 and 3 are tied. Suppose π is a ranking of $\underline{M}$. Define T_π^M on **R** to be the set of all M–tuples $(x_1, x_2, \ldots, x_M)$ so that the greater than relation on $\{x_1, x_2, \ldots, x_M\}$ induces the ranking π on $\underline{M}$. Thus, for instance, if $M = 6$ and π is the ranking 3,6,1-4,2,5, then (51,40,100,51,12,98) is in T_π^6. If $M = 2$ and π is 1,2, then T_π^2 is the $>$ relation. If $M = 2$ and π is 1-2, then T_π^2 is the $=$ relation. The following theorem is the desired characterization.

THEOREM 1 (ROBERTS [1984B]). *If S is an M-ary relation, then (**R**, S) is M–point homogeneous if and only if $S = \phi$ or*

(8) $$S = T_{\pi_1}^M \cup T_{\pi_2}^M \cup \cdots \cup T_{\pi_p}^M$$

for some $p \geq 1$.

It follows Theorem 1 that if $M = 2$, the possible M–point homogeneous M-ary relations (**R**, S) are given by the following list:

$$\phi; \quad T_{1,2}^2 = >; \quad T_{2,1}^2 = <; \quad T_{1-2}^2 = =; \quad T_{1,2}^2 \cup T_{1-2}^2 = \geq;$$
$$T_{1,2}^2 \cup T_{2,1}^2 = \neq; \quad T_{1-2}^2 \cup T_{2,1}^2 = \leq; \quad T_{1,2}^2 \cup T_{1-2}^2 \cup T_{2,1}^2 = \mathbf{R} \times \mathbf{R}.$$

To give another example, $T_{1,2,3}^3 \cup T_{3,2,1}^3$ is 3–point homogeneous. This is the betweenness relation.

What Theorem 1 omits, and what we omit in the study of meaningfulness, is the case of M–point homogeneous N-ary relations where $M \neq N$. The characterization of such relations remains an open problem.

We now present some sample results on the meaningfulness of conclusions $f(a) > f(b)$ when f is a homomorphism from (A, R) into (**R**, S) and the latter is an M–point homogeneous M-ary relation. We shall see that the results depend on combinatorial properties of the set of rankings π_i appearing in the right hand side of (8). To state these results, we use the notion of the *height* $h(\pi)$ of the ranking π, i.e., the number of distinct levels in π. Thus, 3,6,1-4,2,5 has height 5. The first theorem applies to the case $p = 1$ in Theorem 1.

THEOREM 2 (ROBERTS [1984B]). *Suppose $h(\pi) \geq 2$, f is a homomorphism from (A, R) into $(\mathbf{R}, T_\pi^M)$, and $|f(A)| \geq h(\pi)$. Then $f(a) > f(b)$ is meaningful for all a, b in A.*

The next few theorems allow $p > 1$.

THEOREM 3 (HARVEY AND ROBERTS [1988]). *Suppose*

$$S = T_{\pi_1}^M \cup T_{\pi_2}^M \cup \cdots \cup T_{\pi_p}^M$$

and suppose that for all $1 \leq r, s \leq p$ and all $1 \leq i, j \leq M$, i over j in π_r implies i over j in π_s or i, j tied in π_s. Suppose f is a homomorphism from (A, R) into $(\mathbf{R}, S)$. Suppose $h = \min\{h(\pi_r) : 1 \leq r \leq p\}$, $h \geq 2$, and $|f(A)| \geq h$. Then $f(a) > f(b)$ is meaningful for all a, b in A.

Theorem 3 is illustrated by taking $\pi_1 = 1, 2, 3$, $\pi_2 = $1-2,3, and $\pi_3 = 1, 2$-3.

THEOREM 4 (HARVEY AND ROBERTS [1988]). *Suppose $S = T^M_{\pi_1} \cup T^M_{\pi_2}$ and $h(\pi_1) > h(\pi_2)$. Suppose f is a homomorphism from (A, R) into $(\mathbf{R}, S)$ with $|f(A)| \geq h(\pi_1)$. Then $f(a) > f(b)$ is meaningful for all a, b in A.*

Theorem 4 is illustrated by taking $\pi_1 = 1, 2, 3$ and $\pi_2 = 2-3, 1$. Note that the hypothesis $h(\pi_1) > h(\pi_2)$ is needed. For instance, if $\pi_1 = 1,2,3$ and $\pi_2 = 3,2,1$, then $f(a) > f(b)$ is meaningless for all a, b. Simply note that $g(x) = -f(x)$ is another homomorphism.

THEOREM 5 (HARVEY AND ROBERTS [1988]). *Suppose $i \neq j$, π is some ranking of $\underline{M} - \{i, j\}$, π_1 is i, j, π, π_2 is j, i, π, and $S = T^M_{\pi_1} \cup T^M_{\pi_2}$. Suppose f is a homomorphism from (A, R) into $(\mathbf{R}, S)$ with $|f(A)| \geq 2$, $|f(A)| < \infty$. Then $f(a) > f(b)$ is meaningless for all a, b in A.*

Theorem 5 is illustrated by taking $\pi_1 = 5, 2, 4, 1\text{-}3$ and $\pi_2 = 2, 5, 4, 1\text{-}3$.

It is now known (Harvey and Roberts [1988]) when $f(a) > f(b)$ is meaningful in all cases $T^M_{\pi_1} \cup T^M_{\pi_2}$ when $M = 2$ or 3. The general case of S of the form (8) and M arbitrary is still open. Other interesting open questions involve the meaningfulness of $f(a) > f(b)$ for f a homomorphism into $(\mathbf{R}, S)$ for S M–point homogeneous M-ary and into $(\mathbf{R}, S_1, S_2, \ldots, S_p)$ for $S_1, S_2, \ldots, S_p$ M-point homogeneous M-ary. Of course, similar problems for other statements involving scales are also interesting open problems.

4. Analysis of Order and Matching Experiments. A variety of experiments involve judgements such as "I am indifferent between a and b," "a and b are similar," "a and b are the same," or "a and b match." We shall call these *matching experiments*. Related experiments involve judgements like "I prefer a to b," "a is bigger than b," or "a is better than b." We shall call these *order experiments*. A topic of recent interest is the meaningfulness of conclusions like $f(a) > f(b)$ when f somehow measures the results of an order or a matching experiment.

As is observed in the overview paper by Roberts in this volume, observed matching or indifference is often not transitive; moreover, if matching or indifference is not transitive and aRb means that a is ordered higher than (preferred to) b, then there is no function f satisfying (1). This observation led to Luce's [1956] idea of a semiorder, and to Luce's attempt to find conditions on a binary relation (A, R) necessary and sufficient to guarantee that there is a real–valued function f on A so that for all $a, b, \in A$,

(9) $$aRb \leftrightarrow f(a) > f(b) + \delta,$$

where δ is a fixed positive number measuring threshold or just–noticeable difference. It was proved by Scott and Suppes [1958] that if A is finite, there is a function f satisfying (9) if and only if (A, R) defines a semiorder. (See the overview paper Roberts [1989] for the definition of semiorder and see Fishburn [1985], Roberts [1979], or Trotter [1988], for many references on semiorders and the related order relations called interval orders.) Note that this result shows that the particular value of δ is irrelevant. Note also that we can think of a function f satisfying (9) as a homomorphism from (A, R) into $(\mathbf{R}, >_\delta)$, where $x >_\delta y$ holds if and only if $x > y + \delta$. Then the Scott–Suppes Theorem is a representation theorem in the sense of Section 2. There can be no uniqueness theorem in the sense of capturing the class of admissible transformations of f, because f is not a regular scale. To see why, suppose

(10) $$A = \{u, v, w\}, \quad R = \{(w, v), (w, u)\}.$$

Then two functions satisfying (9) with $\delta = 1$ are given by

(11) $\quad f(u) = 0, \; f(v) = 0, \; f(w) = 2; \quad g(u) = .1, \; g(v) = 0, \; g(w) = 2.$

However, there is no admissible transformation which takes f into g. For some uniqueness results in this situation see Roberts [1979].

Corresponding to the representation (9) for order is the representation for matching or indifference which results if we assume that a and b match (or we are indifferent between a and b) if and only if we order neither greater (prefer neither). In symbols, if I means matching or indifference, we assume

$$aIb \leftrightarrow \text{not } aRb \; \& \; \text{not } bRa.$$

In this case, we seek a real–valued function f on A so that for all $a, b \in A$,

(12) $$aIb \leftrightarrow |f(a) - f(b)| \leq \delta.$$

A binary relation (A, I) for which there exists an f satisfying (12) is called an *indifference graph*. We may think of an indifference graph as a graph (vertices correspond to elements of A, edges to pairs in the relation I) for which we can assign numbers to the vertices so that two vertices are adjacent if and only if the corresponding numbers are close. Indifference graphs were introduced and characterized by Roberts [1969]. (See Roberts [1978], Golumbic [1980], and Fishburn [1985] for many references on indifference graphs and the equivalent class of graphs called unit interval graphs.)

We shall be concerned here with the question: If f satisfies (9) or (12), when is the statement $f(a) > f(b)$ meaningful? For indifference graphs, this is obviously meaningless for all a, b. For if f satisfies (12), then so does $-f$. The statement can be meaningless or meaningful for semiorders. For instance, suppose that (A, R) is given in (10). Then f and g given by (11) satisfy (9), but $g(u) > g(v)$ while $f(u)$ is not $> f(v)$. Thus, the statement that $f(u) > f(v)$ is meaningless. Next suppose that $A = \{u, v, w\}$, $R = \{(u, v), (v, w), (u, w)\}$. Then whenever f satisfies (9), we have $f(u) > f(v) + \delta$ and hence $f(u) > f(v)$. Thus, $f(u) > f(v)$ is meaningful. In fact, $f(a) > f(b)$ is meaningful for all $a, b \in A$.

Generalizing this example, let us define a binary relation E on A by taking

$$aEb \leftrightarrow (\forall c)[aRc \leftrightarrow bRc \; \& \; cRa \leftrightarrow cRb].$$

If (A, R) is a semiorder, then (A, E) is an equivalence relation. In (10) above, uEv.

THEOREM 6 (ROBERTS [1984A]). *Suppose f is a homomorphism on (A, R) satisfying (9). Then $f(a) > f(b)$ is meaningful for all $a, b \in A$ if and only if for all $a \neq b \in A$, not aEb.*

Turning to indifference graphs, recall that we cannot expect $f(a) > f(b)$ to be meaningful. However, we might expect betweenness judgements to be meaningful. Let us say that y is *between* x and z if $x < y < z$ or $z < y < x$. If f satisfies (12), is it meaningful to conclude that $f(b)$ is between $f(a)$ and $f(c)$? In answering this question, we observe that (12) implies that aIa for all $a \in A$, i.e., every object matches itself. (In the graph, there is a loop at each vertex.) Assuming aIa for all a, we then define E on A by

$$aEb \leftrightarrow (\forall c)[aIc \leftrightarrow bIc].$$

If (A, I) is an indifference graph, then (A, E) is an equivalence relation. (In terms of the graph, aEb means that a and b are adjacent and have the same closed neighborhoods.)

THEOREM 7 (ROBERTS [1984A]). *Suppose f is a homomorphism on (A, I) satisfying (12) and suppose that $|f(A)| \geq 3$. Then the conclusion that $f(b)$ is between $f(a)$ and $f(c)$ is meaningful for all $a, b, c \in A$ if and only if the following conditions hold:*

(a)   *For all $a \neq b \in A$, not aEb;*
(b)   *The graph (A, I) is connected.*

We conclude that for matching experiments, betweenness conclusions can be drawn, but only under certain conditions.

No other results about betweenness conclusions for other measurement scales are known. They would be of interest, for example, for scales mapping into M-point homogeneous systems $(\mathbf{R}, S)$.

5. Classification of Scales by Properties of Automorphism Groups. As we pointed out in Section 2, a recent trend in the study of the uniqueness of scales of measurement involves classification of scales by considering properties of the group of automorphisms of the image relational system $\mathcal{B}$. This idea has its beginnings in a series of papers initiated by Narens and was subsequently worked on by Alper and Luce. See Alper [1985, 1987], Luce [1986, 1987], Luce and Narens [1983, 1984, 1985, 1986, 1987], and Narens [1981 A,B]. (Luce and Narens [1984, 1986, 1987] are expository papers and Luce and Narens [1983, 1985] primarily include applications of the scale classification ideas.) Alper, Luce, and Narens consider the case where $\mathcal{B}$ is a relational system of the form $(B, \succ, R_2, R_3, \ldots, R_p)$, where $(B, \succ)$ is a strict weak order (a linear order with ties). Such a relational system $\mathcal{B}$ is called *ordered*. Alper, Luce, and Narens classify ordered relational systems by studying two properties of the group of automorphisms: M-point homogeneity and N-point uniqueness. This work is summarized in Luce, et al. [1988, Chapter 20] and in Luce and Narens [1986, 1987]. We shall discuss the resulting classification of scales.

if $(B, \succ)$ is a strict weak order, it is useful to define $\succsim$ by

$$a \succsim b \leftrightarrow \text{not } b \succ a$$

and $\sim$ by

$$a \sim b \leftrightarrow \text{not } a \succ b \ \& \ \text{not } b \succ a.$$

The relation $(B, \sim)$ is an equivalence relation, and we denote by $B*$ the corresponding collection of equivalence classes.

We have already defined M–point homogeneity in a particular context and we now redefine it in this more abstract setting. Suppose $M \leq |B*|$. We say that the ordered relational system $\mathcal{B} = (B, \succ, R_2, R_3, \ldots, R_p)$ is *M–point homogeneous* if whenever

$$x_1 \succ x_2 \succ \ldots \succ x_M \text{ and } y_1 \succ y_2 \succ \ldots \succ y_M,$$

there is an automorphism ϕ of $\mathcal{B}$ so that $\phi(x_i) = y_i$, $i = 1, 2, \ldots, M$. In this sense, as we have already observed above, $\mathcal{B} = (\mathbf{R}, >)$ is M–point homogeneous for every M and $\mathcal{B} = (\mathbf{R}^+, >, +)$ is 1-point homogeneous, but not M–point homogeneous for any $M > 1$. Moreover, $(\mathbf{R}, >, \Delta)$ is 2-point homogeneous and 1-point homogeneous, but not M–point homogeneous for any $M > 2$. The reasoning is the same as that for $(\mathbf{R}, \Delta)$, since $(\mathbf{R}, >, \Delta)$ and $(\mathbf{R}, \Delta)$ have the same automorphisms.

We define the *degree of homogeneity* $h(\mathcal{B})$ to be the largest M so that $\mathcal{B}$ is M–point homogeneous. We take $h(\mathcal{B})$ to be ∞ if $\mathcal{B}$ is M–point homogeneous for arbitrarily large M, and $h(\mathcal{B})$ to be 0 if $\mathcal{B}$ is never M–point homogeneous. Of course, $h(\mathcal{B}) \leq |B*|$. By what we have already observed,

$$h[(\mathbf{R}, >)] = \infty, \ h[(\mathbf{R}^+, >, +)] = 1, \ h[(\mathbf{R}, >, \Delta)] = 2.$$

We say that $\mathcal{B}$ is *N–point unique* if whenever $x_1, x_2, \ldots, x_N$ are distinct elements of B, and ϕ and ϕ' are two automorphisms of $\mathcal{B}$ so that $\phi(x_i) = \phi'(x_i)$, $i = 1, 2, \ldots, N$, then $\phi(x) = \phi'(x)$ for all $x \in B$; that is, if an automorphism is determined by its values on any N points. Some motivation for this definition is provided by the ordered relational system $\mathcal{B} = (Z, >)$, where Z is the set of integers. Note that ϕ is an automorphism of $\mathcal{B}$ if and only if $\phi(x) = x + p$, p an integer. Then an automorphism is determined by its value on any one point, so $\mathcal{B}$ is 1–point unique. Similarly, $(\mathbf{R}^+, >, +)$ is 1-point unique, since an automorphism $\phi(x) = \alpha x$, $\alpha > 0$, is determined by its value on one point. Finally, $(\mathbf{R}, >, \Delta)$ is 2-point unique but not 1-point unique.

We define the *degree of uniqueness* $u(\mathcal{B})$ to be the smallest N so that $\mathcal{B}$ is N–point unique. We take $u(\mathcal{B})$ to be ∞ if $\mathcal{B}$ is never N–point unique. We have

$$u[(Z, >)] = 1, \ u[(\mathbf{R}^+, >, +)] = 1, \ u[(\mathbf{R}, >, \Delta)] = 2, \ u[(\mathbf{R}, >)] = \infty.$$

The problem that Narens posed in general terms is to classify ordered relational systems $\mathcal{B}$ by giving the pair (M, N), where $M = h(\mathcal{B})$ and $N = u(\mathcal{B})$. This pair (M, N) is called the *scale type* of $\mathcal{B}$. He specifically started research on classifying the scale types for ordered relational systems under the special assumption that there is an onto isomorphism (one–to–one homomorphism) from $(B, \succ)$ to $(\mathbf{R}, >)$. Note that for such an ordered relational system, $h(\mathcal{B}) \leq u(\mathcal{B})$.

The following theorem was discovered in pieces.

THEOREM 8 (NARENS [1981A,B], ALPER [1985, 1987]). *Suppose $\mathcal{B}$ is an ordered relational system with $(B, \succ)$ isomorphic onto $(\mathbf{R}, >)$. Then $\mathcal{B}$ can have scale type (M, N) with $1 \leq M \leq N < \infty$ if and only if one of the following holds:*

(a) $M = N = 1;$ (b) $M = N = 2;$ (c) $M = 1, \ N = 2.$

There really is more to the Narens–Alper result. It was also shown that $\mathcal{B}$ is isomorphic to a real relational system whose automorphism group is in case (a) the group of similarities $\phi(x) = \alpha x, \alpha > 0$; in case (b) the group of positive linear transformations $\phi(x) = \alpha x + \beta$, $a > 0$; and in case (c) a proper subgroup of the group of positive linear transformations that contains the group of similarities.

The rather surprising conclusion obtained by Narens and Alper is that in the setting of Theorem 8, there are very few scale types. When one obtains a result like this, one has two contradictory reactions. The first is that the result is exciting and that the classification has probably captured something fundamental about the objects being classified, in this case certain kinds of ordered relational systems. In considering this reaction, one is led to wonder whether the result is truly interesting or whether similar results would be obtained again and again if the classification method were applied in other settings. The latter result would suggest that the classification method is not particularly interesting. However, as we point out below, that is not the case here. The second reaction to a result such as Theorem 8 is that it is too bad that there are not more classes, because a small number of classes does not let us distinguish very much among possible situations. Although a result which leads to a small number of classes is still interesting and useful, one is naturally led to ask if there is a way of refining the classification to distinguish within a class. Thus, one is led to ask whether there are other parameters which could be introduced which would allow us to distinguish among elements which agree on the first two parameters. This will not be possible for the (1,1) and (2,2) cases if the parameters are based solely on the automorphism group of $\mathcal{B}$, since all (i, i) systems in the setting of Theorem 8 have the same automorphism group.

Theorem 8 does not cover the *non–homogeneous cases*, those where $M = 0$. The classification of non–homogeneous cases seems to be rather a complex problem (cf. Alper [1987]). However, many important measurement structures (including most finite relational systems, most semiorders, and probability structures) are non–homogeneous and so an understanding of the ordered relational systems which have different scale types $(0, N)$ would be very helpful.

Let us now address the question: Are there other ordered relational systems where the Narens theory of scale type leads to a classification of homogeneous ($M \geq 1$) cases in which many more types are attainable? It turns out that the answer is affirmative for the special ordered relational systems which correspond to colored digraphs. The results make Theorem 8 all the more interesting and, at the same time, raise a variety of interesting and important graph–theoretical questions.

If one wants to study ordered relational systems, the simplest type to study is the ordered relational system $(B, \succ)$, a strict weak order. The next simplest type is the ordered relational system $\mathcal{B} = (B, \succ, R)$, where R is a binary relation. We can think of (B, R) as a digraph. In trying to organize practical comparison data and to understand the patterns into which it falls, it is natural to consider the case where B is a finite set. It suffices for what follows to assume that $B*$ is finite. In this case, instead of considering $\succ$, we can think of placing a *value* $v(a)$, a real number, on each vertex a of the digraph, with

$$a \succ b \leftrightarrow v(a) > v(b).$$

The function v takes on only finitely many different values. This can always be done if $B*$ is finite. Indeed, if $B*$ is finite, the existence of such a function v is equivalent to $\succ$ defining a strict weak order. (See Roberts [1979].) If R is a binary relation on B and v is a real-valued function on B, the triple (B, R, v) is called a *valued digraph*. In this paper, by our assumption that $B*$ is finite, the function v will take on only finitely many values. Figure 1 shows a valued digraph, with the values indicated as numbers placed next to each vertex. Valued digraphs arise when we make binary comparisons between alternatives which have a hopefully-related natural ordering relation, for instance if we compare the problem-solving abilities of students of differing $I.Q.$'s or the nutritional ranking of cuts of meat of differing grades.[3] We now consider the Narens classification for valued digraphs.

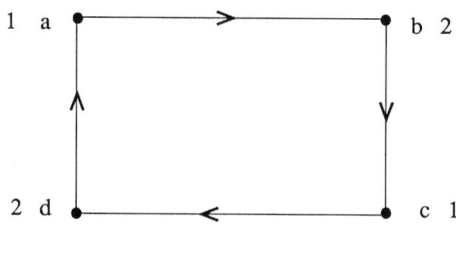

Figure 1

What is an automorphism of a valued digraph? It is a function ϕ from B onto B so that for all $a, b \in B$.

(13) $$aRb \leftrightarrow \phi(a)R\phi(b)$$

and

(14) $$a \succ b \leftrightarrow \phi(a) \succ \phi(b),$$

or equivalently

(15) $$v(a) > v(b) \leftrightarrow v(\phi(a)) > v(\phi(b)).$$

---

[3] R.D. Luce (personal communication) has pointed out that the motivation for valued digraphs is analogous to the idea of system constants in the theory of similar systems in dimensional analysis. (See Causey [1969]; see also Luce [1971] or Krantz, et al. [1971, Chap. 10].)

Now (13) says that ϕ is a digraph automorphism. Since $B*$ is finite, (15) says that $v(\phi(a)) = v(a)$ for all $a \in B$, i.e., ϕ is value–preserving. Hence, the order of the numbers $v(a)$ is irrelevant and we can think of a valued digraph as a colored digraph (colors corresponding to numbers). Then automorphisms are color–preserving digraph automorphisms. In Figure 1, the function ϕ which takes a into c, b into d, c into a, and d into b, is an automorphism of the valued digraph. However, the function ϕ which takes a into b, b into c, c into d, and d into a, is a digraph automorphism which is not a valued digraph automorphism since $v(\phi(a)) \neq v(a)$.

M–point homogeneity for a valued digraph means that if

$$v(x_1) > v(x_2) > \cdots > v(x_M) \quad \text{and} \quad v(y_1) > v(y_2) > \cdots > v(y_M),$$

then there is a valued digraph automorphism ϕ so that $\phi(x_i) = y_i, i = 1, 2, \ldots, M$.

Note that if $v(a) \neq v(b)$ for all $a \neq b$, then $u(\mathcal{B}) = 0$: The only automorphism is the identity. However, $u(\mathcal{B})$ can be interesting if this condition is violated, and we assume that it is. It is also useful to make a technical assumption. We say that $\mathcal{B}$ is *2-tight homogeneous* if

$$v(x_1) \neq v(x_2), v(x_1) = v(y_1), v(x_2) = v(y_2)$$

implies that there is an automorphism ϕ of $\mathcal{B}$ so that $\phi(x_1) = y_1$ and $\phi(x_2) = y_2$. Clearly, 2-point homogeneity implies 2–tight homogeneity. However, it is easy to show that the converse is false.

THEOREM 9 (ROBERTS AND ROSENBAUM [1988]). *Suppose $\mathcal{B}$ is a valued digraph, $v(a) = v(b)$ for some $a \neq b$, and $\mathcal{B}$ is 2-tight homogeneous. Then the possible scale types (M, N) for $\mathcal{B}$ are all possible pairs (M, N) such that $1 < M \leq N$.*

The reader should contrast the number of scale types attainable in this situation to the number attainable in the situation studied by Alper, Luce, and Narens. Dropping the 2-tight homogeneous assumption does not change this conclusion about the *number* of scale types attainable, though the exact scale types possible without this assumption have not been worked out. A result similar to Theorem 9 is obtained by Cameron [1987] in the setting of Theorem 8 when one considers rational relational systems. For other related results, see Roberts and Rosenbaum [1985]. The fact that the Narens classification can lead to a large number of scale types in the context of Theorem 9 and to a small number in the context of Theorem 8 makes Theorem 8 all the more interesting.

Theorem 9 suggests a variety of questions which remain open. First, one would like to characterize the valued digraphs of a given scale type (M, N). Second, one would like to obtain similar results without the assumption of finiteness of $B*$. Third, one would like to obtain similar results for more general ordered relational systems with $B*$ finite. Fourth, one would like to obtain results about the cases $(0, N)$ and $(1, N)$ and a complete understanding of the possible scale types (M, N) if $\mathcal{B}$ is not 2-tight homogeneous. Fifth, one would like to obtain similar results for modifications of the notions of homogeneity and uniqueness. M–tight homogeneity can be defined for arbitrary M in a manner analogous to M–point homogeneity, and a tight degree of homogeneity can then be defined. Several other variants on the notion of homogeneity and similar variants on the notion of uniqueness give rise to a whole collection of parameters, for each of which a similar classification theorem is called for. See Roberts and Rosenbaum [1988] for precise definitions of these variants on the concepts of homogeneity and uniqueness defined here.

6. Concluding Remarks. We have introduced a variety of problems in graph theory and combinatorics which arise in the theory of measurement. The problems we have discussed here have each only been investigated in quite preliminary form. There remain many interesting open questions, quite a few of which have been mentioned in the text.

It is safe to say that the theory of measurement will give rise to similar interesting problems in the future. Moreover, solutions to these problems and new concepts introduced in the course of developing these solutions will be useful for measurement theorists as they continue their attempts to understand the nature of measurement.

Acknowledgements. The author thanks Garth Isaak, Suh–ryung Kim, Wenzhong Li, Duncan Luce, and Barry Tesman for their helpful comments. He also thanks the National Science Foundation for its support under grant number IST–86–04530 to Rutgers University.

REFERENCES

ALPER, T.M., *A Note on Real Measurement Structures of Scale Type* $(m, m+1)$, J. Math. Psychol., 29 (1985), 73–81.

ALPER, T.M., *A Classification of all Order–Preserving Homomorphism Groups of the Reals that Satisfy Finite Uniqueness*, J. Math. Psychol., 31 (1987), 135–154.

CAMERON, P.T., *Groups of Order–Automorphisms of the Rationals with Prescribed Scale Types*, manuscript, 1987.

CAMPBELL, N.R., *Physics: the Elements*, Cambridge University Press, Cambridge, 1920. Reprinted as *Foundations of Science: The Philosophy of Theory and Experiment*, Dover, New York, 1957.

CAMPBELL, N.R., *An Account of the Principles of Measurement and Calculation*, Longmans, Green, London, 1928.

CAMPBELL, N.R., *Symposium: Measurement and its Importance for Philosophy*, Aristotelian Society, Suppl. Vol. 17, Harrison, London, 1938.

CAUSEY, R.L., *Derived Measurement, Dimensions, and Dimensional Analysis*, Phil. Sci., 36 (1969), 252–270.

DEBREU, G., *Topological Methods in Cardinal Utility Theory*, in K.J. Arrow, S. Karlin, and P. Suppes (eds.), *Mathematical Methods in the Social Sciences*, Stanford University Press, Stanford, California, 1960, 16–26.

FALMAGNE, J–C., AND NARENS, L., *Scales and Meaningfulness of Quantitative Laws*, Synthese, 55 (1983), 287–325.

FISHBURN, P.C., *Interval Orders and Interval Graphs*, Wiley–Interscience, New York, 1985.

FISHBURN, P.C., AND ROBERTS, F.S., *Uniqueness in Finite Measurement*, F.S. Roberts (ed.), *Applications of Combinatorics and Graph Theory to the Biological and Social Sciences*, Springer–Verlag, New York, 1989.

GOLUMBIC, M.C., *Algorithmic Graph Theory and Perfect Graphs*, Academic Press, New York, 1980.

HARVEY, L.H., AND ROBERTS, F.S., *On the Theory of Meaningfulness of Ordinal Comparisons in Measurement II*, Annals NY Acad. Sci., 1988, to appear.

HELMHOLTZ, H.V., *Zahlen und Messen*, in *Philosophische Aufsätze*, Fues's Verlag, Leipzig, 1887, pp. 17–52. Translated by C.L. Bryan,"Counting and Measuring," Van Nostrand, Princeton, NJ, 1930.

KRAFT, C.H., PRATT, J.W., AND SEIDENBERG, A., *Intuitive Probability on Finite Sets*, Ann. Math. Stat., 30 (1959), 408–419.

KRANTZ, D.H., LUCE, R.D., SUPPES, P., AND TVERSKY, A., *Foundations of Measurement*, Vol. I, Academic Press, New York, 1971.

LUCE, R.D., *Semiorders and a Theory of Utility Discrimination*, Econometrica, 24 (1956), 178–191.

LUCE, R.D., *Sufficient Conditions for the Existence of a Finitely Additive Probability Measure*, Ann. Math. Stat., 38 (1967), 780–786.

LUCE, R.D., *Similar Systems and Dimensionally Invariant Laws*, Phil. Sci., 38 (1971), 157–169.

LUCE, R.D., *Uniqueness and Homogeneity of Ordered Relational Structures*, J. Math. Psychol., 30 (1986), 391–415.

LUCE, R.D., *Measurement Structures with Archimedean Ordered Translation Groups*, Order, 4 (1987), 165–189.

LUCE, R.D., KRANTZ, D.H., SUPPES, P., AND TVERSKY, A., *Foundations of Measurement*, Vol. III, Academic Press, New York, 1988, in press.

LUCE, R.D., AND NARENS, L., *Symmetry, Scale Types, and Generalizations of Classical Physical Measurement*, J. Math. Psychol., 27 (1983), 44–85.

LUCE, R.D., AND NARENS, L., *Classification of Real Measurement Representations by Scale Type*, Measurement, 2 (1984), 39–44.

LUCE, R.D., AND NARENS, L., *Classification of Concatenation Measurement Structures According to Scale Type*, J. Math. Psychol., 29 (1985), 1–72.

LUCE, R.D., AND NARENS, L., *Measurement: The Theory of Numerical Assignments*, Psychol. Bull., 99 (1986), 166–180.

LUCE, R.D., AND NARENS, L., *Measurement Scales on the Continuum*, Science, 236 (1987), 1527–1532.

LUCE, R.D., AND TUKEY, J.W., *Simultaneous Conjoint Measurement: A New Type of Fundamental Measurement*, J. Math. Psychol., 1 (1964), 1–27.

NARENS, L., *A General Theory of Ratio Scalability with Remarks about the Measurement–theoretic Concept of Meaningfulness*, Theory and Decision, 13 (1981), 1–70. (A).

NARENS, L., *On the Scales of Measurement*, J. Math. Psychol., 24 (1981), 249–275. (B).

NARENS, L., *Abstract Measurement Theory*, MIT Press, Cambridge, MA, 1985.

PFANZAGL, J., *Theory of Measurement*, Wiley, New York, 1968.

ROBERTS, F.S., *Indifference Graphs*, in F. Harary (ed.), *Proof Techniques in Graph Theory*, Academic Press, New York, 1969, 139–146.

ROBERTS, F.S., *Graph Theory and its Applications to Problems of Society*, CBMS–NSF Monograph No. 29, SIAM, Philadelphia, 1978.

ROBERTS, F.S., *Measurement Theory, with Applications to Decisionmaking, Utility, and the Social Sciences*, Addison–Wesley, Reading, MA, 1979.

ROBERTS, F.S., *Applications of the Theory of Meaningfulness to Order and Matching Experiments*, in E. DeGreef and J. Van Buggenhaut (eds.), *Trends in Mathematical Psychology*, North–Holland, Amsterdam, 1984, 283–292. (A).

ROBERTS, F.S., *On the Theory of Meaningfulness of Ordinal Comparisons in Measurement*, Measurement, 2 (1984), 35–38. (B).

ROBERTS, F.S., *Applications of the Theory of Meaningfulness to Psychology*, J. Math. Psychol., 29 (1985), 311–332.

ROBERTS, F.S., *Applications of Combinatorics and Graph Theory to the Biological and Social Sciences: Seven Fundamental Ideas*, in F.S. Roberts (ed.) *Applications of Combinatorics and Graph Theory to the Biological and Social Sciences*, Springer–Verlag, New York, 1989.

ROBERTS, F.S., AND ROSENBAUM, Z., *Some Results on Automorphisms of Ordered Relational Systems and the Theory of Scale Type in Measurement*, in Y. Alavi, et al. (eds.), *Graph Theory and its Applications to Algorithms and Computer Science*, Wiley, New York, 1985, 659–669.

ROBERTS, F.S., AND ROSENBAUM, Z., *Scale Type, Meaningfulness, and the Possible Psychophysical Laws*, Math. Soc. Sci., 12 (1986), 77–95.

ROBERTS, F.S., AND ROSENBAUM, Z., *Tight and Loose Value Automorphisms*, Discrete Appl. Math., 22 (1988), 69–79.

SCOTT, D., *Measurement Models and Linear Inequalities*, J. Math. Psychol., 1 (1964), 233–247.

SCOTT, D., AND SUPPES, P., *Foundational Aspects of Theories of Measurement*, J. Symbolic Logic, 23 (1958), 113–128.

STEVENS, S.S., *On the Theory of Scales of Measurement*, Science, 103 (1946), 677–680.

STEVENS, S.S., *Mathematics, Measurement, and Psychophysics*, in S.S. Stevens (ed.), *Handbook of Experimental Psychology*, Wiley, New York, 1951, 1–49.

STEVENS, S.S., *Measurement, Psychophysics, and Utility*, in C.W. Churchman and P. Ratoosh (eds.), *Measurement: Definitions and Theories*, Wiley, New York, 1959, 18–63.

SUPPES, P., *Measurement, Empirical Meaningfulness and Three–Valued Logic*, in C.W. Churchman and P. Ratoosh (eds.), *Measurement: Definitions and Theories*, Wiley, New York, 1959, 129–143.

SUPPES, P., KRANTZ, D.H., LUCE, R.D., AND TVERSKY, A., *Foundations of Measurement*, Vol. II, Academic Press, New York, 1988, in press.

SUPPES, P., AND ZINNES, J.,, *Basic Measurement Theory*, in R.D. Luce, R.R. Bush, and E. Galanter (eds.), *Handbook of Mathematical Psychology*, Vol. 1, Wiley, New York, 1963, 1–76.

TROTTER, W.T., *Interval Graphs, Interval Orders and their Generalizations*, in R.D. Ringeisen and F.S. Roberts (eds.), *Applications of Discrete Mathematics*, SIAM, Philadelphia, 1988, 45–58.

COMBINATORIAL ASPECTS OF ENZYME KINETICS

PETER H. SELLERS*

Abstract. Two concepts from chemistry, definable in mathematical terms, are the starting point of this paper: A *reaction network* (which is a generalization of a graph) and a *mechanism for a reaction* (which is a generalization of a path from one vertex to another in a graph). Then, as the main result, a statement made in 1964 by P.C. Milner [4] is put into precise terms and proved. To paraphrase Milner's statement, a mechanism for a reaction r in a given network reduces to the superposition of two or more consistently oriented direct mechanisms for r from the same network, where direct mechanisms are capable of no such reduction. This result is the principal justification of an algorithm, described in this paper, for generating a list of all possible direct mechanisms for a given reaction in a given network. Examples are used to show how these ideas apply to enzyme kinetics studies.

1 Introduction. The principal object of chemical kinetics to determine the mechanisms of chemical reactions. The combinatorial aspect of this subject arises from the fact that a chemical mechanism is a flow network whose underlying structure can be pictured as a finite set of polygonal faces, meeting at their edges, as illustrated in figures 6 through 10. This structure is characterized by the incidence relations between its vertices and edges. It is described in geometrical terms only for illustrative purposes.

The main theorem in this paper shows that any mechanism can be decomposed into a sum of irreducible parts, called *direct mechanisms*. The proof is constructive and illuminates the character of direct mechanisms. It supports the idea, proposed by Milner [4], that if a list is required of all possible mechanisms for a given reaction, then it is sufficient to list only the direct mechanisms. Any system has only finitely many direct mechanisms for a given reaction, and, as indicated by Milner, it is a useful matter to list them.

The language and methods of this paper come from algebraic topology, which lends itself to describing higher dimensional networks. Accordingly, a reaction with a certain mechanism will be described as the boundary of a certain *chain*, and a mechanism which produces no reaction as a *cycle*. We shall not use *homology groups*, but an appendix is added showing that they are not necessarily trivial, and thus can be a significant feature of a chemical system.

2 Networks. A chemical system, viewed as a collection of elementary chemical reactions linked by the flow of molecules from one to another, is an example of a kind of network which cannot be represented adequately by a graph or a hypergraph. The expression *reaction network* is well-established in chemistry, but there is nothing about the physical appearance of a mixture of reacting chemicals which fits the word *network* literally, so let us figure out what sort of mathematical structure we can use to describe it.

---

*Department of Mathematics, The Rockefeller University, 1230 York Avenue, NY 10021-6399

It is certainly an oversimplification to take the view that every elementary reaction in a chemical system is an isomerization, where one species of molecule simply changes to another one. Such a reaction is represented in chemistry by an equation

$$a = b.$$

In this case a graphical representation is possible, since every species can be represented by a node and every reaction by an edge or a chain of edges. However, it remains necessary to find a representation of a chemical system which will include isomerizations as a special case, but will extend to the more usual elementary reaction where two species combine to form a third one, or two species result from the dissociation of a third one. Such reactions are represented by equations like

$$a + b = c \quad \text{or} \quad 2a = a_2.$$

If we wish to express the idea that these reactions have proceeded by one unit from left to right, we would write

$$-a - b + c \quad \text{or} \quad -2a + a_2$$

and change signs for the opposite direction.

If it is assumed that an elementary reaction can involve any finite set of species, then it is possible to view it as an edge in a hypergraph and each species as a vertex, but it turns out that the theory of chain complexes, rather than the theory of hypergraphs, provides the appropriate way to generalize the theory of graphs for our purposes. This can be seen by comparing the cycle concept in the two theories.

In the special case of graphs a *cycle* can be defined either as a closed succession of edges or as a chain whose boundary is zero. For instance, a succession AB, BC, CA of graph-theoretic edges constitutes a cycle, because it forms the perimeter of a triangle which is an example of a closed succession of edges. It is also a cycle in the sense that the boundary ∂ of the chain

$$AB + BC + CA$$

in the theory of simplicial complexes is zero:

$$\partial(AB + BC + CA) = \partial(AB) + \partial(BC) + \partial(CA) = -A + B - B + C - C + A = 0$$

Now let us consider an example in which each edge has more than two vertices:

Example 1. Consider a chemical system in which it is observed that the reaction

$$ATP + G = ADP + GP$$

does not occur an an appreciable rate except in the presence of the enzyme E (hexokinase). Let us simplify the above notation and write

$$DP + G = D + GP.$$

If E is present, let us assume (for the sake of this example) that the reaction takes place in four steps:

$$S_1 : DP + E = DEP$$
$$S_2 : DEP + G = DEGP$$
$$S_3 : DEGP = DE + GP$$
$$S_4 : DE = D + E$$

If DP and G are supplied to the system at a steady rate and D and GP carried off at the same rate, then the other species $DEP, DEGP, DE$ and E are recycled at the same rate. Then $S_1 + S_2 + S_3 + S_4$ is called the *mechanism* for the given reaction. What is the combinatorial structure of such a mechanism?

One model is hypergraph with the 8 chemical species

$$D,\ DE,\ DEGP,\ DEP,\ DP,\ E,\ G,\ GP$$

as vertices and the four steps S_1, S_2, S_3 and S_4 as hyperedges. In the theory of hypergraphs, as presented by Berge [1], these steps constitute a cycle and would be portrayed as in Figure 1.

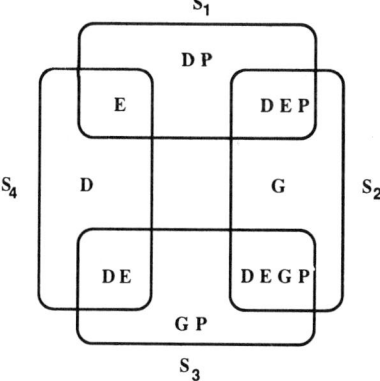

Figure 1. A cyclic hypergraph.

Another model is a chain complex based on the idea of a boundary operator ∂. Here the boundary of a step is defined as "output minus input," which is formalized thus:

$$\partial(S_1) = -DP - E + DEP$$
$$\partial(S_2) = -DEP - G + DEGP$$
$$\partial(S_3) = -DEGP + DE + GP$$
$$\partial(S_4) = -DE + D + E$$

It follows by addition that the *chain* $S_1 + S_2 + S_3 + S_4$ is not a cycle, because its boundary is not zero:

$$\partial(S_1 + S_2 + S_3 + S_4) = -DP - G + D + GP$$

In fact, this example can have no non-zero cycles, because its four elementary reactions $\partial(S_1), \partial(S_2), \partial(S_3)$ and $\partial(S_4)$ are linearly independent. Note that a distinction is maintained between the physical *steps* S_1, S_2, S_3, S_4 and the *elementary reactions* $\partial(S_1), \partial(S_2), \partial(S_3), \partial(S_4)$ which they produce.

The application of chain complexes to this kind of problem has been discussed elsewhere [2, 3]. The emphasis here will be on a problem which seems to have reached the limit of what can be achieved algebraically, so we will try a more combinatorial approach. It is the problem of constructing a list of all the chains of steps by which a given chemical reaction can proceed in a given chemical network. It is analogous to, but not as obvious as, the problem of finding all paths from one vertex to another in a graph-theoretic network.

The word *pathway* is not entirely appropriate in a network in which there are flows that cannot be seen as advancing over one-dimensional curves. Therefore, let us use the terminology of chemistry where it would be said that the chain of steps $S_1 + S_2 + S_3 + S_4$, described in example 1, is a *mechanism* for the reaction

(1) $$DP + G = D + GP.$$

This mechanism is a process which converts one unit of DP and one of G into one of D and one of GP, or vice versa. If we wish to specify the direction of the reaction, say from left to right, we express it as

(2) $$-DP - G + D + GP.$$

Reactions have integral coefficients which give the number of units produced or expended of each species. A reaction, written this way, exists in a chemical system if, and only if, it is equal to the boundary of some mechanism in the system.

Notation (1) for a chemical reaction is called a *chemical equation*, and it holds in a chemical system if the two sides differ by a boundary. This kind of equation conforms exactly with the definition of homological equivalence in a chain complex. Notation (2) is what chemists sometimes call a *reaction vector*, since it displays the direction of a reaction and can be multiplied through by a scalar to indicate the magnitude of the reaction. Ordinarily reactions have integer coefficients, so that strictly speaking they do not constitute a vector space, and, if we want to treat them as such, we must extend the coefficients at least to rational numbers.

3 Direct Mechanisms. In the context of graph theory a mechanism reduces very simply to a path from one vertex to another or several such paths. Let us define a *direct path* as one which goes from one vertex to another and does not intersect itself anywhere. The important thing about such paths, that is, the aspect we would like to generalize to mechanisms in a network, is that they do not contain any unnecessary detours. For instance, the mechanism $S_1 + S_2 + S_3 + S_4$ in example 1 does not contain any superfluous steps in the process of converting $DP + G$ to $D + GP$, but it is meaningless to say that it follows a non-intersecting path, because it does not have the sequential character associated with paths in graphs and hypergraphs. The following definition is applicable to paths in a graph as well as mechanisms. A mechanism is *direct* if no distinct mechanism for the same reaction can be formed from a subset of its steps. This definition applies to cycles, because they are mechanisms of a special kind: Namely, mechanisms whose reactions are equal to zero. But in the case of a cycle we have to say that it is direct if no distinct cycle can be formed from a subset of its steps *except* zero and that a multiple of the cycle is not regarded in this instance as being a *distinct* from it.

Direct mechanisms are analogous to paths in a graph which are not self-intersecting.

THEOREM 1. *A mechanism for a non-zero reaction is direct if, and only if, no cycle can be formed from a subset of its steps.*

Proof. If there were a subset of its steps which formed a cycle, we could subtract the cycle from the mechanism and get a distinct mechanism for the same reaction, which contradicts the definition of directness. Conversely, if a mechanism is not direct, it does yield two mechanisms for the same reaction, whose difference is a cycle.

The next theorem shows that a mechanism for reaction r decomposes into direct submechanisms for r and direct cycles within the context of a given chemical network. For the purposes of this theorem a network is nothing but a homomorphism ∂ from a free Abelian group generated by a finite set of *steps* to a free Abelian group generated by a finite set of *species*. The elements in the domain of ∂ are called *mechanisms*, those in the kernel of ∂ are called *cycles* and those in the image of ∂ are called *reactions*.

THEOREM 2. *Let m be a mechanism for a non-zero reaction r in a given network; then there exist direct mechanisms for r*

$$m_1, m_2, \cdots$$

direct cycles

$$z_1, z_2, \cdots$$

and positive integers

$$\mu = \mu_1 + \mu_2 + \cdots .$$

such that m decomposes as follows:

$$\mu m = \mu_1 m_1 + \mu_2 m_2 + \cdots + z_1 + z_2 + \cdots ,$$

and, whenever a step appears in more than one summand of this decomposition, it appears with the same sign.

In some instances a mechanism can be partitioned into direct mechanisms and direct cycles which are disjoint, in the sense of having no steps in common. What theorem 2 tells us, is that, when such a partition is not possible, we can, nevertheless, choose parts such that where they overlap all steps will be advancing in the same directions. Although, in general, the decomposition is not unique, the set of all direct mechanisms for a given reaction is a unique attribute of a network, as is the set of all direct cycles. The direct mechanisms for a given reaction are of particular interest in practice, so in the remainder of this paper we will concentrate on the problem of listing such mechanisms. The first time, to my knowledge, that this problem was addressed in the context of chemical networks was in a paper on electro-chemistry by P.C. Milner [4] in which the term *direct mechanism* was introduced. The proof of theorem 2, which follows, uses a geometrical point-of-view to make it more intuitive than a purely algebraic proof would be. This geometrical method was introduced in reference [5].

Proof. Without loss of generality in proving theorem 2, we can allow the mechanisms and reactions to have rational coefficients. Then the network under consideration is a linear transformation ∂ from the vector space $\mathcal{M}$ of all linear combinations of steps to the vector space $\mathcal{N}$ of all linear combinations of molecular species. A mechanism m for a non-zero reaction r is given. In other words, $m \in \mathcal{M}$ and $r = \partial(m) \neq 0$. Since we wish to reduce m to a linear combination of certain cycles and certain mechanisms for r, let us begin by noticing that the set of all cycles in $\mathcal{M}$ equals $ker \partial$, and the set of all mechanisms for r is equal to the coset $(m + ker \partial)$. By definition $z \in ker \partial$ if $\partial(z) = 0$. A typical element $m + z$ of the coset is a mechanism for r, because

$$\partial(m + z) = r.$$

Conversely, any mechanism m' for r can be written as

$$m + (m' - m)$$

This is in the coset, because $m' - m$ is a cycle.

Let $x_1, x_2, \ldots, x_C$ be a basis for $ker \, \partial$; then any mechanism for r can be written as

(3) $$m + \phi_1 x_1 + \phi_2 x_2 + \cdots + \phi_C x_C.$$

If $S_1, S_2, \cdots, S_M$ are all the steps, then (3) can be rewritten as

(4) $$\sum_{I=1}^{M} L_I(\phi_1, \phi_2, \cdots, \phi_C) S_I$$

where L_I is an affine transformation.

It is clear from (3) that every point $(\phi_1, \cdots, \phi_C)$ in a C-dimensional vector space $\mathcal{K}$ represents a unique mechanism for r and that all such mechanisms are so represented. It is clear from (4) that $\mathcal{K}$ has a family of $(C-1)$-dimensional subspaces $\mathcal{L}_1, \cdots, \mathcal{L}_C$ defined as follows:

$$\mathcal{L}_I = \left\{ (\phi_1, \cdots, \phi_C) \in \mathcal{K} | L_I(\phi_1, \cdots, \phi_C) = 0 \right\}$$

$\mathcal{L}_1$ represents all mechanism for r which do not involve the step S_1 and partitions $\mathcal{K}$ into three parts: The C-dimensional region in which S_1 has a positive coefficient, the C-dimensional region where S_1 has a negative coefficient, and finally the $(C-1)$-dimensional region $\mathcal{L}_1$ which separates the first two. $\mathcal{L}_2$ creates another partition separating each of the above regions into at most 3 parts, with the result that $\mathcal{K}$ is partitioned by $\mathcal{L}_1$ and $\mathcal{L}_2$ into at most 9 parts, whose dimensions are $C, C-1, C-2$. Continuing thus up to $\mathcal{L}_C$, we arrive at a partition of $\mathcal{K}$ into a set of convex regions of all dimensions from 0 to C, some of them bounded and some unbounded.

Notice that each point $(\phi_1, \cdots, \phi_C)$ in $\mathcal{K}$ has a signature (sign $\phi_1, \cdots,$ sign ϕ_C) where sign ϕ equals $+, 0$ or $-$. Two points are in the same region of $\mathcal{K}$ if, and only if, they have the same signature. Furthermore, if two points in the same region are joined by a line segment, the equation of such a line tells us that all intermediate points have the same signature as the end-points, from which it follows that every region is convex.

In every region of dimension greater than zero there are infinitely many points, any two of which represent mechanisms for r involving exactly the same set of steps. Each zero-dimensional region is an isolated point representing mechanism for r which does not have the same set of steps as any other mechanism for r. Therefore, each of these isolated points represents a direct mechanism for r and, no other point in $\mathcal{K}$ does so.

For any $k \in K$ let $m(k)$ denote the mechanism represented by k. Therefore, if $k_1, k_2, \cdots$ are the elements of the zero-dimensional regions, then $m(k_1), m(k_2), \cdots$ are the direct mechanisms for r.

Now let us turn our attention to the mechanism m for r which we wish to decompose into direct mechanism for r and direct cycles. Let k be the point in $\mathcal{K}$ which represents m i.e.,

$$m(k) = m$$

and let I be the dimension of the region which contains k. This region will have faces of all dimensions on its boundary, including at least $I + 1$ zero-dimensional faces consisting of points from among $k_1, k_2, \cdots$. The closure of the region is a convex I-dimensional polyhedron $\mathcal{P}$. If it has $I+1$ vertices, then it is an I-simplex. Otherwise, we can choose a subset of the vertices of $\mathcal{P}$ to determine a simplex in $\mathcal{P}$ which contains k and has dimension I or less.

Given points $l_0, \cdots, l_I$ which span an I-dimensional space, the I-simplex with these points as vertices is the set

$$\{(\lambda_0 l_0 + \cdots + \lambda_I l_I)\}$$

where $\lambda_0, \cdots, \lambda_I$ are non-negative coefficients, known as barycentric coordinates, such that

$$\lambda_0 + \cdots + \lambda_I = 1.$$

Any $(J+1)$-element subset of the vertices determine a J-simplex defined as J-dimensional face of the I-simplex.

Making use of the convexity of all regions, we can construct the desired simplex by choosing its vertices thus: First, take any vertex l of $\mathcal{P}$ and run a line from l through k to a point k' in some face $\mathcal{P}'$ of $\mathcal{P}$. Second, take any vertex l' of $\mathcal{P}'$ and run a line from l' through k' to a point k'' in some face $\mathcal{P}''$ of $\mathcal{P}'$. Continuation of this process leads to a succession of vertices $l, l', l'', \cdots$, which determine a simplex within $\mathcal{P}$ containing k. Then, using the barycentric coordinate system of this simplex, we can express k as follows with positive coefficients

$$k = \lambda l + \lambda' l' + \lambda'' l'' + \cdots$$

where $\lambda + \lambda' + \lambda'' + \cdots = 1$. Therefore,

$$m = m(k) = \lambda m(l) + \lambda' m(l') + \lambda'' m(l'') + \cdots,$$

where $m(l), m(l'), \cdots$ are direct mechanisms.

Therefore, we may conclude that, if k is in one of the bounded regions of $\mathcal{K}$, then the mechanism $m(k)$ decomposes into direct mechanisms, and no cycles have to be included in its decomposition. These are the mechanisms which are of principal interest in chemical applications. Other mechanisms, having cycles in their decomposition, are generalizations of the paths in a graph-theoretic network which are self-intersecting. They contain loops, and it is intuitively obvious if you remove loops from a path that you make it more "direct."

To conclude this proof, we start with a point k in one of the unbounded regions and show that there exists l in one of the bounded regions such that the cycle $m(k) - m(l)$ is decomposable into direct cycles. This, together with the fact that $m(l)$ is decomposable into direct mechanisms, proves that $m(k)$ is decomposable as stated in the theorem, because

$$m(k) = m(l) + \bigl(m(k) - m(l)\bigr).$$

Let $\mathcal{P}$ be the unbounded polyhedron equal to the closure of the region of k. Follow a line through k parallel to an unbounded edge of $\mathcal{P}$ until it strikes a face $\mathcal{P}'$ of $\mathcal{P}$ at k'. If $\mathcal{P}'$ is unbounded, follow a line through k' parallel to an unbounded edge of $\mathcal{P}'$ until it strikes a face $\mathcal{P}''$ of $\mathcal{P}'$ at k''. Continue thus, constructing the sequence $k, k', k'', \cdots$ until a point l is reached which lies in a bounded face. These points determine the following decomposition of $m(k-l)$:

$$m(k - k') + m(k' - k'') + \cdots.$$

The proof is completed by showing that these summands are direct cycles. It is clear that they are cycles and that, being within $\mathcal{P}$, any step which appears in two of them will have the same sign in both.

The points k and k' lie on a straight line parallel to a one-dimensional region in $\mathcal{K}$. These regions are in turn parallel to new one-dimensional regions in $\mathcal{K}$ formed as follows: Modify formula (3) to express any cycle in the form

(5) $$\phi_1 x_1 + \phi_2 x_2 + \cdots + \phi_C x_C.$$

Formula (4) becomes

(6) $$\sum_{I=1}^{M} L'_I(\phi_1, \phi_2, \cdots, \phi_C)$$

where L'_1 is a homogeneous transformation. Let

$$\mathcal{L}'_I = \{(\phi_1, \cdots, \phi_C) \in \mathcal{K} | L'_I(\phi_1, \cdots, \phi_C) = 0\}.$$

$\mathcal{L}'_I$ is merely translation of $\mathcal{L}_I$ such as to put it through the origin. Accordingly, the partition of $\mathcal{K}$ determined by $\mathcal{L}'_1, \mathcal{L}'_2, \cdots \mathcal{L}'_M$ has one-dimensional regions which radiate from the origin. Each of these radial lines represents a direct cycle. Distinct points on the same radial line represent distinct multiples of the same direct cycle and, likewise, their difference is a multiple of that direct cycle. Since k and k' lie on a line parallel to one of these radial lies, $m(k - k')$ is a multiple of a direct cycle. It is a positive multiple because k is beyond k' as their line goes toward infinity. Q.E.D.

Example 2. Suppose we do not know the mechanism for the reaction discussed in example 1, but that it could involve any of the steps S_1 through S_9, whose elementary reactions are as follows:

$\partial(S_1) = -E - G + EG$ $\partial(S_6) = -DE - GP + DEGP$
$\partial(S_2) = -DP - EG + DEGP$ $\partial(S_7) = -E - GP + EGP$
$\partial(S_3) = -D - EGP + DEGP$ $\partial(S_8) = -DEP - G + DEGP$
$\partial(S_4) = -EP - G + EGP$ $\partial(S_9) = -D - E + DE$
$\partial(S_5) = -DP - E + DEP$

Let us find every direct mechanism m, such that

$$\partial(m) = -DP - G + D + GP,$$

First, we must find one mechanism for this reaction and a complete set of independent cycles.

Write all the elementary reactions as row vectors in a matrix, as shown in Figure 2, putting the columns which correspond to the species DP, G, D and GP on the right. Then use row operations to reduce the matrix to row-echelon form, as shown in Figure 3.

Figure 2. A tabulation of the elementary reactions in the system of example 2.

Figure 3. A diagonalization of the matrix in figure 2.

Looking at the first seven rows of Figure 3, we see that there is only one possible reaction involving DP, G, D and GP,

$$DP + G = D + GP$$

and that

$$S_1 + S_2 - S_3 - S_7$$

is a mechanism for $-DP - G + D + GP$. The last two rows show that there are two independent cycles:

$$S_5 + S_8 - S_1 - S_2 \quad \& \quad S_6 + S_9 - S_3 - S_7 \ .$$

Therefore, a general expression for any mechanism for $-DP - G + D + GP$ is as follows:

$$m(\phi, \psi) = S_1 + S_2 - S_3 - S_7 + \phi(S_5 + S_8 - S_1 - S_2) + \psi(S_6 + S_9 - S_3 - S_7)$$

Collecting terms, we get

$$m(\phi,\psi) = (1-\phi)(S_1 + S_2) - (1+\psi)(S_3 + S_7) + \phi(S_5 + S_8) + \psi(S_6 + S_9),$$

This expression for any mechanism $m(\phi,\psi)$ allows us to plot four lines on the (ϕ,ψ)-plane:

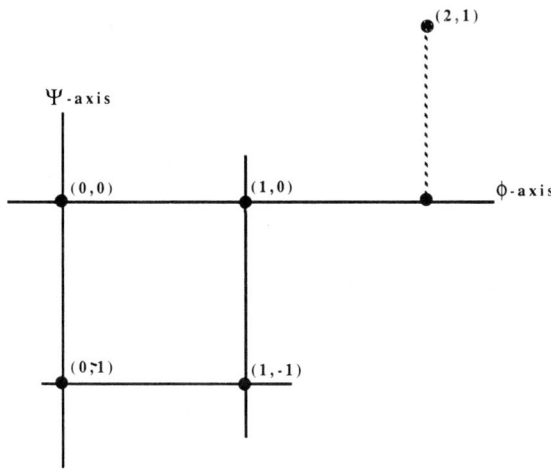

Figure 4. Each point in the (Φ,Ψ) - plane represents a mechanism for DP+G=D+PG in the chemical system of example 2. The points where two solid lines intersect represent direct mechanisms.

The line $1 - \phi = 0$ represents mechanisms without S_1 or S_2, the line $1 + \psi = 0$ represents mechanisms without S_3 or S_7, the line $\phi = 0$ represents mechanisms without S_5 or S_8, and the line $\psi = 0$ represents mechanism without S_6 or S_9. As seen in Figure 4, these lines partition the (ϕ,ψ)-plane (called $\mathcal{K}$ in the proof of theorem 2) into 9 two-dimensional regions, 12 one-dimensional regions, and 4 zero-dimensional regions. The last 4 regions are the isolated points $(1,0), (1,-1), (0,0)$ and $(0,-1)$, which represent mechanisms with minimal sets of steps. Therefore, the system has 4 direct mechanisms:

$$m(1,0) = S_5 + S_8 - S_3 - S_7$$
$$m(1,-1) = S_5 + S_8 - S_6 - S_9$$
$$m(0,0) = S_1 + S_2 - S_3 - S_7$$
$$m(0,-1) = S_1 + S_2 - S_6 - S_9$$

Each line in Figure 4 represents a direct cycle, which can be determined by taking the difference of any two points on the line. However, two parallel lines represent the same direct cycle, so, using just lines through the origin we can fine all the direct cycles:

$$m(1,0) - m(0,0) = S_5 + S_8 - S_1 - S_2$$
$$m(0,0) - m(0,-1) = S_6 + S_9 - S_3 - S_7$$

Let us conclude this example by illustrating the decomposition of an arbitrarily chosen mechanism $m(2,1)$ into direct cycles and direct mechanisms:

$$m(2,1) = -S_1 - S_2 - 2S_3 + 2S_5 + S_6 - 2S_7 + 2S_8 + S_9$$

$$= \begin{cases} & -S_3 & +S_6 \; -S_7 & +S_9 \\ -S_1 \; -S_2 & +S_5 & +S_8 \\ & -S_3 \; +S_5 & -S_7 \; +S_8 \end{cases}$$

The first two rows of this decomposition are direct cycles and the last is a direct mechanism. As stated in theorem 2, every time a step appears in a member of the decomposition it has the same sign. The method used to construct a decomposition in the proof of theorem 2 is based on the existence of a chain of line segments, which are illustrated in Figure 4 by a broken line from (2,1) to (2,0) to (1,0). These define the decomposition of $m(2,1)$ into the three parts whose values have been given above.

$$m(2,1) - m(2,0) = -S_3 + S_6 - S_7 + S_9$$
$$m(2,0) - m(1,0) = -S_1 - S_2 + S_5 + S_8$$
$$m(1,0) \qquad\quad = -S_3 + S_5 - S_7 + S_8$$

4 A Diagram For A Mechanism. In a graph-theoretic network a direct mechanism is a sequence of edges which connect two points, every intermediate point being incident to exactly two edges. A four-step direct mechanism is illustrated thus:

In this context a direct mechanism is so simple that it gives us no hint of the general nature of direct mechanisms. In a network, as we have defined it, a direct mechanism can have more than two terminals, more than two steps meeting at an intermediate point and more than one step at each terminal. It can have branch points and be multiply connected. The easiest way to appreciate these possibilities is to introduce a diagram for mechanisms and display some examples of direct mechanisms. These diagrams do not enter into the proofs of theorems, but they give an idea of the range of objects embraced by the theorems.

To generalize the graph-theoretic network of line segments meeting at their end points, let us define a network as polygonal surfaces meeting at their edges. In example 1 each of the four steps is a triangle, because it has three terms in its boundary. Figure 5 shows the four steps.

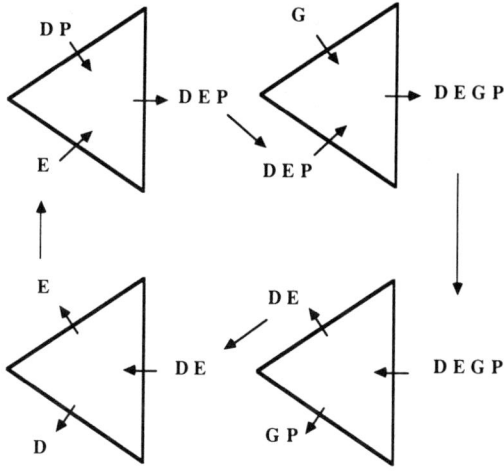

Figure 5. Diagrams for the four steps in example 1.

When the step changes sign, all the arrows are reversed. The whole mechanism of example 1 is shown by the diagram in Figure 6. The boundary of this diagram is

$$DP + G = D + GP$$

which exactly describes the overall reaction of the mechanism.

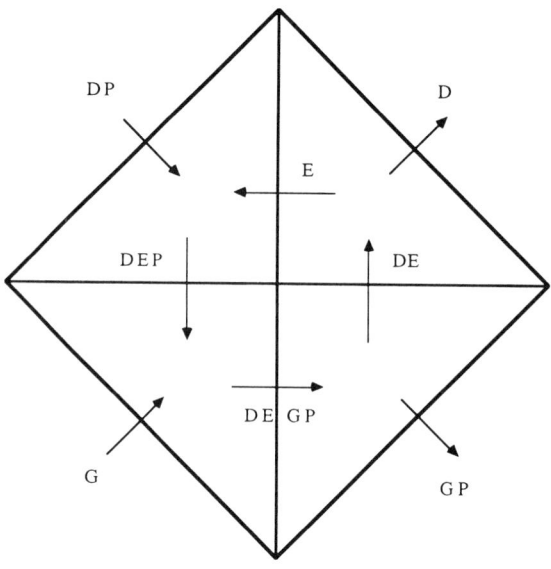

Figure 6. Diagram for the mechanism in in example 1.

Now let us use this kind of diagram to display some direct mechanisms, so that we can see how fundamentally they differ from the trivial graph-theoretic mechanisms. All the examples we will look at are direct mechanisms for the reaction

$$DP + G = D + GP$$

in a chemical system, which has 27 elementary reactions, given by Hammes and Kochavi [6] as possible steps in the action of the enzyme hexokinase and coenzyme Mg^{++}. These two species will be represented by E and M, respectively. The mechanism in Figure 6 is a direct mechanism in this context, but it does not involve M, which is known to be involved in the action of hexokinase. The mechanisms to be shown may not actually exist in nature, but, if the 27 steps given by Hammes and Kochavi are possible, then any mechanism using only those may be considered as a possibility.

In the hexokinase system, as defined in reference[6], there are 6 possible 4-step mechanisms, one of which is illustrated in Figure 6. They are all direct mechanisms. There are no direct mechanisms with 5 steps. There are altogether 55 direct mechanisms with 6 steps, one of which is illustrated in Figure 7.

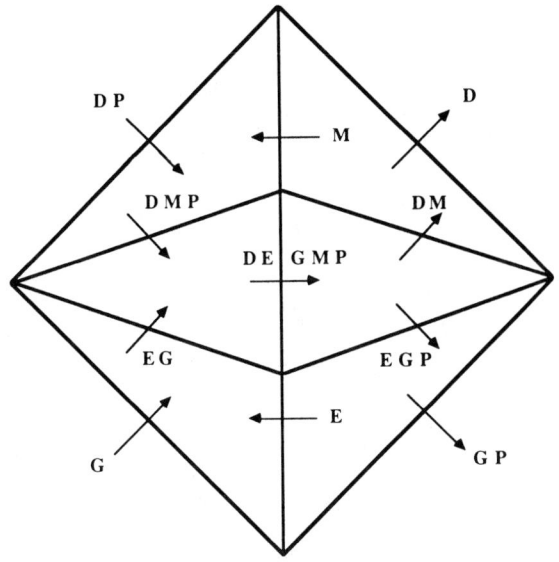

Figure 7. A direct mechanism with 11 species and 6 steps.

The procedure for determining all the direct mechanisms in a given system will be discussed further in section 5. Now let us consider some examples of direct mechanisms to illustrate how they differ from the trivial direct mechanisms in the graph theoretic case.

First let us consider a direct mechanism where there is branching. Figure 8 shows the possibility of 3 steps meeting at a common edge. There is one step (shaded) which appears in two places in the diagram. They are drawn separately to simplify the picture, but they represent a single step which advances at twice the rate of the other steps.

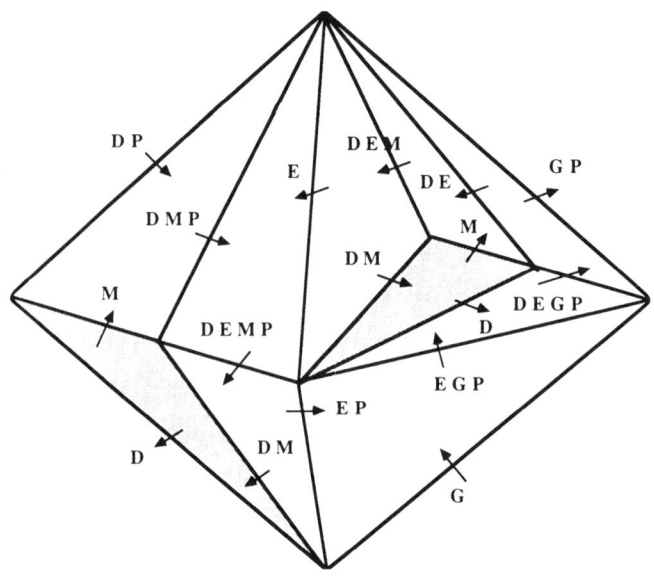

Figure 8. A direct mechanism with 14 species and 9 steps. The shaded triangles represent one step advancing twice.

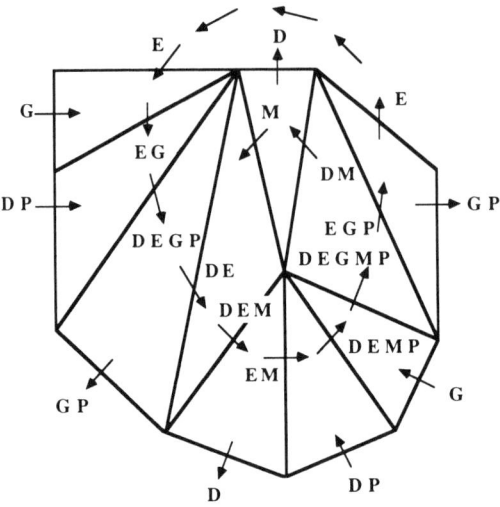

Figure 9. A direct mechanism with 15 species and 10 steps for the reaction 2DP + 2G = 2 D + 2GP. The two edges marked E must be understood as being identified, but written separately to allow a 2-dimensional representation.

Next let us consider a direct mechanism whose unusual character depends on the fact that reactions have integer coefficients. Figure 9 shows a mechanism with 10 steps in it, no combination of which will produce the reaction

$$DP + G = D + GP.$$

However, they produce the reaction

$$2DP + 2G = 2D + 2GP.$$

Finally, let us consider a direct mechanism which has unusually many chemical components. The number of chemical components in a system is the number of steps minus the number of independent reactions. If we look at each of our examples as a chemical system with no other reactions possible, the number of components in Figure 6 is 4, the number of components in Figures 7, 8 and 9 is 5, and the number of components in Figure 10 is 6. Notice that the number of components in each one except the last is equal to the number of letters used in the chemical formulas. Figure 10 is an example of the unusual case where the number of components exceeds the number of atoms and stable radicals in a system with balanced chemical reactions.

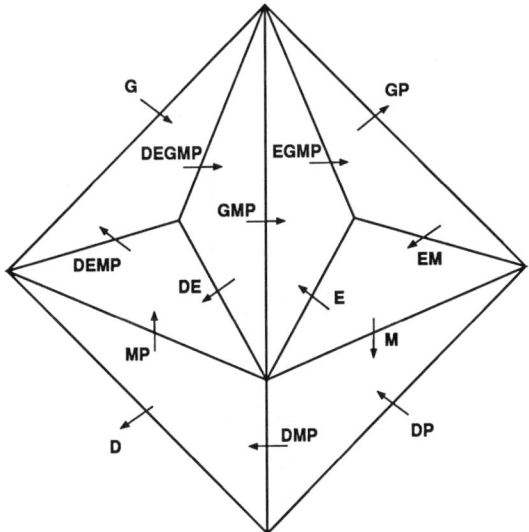

Figure 10. A direct mechanism with 14 species and 8 steps.

The above examples show that direct mechanisms are extremely varied, which suggests the need for a classification system. In references [7] and [8] preliminary efforts were made at an algebraic classification. In the appendix to this paper I have shown how the first homology group of a chemical system can be defined.

5 Applications. In reference [9] an algorithm was presented which would determine all the direct mechanisms which would produce a given overall reaction in a given chemical system. The algorithm requires two kinds of input: a list of all the species in the overall reaction known as the *terminal species*, and a list of all elementary reactions in the system. It has been programmed to run on a personal computer, and the way it has been used is to postulate a list of elementary reactions for a chemical system, including not only those which are known to occur but others which might occur, and then to regard the output as a list of possible mechanisms, which may require further screening. The algorithm determines what overall reactions are possible, involving the given terminal species, and finds the direct mechanisms for each overall reaction in a complete independent set of them.

The algorithm uses the same method of constructing direct mechanisms as was used in the proof of theorem 2 and as illustrated in example 2. It will be remembered that each direct mechanism was determined by the intersection of C hyperplanes selected from a possible set of M hyperplanes. The algorithm considers each of the $\binom{M}{C}$ ways of choosing C hyperplanes. Each choice corresponds to a system of C linear equations. The system has a solution if, and only if, the C hyperplanes intersect at a point. This point represents a direct mechanism, whose coefficients are determined by the solution of the equations. Therefore, the running time of the algorithm is equal to $\binom{M}{C}$ multiplied by the average time it takes to invert a matrix or determine that it is not invertible. Though this running time increases rapidly with the size of the system, the number of direct mechanisms in the solution also increases. So, if we want to examine all the direct mechanisms in the output, we would be limited to reasonably small systems.

Let me conclude by describing the results I have obtained, so far, with the hexokinase system. Hammes and Kochavi [6] studied 8 mechanisms for $DP + G = D + GP$ using a total of 27 steps or elementary reactions. All 8 were direct, and among them there were 4 with 4 steps and 4 with 6 steps. Therefore, it is natural to ask what are all the direct mechanisms, using only the 27 steps which were proposed as being possible.

There are some steps which must occur together, so the system can be reduced to 24 steps, and there are 12 cycles. Therefore, there is a factor of $\binom{24}{12}$ in the running time of the algorithm. However, if we confine ourselves to mechanism with 4 or 6 steps, it is clear that we only have to look at $\binom{24}{4} + \binom{24}{6}$ cases at most. Using further simplifications, I was able to construct by hand a complete set of all direct mechanisms for

$$DP + G = D + GP$$

using only the given steps and found 6 mechanism with 4 steps and 55 with 6 steps. From a chemical viewpoint, listing the possible mechanisms in a system provides a basis for further investigation.

6 Appendix. The First Homology Group of a Chemical System. So far, a network describing a chemical system consists of a boundary homomorphism $\partial : \mathcal{M} \to \mathcal{N}$ where $\mathcal{M}$ and $\mathcal{N}$ are free Abelian groups generated by steps and species respectively, and, if step S is a process by which a collides with b to form ab then,

$$\partial(S) = a - b + ab.$$

This expression is the reaction produced by S. To express the idea that reaction is *balanced* we define a new boundary homomorphism $\partial : \mathcal{N} \to \mathcal{A}$, where $\mathcal{A}$ is a free Abelian group generated by the atoms or stable radicals which are the constituents of the species, where ∂ separates a species into its constituents:

$$\partial(ab) = a + b.$$

A *balanced reaction* has a boundary of zero. For instance,

$$\partial\partial(S) = \partial(-a - b + ab) = -a - b + a + b = 0.$$

These two homomorphisms are familiar to chemists as matrices: The *step-by-species* matrix and the *species-by-atom* matrix. If a chemical system has M elementary reactions, N species and A atoms, then the incidence relations between elementary reaction steps and species are characterized by an $M \times N$ matrix, and the incidence relations between species and atoms are characterized by an $N \times A$ matrix. If all chemical reactions are balanced, then the product of these matrices is zero. Algebraically these matrices define homomorphisms

$$\mathcal{M} \longrightarrow \mathcal{N} \longrightarrow \mathcal{A}$$

between the free Abelian groups generated by the M steps, N species and A atoms such that $\partial\partial = 0$ or, equivalently, $im\partial \subset ker\partial$. This allows us to define the *homology group* $\dfrac{ker\partial}{im\partial}$ of the system.

The homology groups of the chemical systems defined by Figures 6, 7 and 8 are trivial. Figure 9 defines a system in which

$$(-DP - G + D + GP + im\partial).$$

is an element of order 2 in the homology group, and Figure 10 defines a system in which the homology group is a free Abelian group of rank 1.

REFERENCES

[1] C. BERGE, *Graphs and Hypergraphs*, North Holland Publishing Co., Amsterdam-London (1973).

[2] P.H. SELLERS, *Algebraic Complexes Which Characterize Chemical Networks*, SIAM J. Appl. Math., 15 (1) (1967), pp. 13–68.

[3] P.H. SELLERS, *Combinatorial Complexes*, D. Reidel Publishing Co., Dordrecht, Holland (1979).

[4] P.C. MILNER, *The Possible Mechanisms of Complex Reactions Involving Consecutive Steps*, J. Electrochem Soc., III (1964), pp. 228–232.

[5] P.H. SELLERS, *The Classification of Chemical Mechanisms From a Geometric Viewpoint, Chemical Applications of Topology and Graph Theory*, R.B. King, Ed., Elsevier, N.Y. (1983), pp. 420–429.

[6] C.G. HAMMES AND D. KOCHAVI, *Studies of the Enzyme Hexokinase*, J. Am. Chem. Soc., 84 (1962), pp. 2069–2079.

[7] P.H. SELLERS, *An Introduction to a Mathematical Theory of Chemical Networks I & II*, Archive for Rational Mechanics and Analysis, 44 (1) (1971), pp. 23–40 and 44 (5) (1972), pp. 376–386.

[8] P.H. SELLERS, *Combinatorial Classification of Chemical Mechanisms*, SIAM J. Appl. Math., 44 (4) (1984), pp. 784–792.

SPATIAL MODELS OF POWER AND VOTING OUTCOMES

PHILIP D. STRAFFIN JR.*

Introduction. This article is a brief, subjective guided tour into an area of active, recent work in political science—spatial models of voting. We will be concerned with two main questions. The first is how we can measure the *power* of voters in an asymmetric voting situation. The second is how we can predict or judge the *outcome* of a voting situation.

The common theme as we treat these two questions will be the interplay between combinatorial ideas and a geometric framework. For each question, the first answers given were combinatorial, and in general they ignored the ideological positions of voters. Geometry was brought in to model the effects of ideology. The resulting fusion of combinatorics and geometry has been producing some surprising and elegant results. At the end, we will see how a recent geometric theorem ties together our two questions very neatly.

Because of its restricted focus on power and outcomes, and its emphasis on the relation between combinatorial and geometric ideas, this article should not be taken as a survey of the use of spatial models in political science. It is meant to be illustrative, but it cannot pretend to completeness.

I. Voting Power. The classical combinatorial measure of voting power is the Shapley–Shubik power index [Shapley and Shubik, 1954]. To illustrate how it works, consider the weighted voting situation described by

$$[\,5;\ 3,\ 2,\ 2,\ 1\,]$$
$$A\ \ B\ \ C\ \ D$$

Here "5" is the quota of votes needed to pass a bill, and the other numbers tell how many votes are cast by each of the four voters A, B, C and D. In applications, these voters could be legislators casting weighted votes (as in New York county boards), countries casting weighted votes (as in UN agencies or the European Economic Community), or political parties in a multi-party legislature.

Shapley and Shubik imagine a coalition forming in favor of some bill. At some point, a voter joining the coalition changes it from a losing coalition (fewer than 5 votes) to a winning coalition (5 or more votes). That voter—the *pivotal voter* for that ordering—has considerable power. Hence we can measure the power of a voter by the proportion of orderings in which that voter is pivotal. The 4! orderings of the voters in our example are listed below, with the pivotal voter underline in

---

*Department of Mathematics and Computer Science, Beloit College, Beloit, Wisconsin 53511

each.

A	$\underline{B}$	C	D	B	$\underline{A}$	C	D	C	$\underline{A}$	B	D	D	A	$\underline{B}$	C
A	$\underline{B}$	D	C	B	$\underline{A}$	D	C	C	$\underline{A}$	D	B	D	A	$\underline{C}$	B
A	$\underline{C}$	B	D	B	C	$\underline{A}$	D	C	B	$\underline{A}$	D	D	B	$\underline{A}$	C
A	$\underline{C}$	D	B	B	C	$\underline{D}$	A	C	B	$\underline{D}$	A	D	B	$\underline{C}$	A
A	D	$\underline{B}$	C	B	D	$\underline{A}$	C	C	D	$\underline{A}$	B	D	C	$\underline{A}$	B
A	D	$\underline{C}$	B	B	D	$\underline{C}$	A	C	D	$\underline{B}$	A	D	C	$\underline{B}$	A

The Shapley–Shubik indices for the voters are

$$\phi_A = 10/24 = .42 \qquad \phi_B = \phi_C = 6/24 = .25 \qquad \phi_D = 2/24 = .08$$

In a general situation with n voters, the Shapley–Shubik index for voter i will be

$$\phi_i = \frac{\text{number of orders in which } i \text{ is pivotal}}{n!}$$

One can think of this as the probability that i will be the pivotal voter on a bill, assuming that all orders of coalition formation are equally likely.

The Shapley–Shubik power index has been widely used, mostly at the constitutional level where it is natural to assume that we have no information about the beliefs of individual voters. See [Lucas, 1983] and [Straffin, 1983] for surveys. However, in any real voting situation it is clear that ideological concerns of voters would make certain orders of coalition formation more likely than others. One familiar model of voter ideology is the left-right, liberal-conservative continuum. Figure 1, for example, shows a left-right placement of United States Supreme Court justices from [Rohde and Spaeth, 1976]. One can imagine a liberal coalition building from left to right, starting with Douglas and becoming winning with White, or a conservative coalition forming from right to left beginning with Rehnquist and becoming winning with White. If we think of these two spatial orderings as being more likely than others, Justice White in the median position will have more power than the other justices. Indeed, it is part of common political wisdom that centrist voters have more power than extremists.

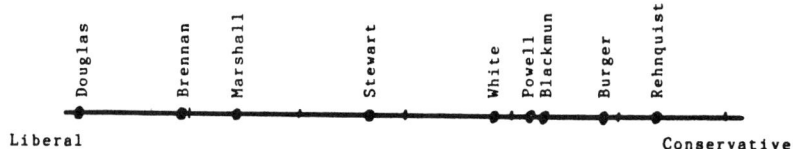

FIGURE 1. THE U.S. SUPREME COURT IN 1976.
FROM [ROHDE AND SPAETH, 1976]

For many–perhaps most–political purposes, the one-dimensional liberal-conservative continuum is a very crude model. One can make it more sophisticated by placing voters in a space of more than one dimension. Such *spatial models* of voter ideology have become increasingly important in the past twenty years. Figure 2 shows a two-dimensional spatial model of states in United States presidential elections 1944–1980, from [Rabinowitz and Macdonald, 1986]. The placements of states were obtained by subjecting voting data to principal component analysis. The first component (vertical, explaining 48% of the variance) is identifiable as liberal-conservative. The second component (horizontal, explaining an additional 31% of the variance) has to do with traditional Republican-Democratic party identification. Since the third component explains only an additional 7% of the variance, the two-dimensional spatial plot captures most of the ideological information in the voting data. Notice that the analysis also gives the projection of the ten election axes onto the principal component plane.

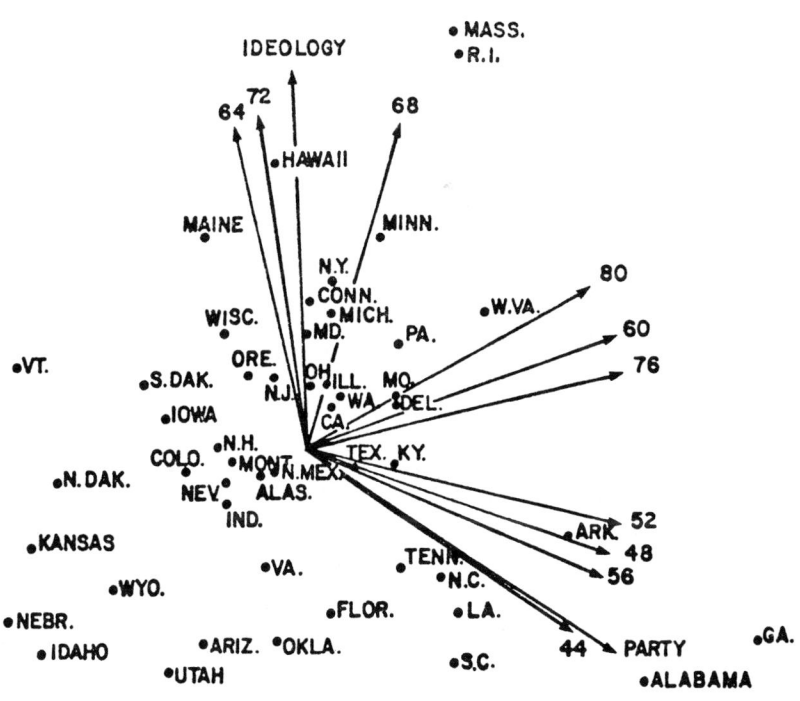

FIGURE 2. FIFTY STATES AND ELECTIONS WITH EXTERNAL AXES
FROM [RABINOWITZ AND MACDONALD, 1986]

In the 1970's Owen [1971] and Shapley [1977] proposed adapting the Shapley-Shubik power index to spatial voting models in R^k. In this paper we will only consider the easily picturable case $k = 2$. Figure 3 shows a two-dimensional spatial placement of the voters in

$$[\,3;\ 1,\ 1,\ 1,\ 1\ \ 1\,]$$
$$A\ \ B\ \ C\ \ D\ \ E$$

The classical Shapley-Shubik index would, of course, give each voter 1/5 of the voting power. On the other hand, common wisdom would expect centrist voter D to have more power than the other voters.

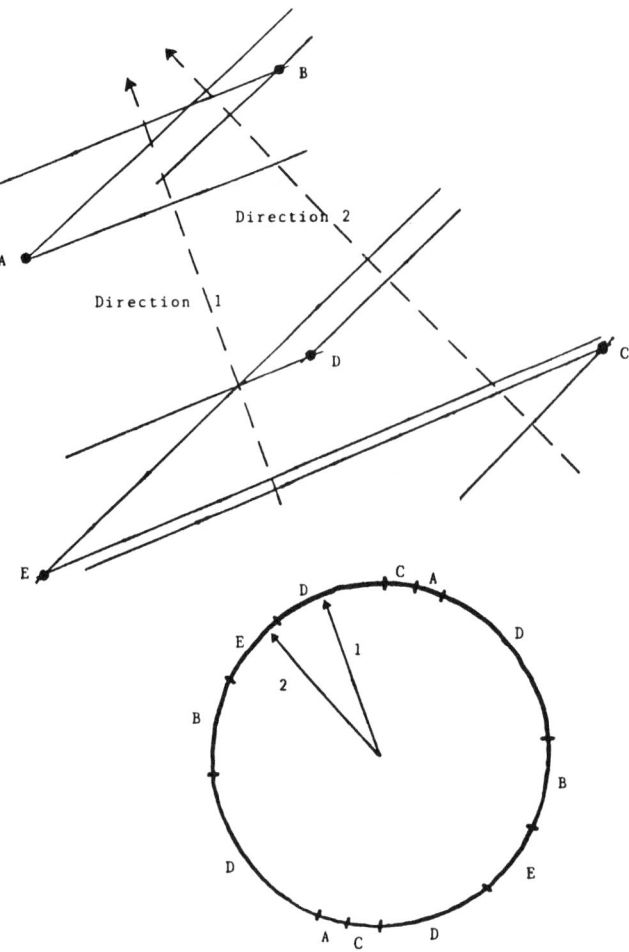

FIGURE 3. COALITION FORMED BY SWEEPING LINES.

In this context we picture a forming coalition as a line sweeping through the plane in a fixed direction, picking up the voters as it moves. Thus a line sweeping in direction 1 picks up voters in the order CEDAB with D pivotal, while a line sweeping in direction 2 picks them up as CDEBA with E pivotal. Label each point on a reference circle by the voter who is pivotal in the ordering given by a line sweeping in the direction of a vector from the circle's center to that point. A voter's Shapley-Owen power index is the proportion of the circle labeled by that voter. In the example of Figure 3 the indices are

$$\psi_A = \frac{20°}{360°} = .06 \quad \psi_B = \frac{62°}{360°} = .17 \quad \psi_C = \frac{22°}{360°} = .06$$
$$\psi_D = \frac{204°}{360°} = .57 \quad \psi_E = \frac{52°}{360°} = .14$$

Voter D is indeed powerful. In general, one can think of a voter's Shapley-Owen power index as the probability that voter will be pivotal, given that coalitions form by this sweeping line method, with all directions of sweep equally likely.

In two dimensions, the Shapley-Owen power index is easily calculable by a "rotation" algorithm. I will explain it in the context of a voting situation with an odd number of votes and a simple majority needed to win, although the method is applicable more generally. Define a *median line* to be a line L such that each open half plane H_1 and H_2 of $R^2 - L$ contains less than a majority of votes. It follows that

i) every median line L contains at least one voter point, and each closed half plane $H_1 \cup L$ and $H_2 \cup L$ contains a majority of votes;

ii) there is exactly one median line in each direction;

iii) voter i's Shapley-Owen power index is the sum of the angles swept out by median lines passing through point i, divided by 180° (if we think of the lines as undirected).

To calculate the indices, start with any median line, say passing through point A in Figure 4. Imagine rotating it counterclockwise until it passes through another voter point (C in this example), keeping track of the rotation angle, and assigning that angle to voter A. As we rotate past AC, the median line will either continue to pass through A or switch to passing through C (here it switches). Continue rotating, always assigning the angle of rotation to the voter through whose point the median line passes, until the median line has rotated through 180° and returns to its original position. In practice, of course, the algorithm is discrete, since the median line can only change its voter point at those directions (finite in number) when it passes through two or more voter points.

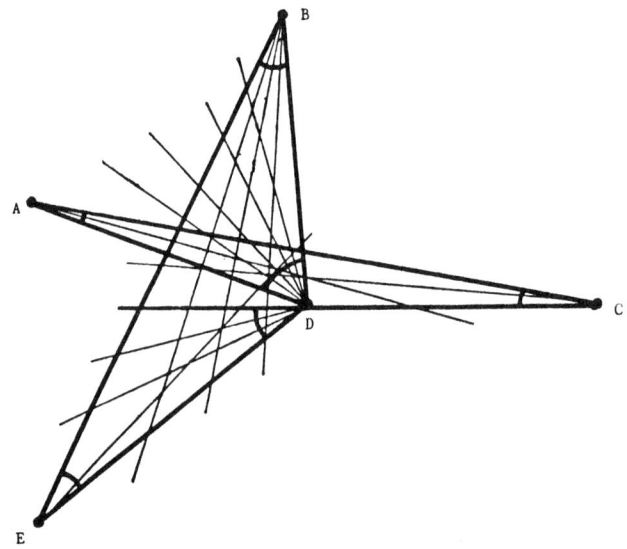

FIGURE 4. ROTATING THE MEDIAN LINE.

The Shapley-Owen indices of the voters in our example are the measures of the angles marked in Figure 4, divided by 180°. Figures 5a and 5b show two common elementary cases. If the voters in [2; 1, 1, 1] are at the vertices of a triangle, their Shapley-Owen power is in proportion to the angles of the triangle. If the voters in [3; 1, 1, 1, 1, 1] are at the vertices of a convex pentagon, their power is in proportion to the angles of the corresponding star-pentagon.

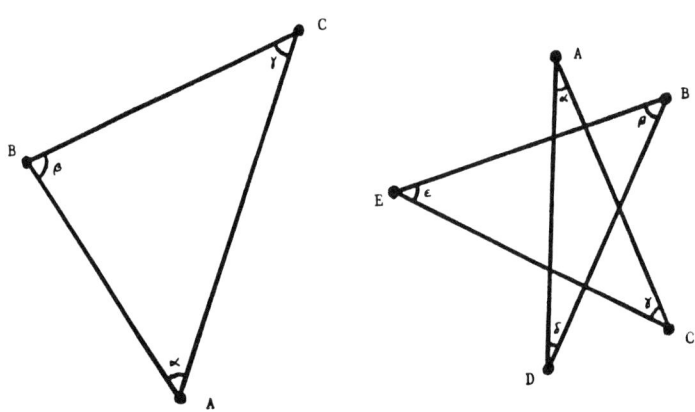

FIGURE 5A.   FIGURE 5B.
SIMPLE CASES OF THE SHAPLEY-OWEN POWER INDEX.

Rabinowitz and Macdonald calculated the Shapley-Owen indices of the states in Figure 2. However, they noted that the directions corresponding to presidential elections 1944-1980 lie within a sector of about 140°, and those corresponding to elections 1964-1980 like within a sector of about 90°. Hence to get a measure of electoral power applicable to the modern period, they only considered directions in this sector. The results are shown in Table 1, which is taken from [Rabinowitz and Macdonald, 1986]. The modified Shapley-Owen power indices are given in column 3, and are translated into effective electoral votes in column 5. Column 6 gives the classical Shapley-Shubik indices in the same form. The difference between these figures, in column 7, is a measure of the extent to which a state's ideological position increases or decreases its effective electoral power. Notice that the states which benefit most heavily in proportion to their number of votes–California, Texas, Illinois, Ohio and Washington–are all centrally located in Figure 2. States which lose most because of ideology are the outlying states–Massachusetts, Rhode Island, Idaho, Nebraska and Utah.

Table 1. Power in the Modern Sector

Rank	State	Power %[a]	Electoral Votes	Power Electoral Votes[b]	Shapley Electoral Votes	Power-Shapley[c]
1	California	12.02	47	64.64	49.91	+14.74
2	Texas	6.99	29	37.60	29.60	+ 8.00
3	New York	6.70	36	36.02	37.29	− 1.27
4	Illinois	5.71	24	30.74	24.25	+ 6.50
5	Ohio	5.47	23	29.42	23.19	+ 6.23
6	Pennsylvania	4.96	25	26.68	25.31	+ 1.37
7	Michigan	3.97	20	21.37	20.05	+ 1.33
8	New Jersey	3.66	16	19.68	15.92	+ 3.77
9	Florida	3.36	21	18.09	21.09	− 3.00
10	North Carolina	2.53	13	13.63	12.86	+ 0.77
11	Missouri	2.41	11	12.95	10.84	+ 2.11
12	Wisconsin	2.34	11	12.57	10.84	+ 1.73
13	Washington	2.33	10	12.52	9.84	+ 2.68
14	Tennessee	2.22	11	11.93	10.84	+ 1.08
15	Indiana	2.19	12	11.78	11.85	− 0.07
16	Maryland	2.16	10	11.62	9.84	+ 1.78
17	Kentucky	2.02	9	10.86	8.84	+ 2.02
18	Virginia	1.91	12	10.29	11.85	+ 1.56
19	Louisiana	1.85	10	9.95	9.84	− 0.11
20	Connecticut	1.60	8	8.62	7.84	+ 0.78
21	Iowa	1.56	8	8.37	7.84	+ 0.52
22	Oregon	1.53	7	8.22	6.85	+ 1.37
23	Colorado	1.48	8	7.93	7.84	+ 0.10
24	Georgia	1.47	12	7.89	11.85	− 3.96
25	Minnesota	1.32	10	7.10	9.84	− 2.74
26	South Carolina	1.26	8	6.78	7.84	− 1.06
27	Alabama	1.26	9	6.77	8.84	− 2.07
28	Arkansas	1.07	6	5.77	5.86	− 0.09
29	New Mexico	1.06	5	5.69	4.87	+ 0.82
30	Oklahoma	0.95	8	5.08	7.84	− 2.76
31	West Virginia	0.85	6	4.57	5.86	− 1.29
32	New Hampshire	0.83	4	4.45	3.89	+ 0.56
33	Montana	0.82	4	4.39	3.89	+ 0.50
34	Mississippi	0.78	7	4.17	6.85	− 2.68
35	Nevada	0.76	4	4.06	3.89	+ 0.17
36	Maine	0.73	4	3.91	3.89	+ 0.02
37	Delaware	0.65	3	3.51	2.91	+ 0.60
38	Kansas	0.63	7	3.41	6.85	− 3.44
39	Alaska	0.62	3	3.34	2.91	+ 0.43
40	Arizona	0.62	7	3.33	6.85	− 3.52
41	South Dakota	0.57	3	3.07	2.91	+ 0.16
42	Hawaii	0.52	4	2.77	3.89	− 1.12
43	Vermont	0.44	3	2.39	2.91	− 0.53
44	North Dakota	0.37	3	2.00	2.91	− 0.92
45	Massachusetts	0.33	13	1.78	12.86	−11.08
46	Utah	0.32	5	1.75	4.87	− 3.12
47	Wyoming	0.27	3	1.46	2.91	− 1.46
48	Nebraska	0.26	5	1.42	4.87	− 3.45
49	Idaho	0.19	4	1.02	3.89	− 2.87
50	Rhode Island	0.12	4	0.64	3.89	− 3.25

[a] Percentage of times the state occupies the pivotal position.
[b] Percentage of times the state occupies the pivotal position multiplied by the total number of electoral votes (538).
[c] Difference between power measures.

from [Rabinowitz and Macdonald, 1986]

The Shapley-Owen power index is a useful tool in political analysis. Moreover, we will see it appearing in a surprising way in a different context at the end of the last section of this paper.

II. Voting Outcomes: The Discrete Case. The second problem we will consider is the problem of predicting or judging voting outcomes. Given a set of alternatives to be voted on, a set of voters, and a transitive preference list of alternatives for each voter, which alternative should win? For example, suppose there are four alternatives w, x, y, z and three voters A, B, C with preference lists

$\underline{A}$	$\underline{B}$	$\underline{C}$
x	w	y
w	z	x
z	y	w
y	x	z

When the set of alternatives is finite, situations like this have been modeled as tournaments. The vertices are the alternatives, and we draw a directed edge from alternative i to alternative j if i is preferred to j by a majority of the voters (so i would beat j in a two-way election). The tournament for the above example is

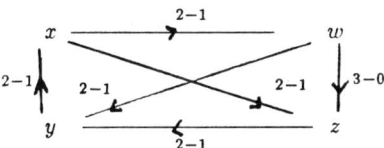

I have included the strengths of the two-way votes. The tournament makes evident a number of anomalies in traditional voting processes. For example, consider a traditional "amendment process" in which two alternatives are paired against each other, the winner goes on to face a third alternative, and so on until a final winner is chosen. If cycles are present in the tournament, the *agenda*, i.e. the order in which the alternative are presented, can affect the outcome of the process. In our example

are agendas which produce z and y as respective winners, and other agendas (see them?) will make x or w win. This *agenda effect* means that in legislative politics control of the agenda can be important, a fact well known to politicians.

A second, disturbing anomaly is illustrated in our example by the fact that alternative z can win. It seems nonsensical to choose an alternative like z when there is another alternative, w, which the voters *unanimously* prefer to z. Yet the amendment procedure can choose such *Pareto inferior* outcomes. See [Straffin, 1980] for further discussion of these issues.

Given a voting situation represented as a tournament, a fundamental problem is to specify an alternative which should be the winner or, failing that, to specify a set of alternatives within which the winner should lie. I will consider four possible solutions to this problem for the discrete case, and then we will investigate how these solution concepts fare in spatial models of voting.

CONDORCET WINNER

In 1785 the Marquis de Condorcet proposed that if there is an alternative which beats all others in pairwise contests, that alternative should win. Such an alternative is called a Condorcet winner. Unfortunately, as our example above shows, there may not be a Condorcet winner.

TOP CYCLE SET

If there is no alternative which beats all other alternatives directly, we could consider the set of alternatives which beat all other alternatives either directly or indirectly, i.e. via a chain (i beats j, j beats $k,\ldots,l$ beats m). This is called the top cycle set. It is the initial strong component of the tournament digraph. The top cycle set can also be characterized as the set of all possible amendment agenda winners. It would certainly seem reasonable to require that the winner be in the top cycle set.

Unfortunately, the top cycle set may be large. In our example, w, x, y and z are all in the top cycle set. Indeed, Moon [1968] proved that in a random tournament of size n, the probability that *every* alternative is in the top cycle set, goes to one as n becomes large.

LANDAU SET

If the top cycle set is too large, we could try to restrict it by including only those alternatives which beat all other alternatives either directly or indirectly by a chain of length two (i beats j, j beats k). This set is called the Landau set after [Landau, 1953]. See [Maurer, 1980] for a clear and entertaining exposition. In our example the Landau set is $\{w, x, y\}$, since

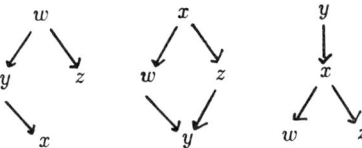

and z is not in the Landau set since z cannot beat w in two steps. This last fact illustrates a nice general property.

THEOREM 1. *The Landau set cannot contain a Pareto inferior alternative.*

Proof. If an alternative w is unanimously preferred to an alternative z, then z cannot beat w in two steps. For if a majority M_1 of voters prefers z to y, and a majority M_2 of voters prefers y to w, then voters in the (necessarily non-empty) intersection $M_1 \cap M_2$ must, since their preferences are transitive, prefer z to w. This would contradict the supposed unanimous preference for w over z.

The Landau set has a number of other good properties. For example, it is non-empty (see below), and it is obviously contained in the top cycle set. Moreover, Miller [1980, 1983] showed it can be characterized game theoretically. Suppose in our example that two politicians P and Q must each choose one of the alternatives w, x, y, z as his "platform." P will beat Q and get a payoff $+1$ precisely if a majority of voters prefers P's platform to Q's platform. The corresponding zero-sum game is

		Q chooses			
		w	x	y	z
P chooses	w	0	-1	1	1
	x	1	0	-1	1
	y	-1	1	0	-1
	z	-1	-1	1	0

(Payoffs to P)

Notice that, for either player, the strategy of choosing z is dominated by the strategy of choosing w, since the payoff for w is in all cases as favorable or more favorable than the payoff for z. On the other hand, w, x and y are all undominated strategies. Miller's result is

THEOREM 2. *The Landau set is exactly the set of undominated platforms.*

Proof. A platform w dominates a platform z if w beats z, and whenever y beats w, y also beats z. Thus w dominates z precisely when z does not beat w either directly or by a chain of length two.

Miller called the relation of domination *covering*, and the Landau set the *uncovered set*.

With all these nice properties, it is unfortunately true that the Landau set can still be large. In fact, Moon [1968] also proved that in a random tournament of size n, the probability that every alternative is in the Landau set, goes to one as n gets large. See [Maurer, 1980].

COPELAND WINNER

Our last discrete solution concept was proposed explicitly by A.H. Copeland in 1950, although the idea is implicit in earlier work. See [Straffin, 1980]. Define the *score* of an alternative i as the number of alternatives which i beats. The Copeland winner is the alternative with the highest score. In our example x and w have score we can get our small society A, B, C to accept z, or any other alternative in the plane. In a spatial voting situation with $k \geq 2$, in the generic case where there is no Condorcet winner, an agenda controller has complete power!

Figure 6 illustrates how to construct the McKelvey chain of x_i's for this example. Notice that AB, AC and BC are all median lines. By the result of Figure 7, if we reflect w across AC and move it in slightly to get x, then x beats w. Now reflect x across BC and move it in slightly to get y, which beats x. Now reflect y across AB and move it in slightly to get z. We have found the required chain w, x, y, z.

The remarkable thing is that this simple example has in it the proof of the general theorem for $k = 2$. For first suppose that we were trying to get to some other point t. Just keep repeating the three reflections in order, obtaining points $x_1, y_1, z_1, x_2, y_2, z_2, \ldots$ as far away from w as we need. When we get some z_m far enough away beyond t, t will be unanimously preferred to z_m and we will have our chain. Second, suppose we have any configuration of any number of voters, but no Condorcet winning alternative. Then by Theorem 3a there must be three median lines which do not pass through a common point and hence bound a triangle. Let this triangle play the role of ABC in the above argument.

McKelvey's beautiful result, with its implications of the inherent instability of majority rule and the power of agenda control, has generated a large literature in political science. For generalizations, see [Cohen, 1979], [Schofield, 1978] and [McKelvey, 1979]. For a thoughtful discussion, see [Riker, 1980].

LANDAU SET (UNCOVERED SET)

In the spatial context, it is convenient to use Miller's alternative formulation of the Landau set.

DEFINITION. For a point x in R^k, *DEF(x)* is the set of all points which beat x under majority rule.

DEFINITION. For points x and y, x *covers* y if x beats y, and $DEF(x) \subseteq DEF(y)$. See Figure 9.

DEFINITION. The *uncovered set* (Landau set) is the set of all points which are not covered by any other point.

Spatial models of voting have been increasingly studied by political scientists since the 1950's. A good survey is [Enelow and Hinich, 1984]. We will examine how the four solution concepts from Section II apply in this spatial context. For simplicity, we will continue to consider only voting by majority rule with an odd number of votes. We will assume that all points in R^2 are possible alternatives to be voted on. We thus have an infinite tournament. On the other hand, the spatial placement of the alternatives provides additional structure which we can exploit.

CONDORCET WINNER (CORE)

In the spatial voting literature the set of Condorcet winners is known by its game theoretic name: the *core*. It is the set of alternatives which beats (or ties) all other alternatives. In a one-dimensional spatial model, the core is the ideal point of the median voter, so the core is always non-empty. For higher dimensions the core will be non-empty if and only if the voter ideal points satisfy a restrictive symmetry condition first given by Plott [1967]. To derive this condition for $k = 2$, recall the idea of a median line from Section I. Notice that if L is a median line, and x is a point not on L (see Figure 7), then x will lose under majority voting to x', which is the reflection of x across L, moved slightly towards L. This is because x' is preferred to x by all voters to the right of the perpendicular bisector of the line segment xx' (shown dotted in Figure 7). This includes all voters in $L \cup H_2$, which is a majority by the definition of a median line.

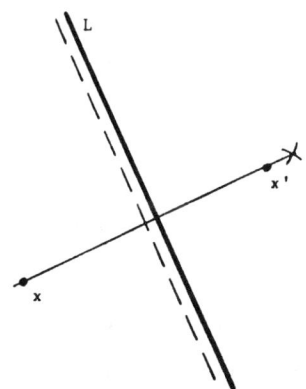

FIGURE 7. REFLECTING ACROSS A MEDIAN LINE.

Since x can be beaten if there is some median line which does not contain x, an alternative x can be a Condorcet winner only if it lies on all median lines. Hence we have shown

THEOREM 3A. *The core will be non-empty if and only if all median lines pass* through a single point. If they do, that point is the unique Condorcet winner. [Davis et. al., 1972] [Feld and Grofman, 1987]

A bit of geometry shows that this is equivalent to

THEOREM 3B. *The core will be non-empty if and only if the voter ideal points all lie on a collection of lines L_i such that*

  i) *the lines L_i all pass through the ideal point of a voter A, and*
  ii) *A is the median voter on each of the lines L_i.* See Figure 8. [Plott, 1967]

This condition is clearly structurally unstable–small perturbations of voter ideal points will destroy it. Hence

THEOREM 3C. *Generically, spatial voting situations in dimensions two or larger do not have a Condorcet winner.*

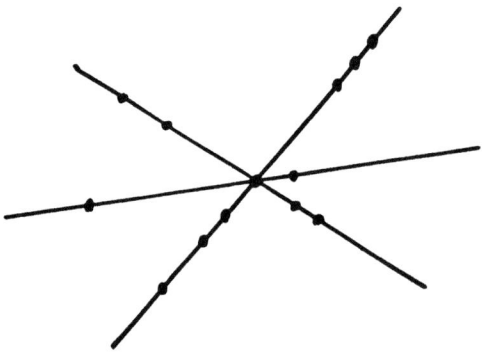

FIGURE 8. THE PLOTT CONDITION FOR A NON-EMPTY CORE.

TOP CYCLE SET

When there is no Condorcet winner, what is the top cycle set? The following result caused astonishment when it was first announced.

THEOREM 4. *If there is no Condorcet winner, the top cycle set of a spatial voting situation in R^k is all of R^k.* [McKelvey, 1976]

To see what this means, suppose we have three voters A, B, C as in Figure 6, and suppose we start with alternative w. Choose any other point in the plane, perhaps one far away like z. McKelvey's theorem says that there must be a chain of alternatives $w = x_0, x_1, x_2, \ldots, x_m = z$ such that x_{i+1} beats x_i by majority rule ($i = 0$ to $m - 1$). If we control the agenda and present alternatives x_i in order,

2, while y and z have score 1, so x and w tie as Copeland winners. Landau proved that any Copeland winner must be in the Landau set (which is why the Landau set must be non-empty). This follows immediately from Theorem 2 and the observation that an alternative with maximal score cannot be dominated by another alternative.

III. Voting Outcomes: The Spatial Case.

The spatial model which we used to introduce ideology into the discussion of voting power can be used to study voting outcomes. The alternatives to be voted on are points in a k-dimensional *issue space* R^k. We will continue to consider only $k = 2$. Voters are also represented by points in this same issue space, their *ideal points*. In judging between two alternatives x and y, voter A will prefer the one which is closer to her ideal point. Thus in Figure 6, voter A will prefer x to w, w to z, and z to y. In fact, the configuration of voters and alternatives in Figure 6 produces preferences which are exactly those in the tournament example in Section II.

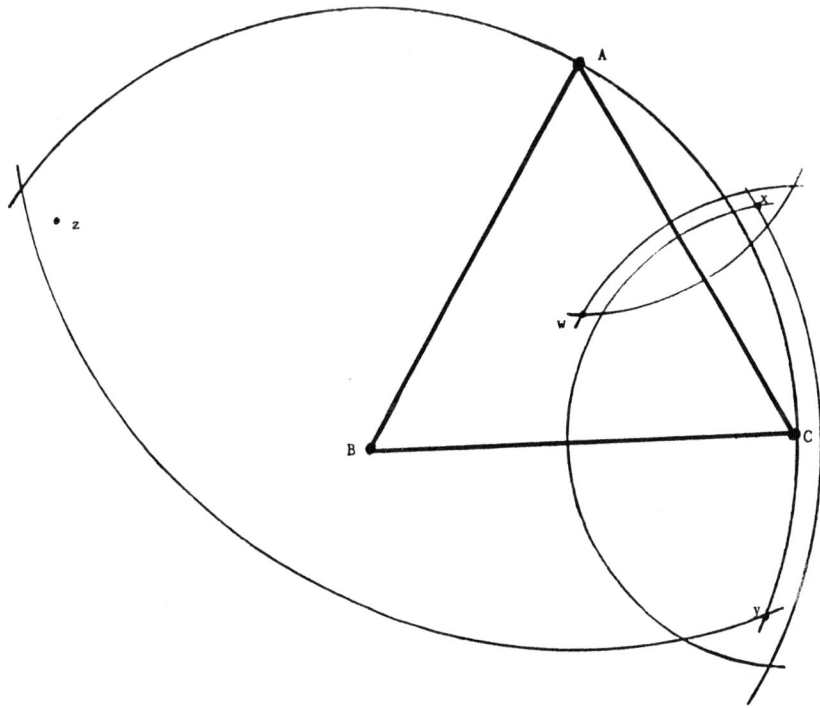

FIGURE 6. SPATIAL VOTING.

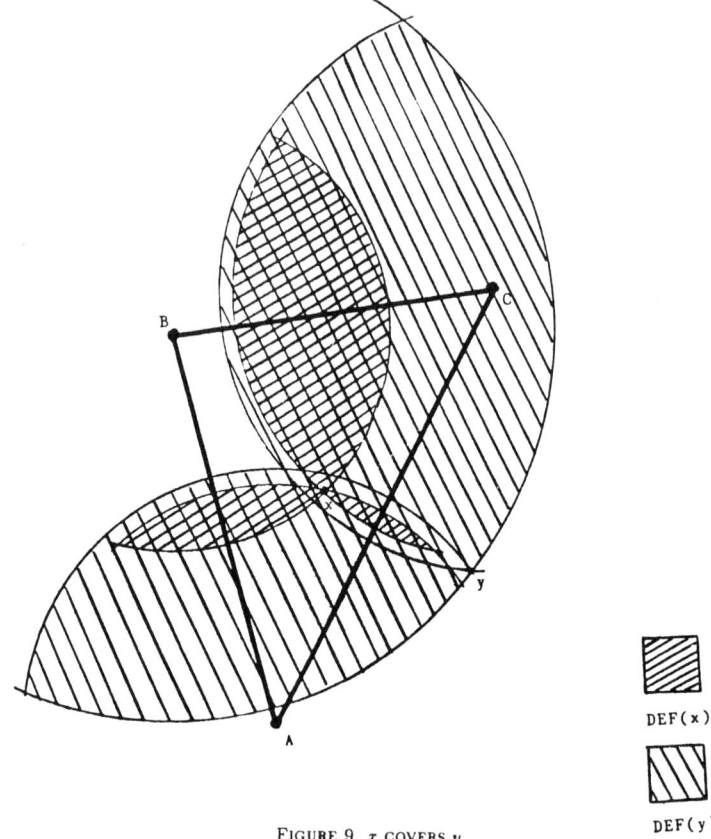

FIGURE 9. x COVERS y.

If the top cycle set is all of R^k, and in the discrete case the Landau set is not much smaller, one might be pessimistic about the size of the uncovered set in a spatial voting situation. However, it is easy to show that the Landau set cannot be all of R^k. First, in a spatial voting situation the set of non-Pareto-inferior alternatives ("Pareto optimal alternatives") is exactly the *convex hull* $\mathcal{C}$ of the voter ideal points. The argument of Theorem 1 goes through to show that the uncovered set cannot contain any Pareto inferior alternatives, hence must be contained in $\mathcal{C}$.

In fact, recent results show that the uncovered set is often considerably smaller than $\mathcal{C}$.

DEFINITION. The *yolk* of a spatial voting situation in R^2 is the smallest disk $B(x,r)$ which intersects all median lines. Here x is the center of the disk, and r is its radius. See Figure 10.

THEOREM 5. *The uncovered set is contained in $B(x, 4r)$, the disk with the same center as the yolk, but radius four times as large.* [McKelvey, 1986]

One consequence of this theorem is that if a voting situation *almost* has a Condorcet winner (so the yolk is small), the uncovered set will be small. Hence the uncovered set looks like an attractive solution concept in spatial voting, although its precise geometry is not yet well understood. The importance of the uncovered set is increased by a beautiful result of Shepsle and Weingast [1984] which characterizes it as the set of alternatives which can be reached from any other point by a chain of majority votes when the voters vote *sophisticatedly* (taking into account what other voters will do and adjusting their vote accordingly) rather than sincerely.

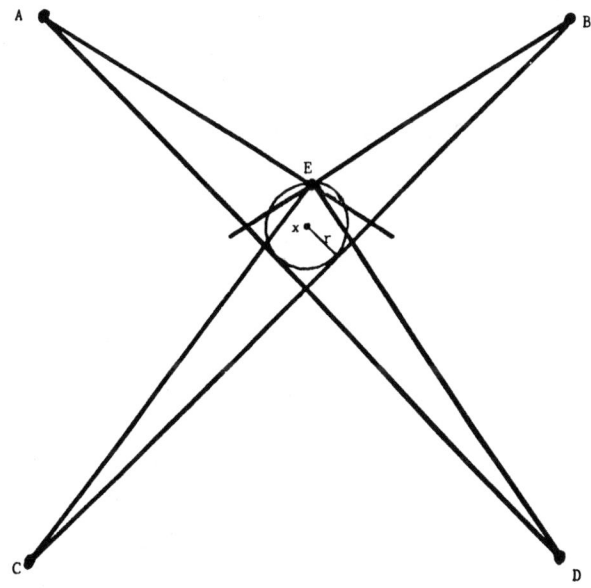

FIGURE 10. THE YOLK OF A FIVE PERSON VOTING SITUATION.

COPELAND WINNER (STRONG POINT)

In the two-dimensional spatial context, the appropriate definition of the Copeland winner is that alternative x which minimizes the area of $DEF(x)$. Thus the Copeland winner exists, and it is not hard to show that it is unique (unlike the tournament case). If there is a Condorcet winner, the Copeland winner is the Condorcet winner. A version of Landau's original argument goes through to show that the Copeland winner is in the uncovered set. In the spatial voting literature the Copeland winner has been called the *strong point*. [Grofman et. al., 1987]

If we are looking for a single point solution for the outcome of a spatial voting situation, the Copeland winner seems very attractive. One possible drawback might be that it could be difficult to compute. That is not so, due to a remarkable recent result of Owen and Shapley.

THEOREM 6. *The Copeland winner in a two-dimensional spatial voting situation is the weighted average of the voter ideal points, where the weights are the voter's Shapley-Owen power indices:*

$$X = \sum_{i=1}^{n} \psi_i V_i \quad \text{where } V_i \text{ is voter } i\text{'s ideal point.}$$

I refer you to [Owen and Shapley, 1988] for the general proof of this theorem, but I would like to illustrate in a simple case the elegant geometric insight on which it is based. In Figure 11 we would like to find a convenient formula for the area of $DEF(X)$, which in this case is the union of six segments of circles, one of which, $XSTU$, is shaded in the figure. The area of this segment is twice the difference between the circular sector BSX and the triangle BUX. Adding the six such areas together, and measuring angles in radians, we get

$$\text{Area of } DEF(X) = 2\left[\frac{\alpha}{2\pi}\pi\overline{AX}^2 + \frac{\beta}{2\pi}\pi\overline{BX}^2 + \frac{\gamma}{2\pi}\pi\overline{CX}^2 - \triangle ABC\right]$$
$$= \alpha\overline{AX}^2 + \beta\overline{BX}^2 + \gamma\overline{CX}^2 - 2\triangle ABC.$$

Since the area of triangle ABC is a constant not depending on X, to minimize the area of $DEF(X)$ we must choose X to minimize $\alpha\overline{AX}^2 + \beta\overline{BX}^2 + \gamma\overline{CX}^2$. But it is well known that this weighted sum of squared distances is minimized by taking

$$X = \frac{\alpha A + \beta B + \gamma C}{\alpha + \beta + \gamma} = \frac{\alpha}{\pi}A + \frac{\beta}{\pi}B + \frac{\gamma}{\pi}C.$$

Recall (Figure 5a) that $\frac{\alpha}{\pi}, \frac{\beta}{\pi}$ and $\frac{\gamma}{\pi}$ are the Shapley-Owen power indices of voters A, B and C.

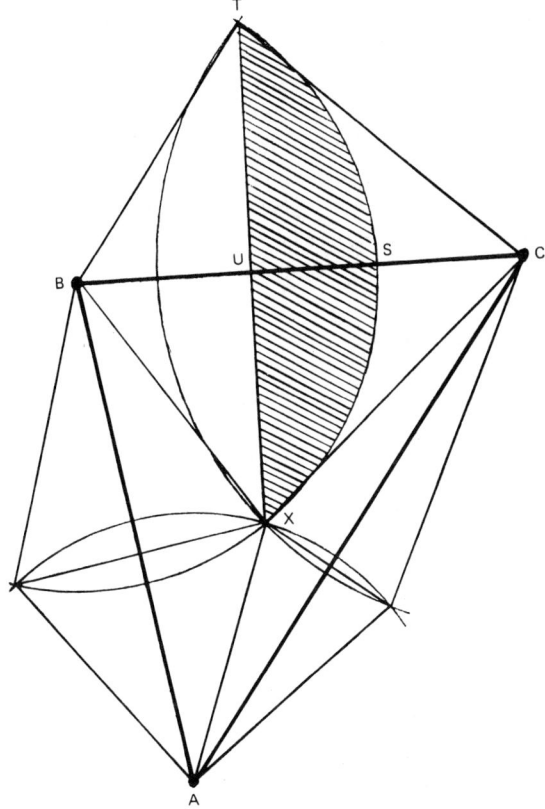

FIGURE 11. DERIVING A FORMULA FOR THE AREA OF $DEF(X)$.

Of course in a more general situation the geometry becomes more complicated, but the rotation algorithm for the Shapley-Owen index carries the basic geometric argument through. I should remark that while the other theorems in this section generalize easily to dimensions greater than two, the proof of Theorem 6 depends crucially on the rotation algorithm and how triangles fit together, and hence is two-dimensional. I have a counter-example which shows that the conclusion of Theorem 6 does not hold in dimensions greater than two.

Conclusion. Where has this excursion into mathematical models of voting situations taken us?

First, we have met a number of political ideas along the way, given precise forms within spatial models of voting. Centrist voters are more influential than extremists, and we can quantify this in the Shapley-Owen power index. When issues have several ideological dimensions, majority rule can be highly unstable, and agenda control can powerfully affect voting outcomes. In the midst of this instability, however, centrist outcomes are least unstable, and we can make this idea precise through the strong point. Spatial models lead to real and important political insights.

Second, we have seen that ideas from combinatorics and graph theory can find new meanings, relationships, and domains of applicability within the geometric context of spatial models. There are elegant mathematical insights here as well.

REFERENCES

COHEN, L. (1979), *Cyclic sets in multidimensional voting models*, Journal of Economic Theory 20: 1–12.

DAVIS, O., DEGROOT, M., AND HINICH, M. (1972), *Social preference orderings and majority rule*, Econometrica 40: 147–157.

ENELOW, J. AND HINICH, M. (1984), *The Spatial Theory of Voting: An Introduction*, Cambridge University Press, Cambridge.

FELD, S. AND GROFMAN, B. (1987), *Necessary and sufficient conditions for a majority winner in n-dimensional spatial voting games: an intuitive geometric approach*, American Journal of Political Science 31: 709–728.

GROFMAN, B., OWEN, G., NOVIELLO, N. AND GLAZER, G. (1987), *Stability and centrality of legislative choice in the spatial context*, American Political Science Review 81: 539–552.

KRAMER, G.H. (1977), *A dynamic model of political equilibrium*, Journal of Economic Theory 16: 310–334.

LANDAU, H.G. (1953), *On dominance relations and the structure of animal societies III: the conditions for a score structure*, Bull. Math. Biophysics 15: 143–148.

LUCAS, W. (1983), *Measuring power in weighted voting systems*, Chapter 9 in Brams S., Lucas W. and Straffin P., eds., Political and Related Models, Springer-Verlag, New York.

MAURER, S. (1980), *The king chicken theorems*, Mathematics Magazine 53: 67–80.

MCKELVEY, R. (1976), *Intransitivities in multidimensional voting models and some implications for agenda control*, Journal of Economic Theory 12: 472–482.

MCKELVEY, R. (1979), *General conditions for global intransitivities in formal voting models*, Econometrica 47: 1085–1112.

MCKELVEY, R. (1986), *Covering, dominance and institution-free properties of social choice*, American Journal of Political Science 30: 283–314.

MILLER, N. (1980), *A new solution concept for tournaments and majority voting*, American Journal of Political Science 24: 68–96.

MILLER, N. (1983), *The covering relationship in tournaments: two corrections*, Amer. J. Pol. Sci. 27: 382–385. See also Errata, 28: 434.

MOON, J.W. (1968), *Topics on Tournaments*, Holt, Rinehart and Winston, New York.

OWEN, G. (1971), *Political games*, Naval Research Logistics Quarterly 18: 345–354.

OWEN, G. AND SHAPLEY, L. (1989), *The Copeland winner and the Shapley value in spatial voting games*, International Journal of Game Theory, to appear.

PLOTT, C. (1967), *A notion of equilibrium and its possibility under majority rule*, American Economic Review 57: 787–806.

RABINOWITZ, G. AND MACDONALD, S. (1986), *The power of the states in U.S. presidential elections*, American Political Science Review 80: 65–87.

RIKER, W.H. (1980), *Implications from the disequilibrium of majority rule for the study of institutions*, American Political Science Review 74: 432–446.

ROHDE, D. AND SPAETH, H. (1976), *Supreme Court Decision Making*, Freeman, San Francisco.

SCHOFIELD, N. (1978), *Instability of simple dynamic games*, Review of Economic Studies 45: 575–594.

SHAPLEY, L. (1977), *A comparison of power indices and a non-symmetric generalization*, RAND paper P-5872, Rand Corporation, Santa Monica.

SHAPLEY, L. AND SHUBIK, M. (1954), *A method for evaluating the distribution of power in a committee system*, American Political Science Review 48: 787–792.

SHEPSLE, K. AND WEINGAST, B. (1984), *Uncovered sets and sophisticated voting outcomes with implications for agenda institutions*, American Journal of Political Science 28: 49–74.

STRAFFIN, P. (1980), *Topics in the Theory of Voting*, Birkhauser, Boston.

STRAFFIN, P. (1983), *Power indices in politics*, Chapter 11 in Brams, S., Lucas, W. and Straffin, P., eds., Political and Related Models, Springer-Verlag, New York.

SOME MATHEMATICS FOR DNA RESTRICTION MAPPING*

MICHAEL S. WATERMAN†

Abstract. DNA sequences are finite sequences over a four letter alphabet. Molecular biologists have available restriction enzymes which cut the DNA sequence at short patterns, specific to the enzyme. By using restriction enzymes singly and in combination, the biologist constructs a map of the location of the enzyme cut sites. It is shown that the simplest such problem is in the class of NP complete problems, and a simulated annealing algorithm for restriction mapping is studied. Under a simple probability model, the mapping problem is shown to have an exponentially increasing number of solutions.

Key words. DNA, restriction mapping, NP complete, stochastic annealing, subadditive ergodic theory

Introduction. In the last dozen or so years a major revolution has occurred in the biological sciences. The genetic information of an organism is encoded in DNA in the form of a linear sequence of nucleotides along either strand of the double helix. The nucleotides are adenine(A), cytocine(C), guanine(G), and thymine(T). DNA is double-stranded and the two strands are the Watson–Crick complements of each other: A forms hydrogen bonds with T on the opposite strand as C does with G. The nucleotides of each strand are arranged along a sugar–phosphate backbone. Each strand has a chemically determined direction determined by the necessity for the base pairs to fit properly; the sugars of one strand must all be in the opposite orientation of the sugars in the other strand. Thus it is justified for our purposes to simply view the information in a DNA molecule to be contained in one strand. This means that we will consider DNA to be a word over an alphabet of four letters {A,C,G,T}. The biological revolution has been created by experimental advances that allow the biologist to read and manipulate DNA. Lewin [12] provides an excellent general reference.

Currently the longest continuous sequence to be determined is that of Epstein Barr virus which is 172,282 nucleotides long. Segments of sequence from a large variety of organisms have been determined for a total of over 20×10^6 nucleotides in the spring of 1988. The average segment length is approximately 1000 nucleotides. The sequence of a DNA segment is fine scale information. If all the DNA of an organism, called the genome, is viewed as a very long street, then a segment of sequenced DNA corresponds to an ordered list of the houses along that segment. As is the case with street maps, it is very useful to have less detailed maps that allow the biologists to find their way about. The larger scale maps which are discussed below are fundamental to organizing unknown DNA sequences for further analysis and experiments. Maps at various levels of resolution are, as in a cross

---

*This research was supported by the National Institutes of Health (GM-36230) and by the System Development Foundation.

†Departments of Mathematics and Molecular Biology, University of Southern California, Los Angeles, California 90089-1113

continent drive to visit a friend in a large city, all useful. It should be no surprise that the first maps made are usually course scale maps.

Relative locations of genes along chromosomes have long been studied by a technique known as linkage analysis. Until recently it was required that the genes have observable mutations to serve as genetic markers for linkage analysis. The data for linkage mapping results from recombination of chromosomes: genes that are located near to one another will usually remain together after recombination, while those farther apart will often be separated by recombination. Thus fruit flies, bacteria and yeast have all had the relative locations of many genes mapped. In 1980 [2] it was proposed that measurable changes in the DNA itself could be used for linkage mapping. These changes are in the length of restriction fragments which are to be discussed below. Restriction fragment lengths are easily measured characteristics of DNA. Since 1980 there has been rapid progress in localizing genes associated with some major human diseases [11]. Still, in linkage mapping, the map distances do not usually correlate in a linear way with physical distances, that is with number of nucleotides. In this paper physical mapping will refer to maps of DNA where the distances are expressed in numbers of nucleotides.

Physical maps are also at more than one scale. Recently there has been much discussion of the physical mapping of genomes. One of these approaches can easily be outlined. The biologist constructs a library or sample space of randomly selected segments of DNA. The lengths of the segments, depending on the experimental approach, are restricted to some interval. In one popular approach the segment lengths are approximately 15×10^3 nucleotides. The biologist then randomly samples the segments and determines a "fingerprint" for each segment. Segments that have enough overlap will have sufficiently similar fingerprints for the experimentalist to detect the overlap. The goal is cover or map the genome with overlapping segments. $E.\ coli$ has just been mapped in this manner and with 4.7×10^6 nucleotides it is the largest genome to be physically mapped [9].

Restriction enzymes are fundamental to almost all the recent mapping techniques and to many experimental advances in molecular biology. In 1970 [14] site specific restriction enzymes were discovered in bacteria and over 300 have been isolated and are commercially available. Restriction enzymes cut double–stranded DNA at short specific patterns that are usually four to six letters in length. For example the enzyme EcoRI cuts DNA at GAATTC and HhaI cuts at GCGC. Biologists soon learned to map the approximate locations of the restriction sites. The resulting map is known as a restriction map and usually covers a few thousand nucleotides of DNA. Mathematical aspects of the construction of restriction maps is the topic of this paper; many of the basic results are from [8]. For a treatment of restriction maps as graphs see [15].

2. The double digest problem. The generic restriction mapping problem has two enzymes X and Y which can be applied to a segment of linear DNA of length L. Experimentally it is possible to apply the enzymes, alone or in combination, and to obtain approximate measurements of the resulting fragment lengths. In the interest of simplifying the problem as much as possible, it is assumed here that the measurements are exact. Of course the order of the fragments is unknown even after their measurement; when this order is determined the DNA is said to be mapped. From digestion with enzyme X, the set of length measurements is given by

$$\underline{X} = \{x_1, x_2, \ldots x_n\},$$

where by convention $x_1 \leq x_2 \leq \cdots \leq x_n$. From digestion with enzyme Y, the set of measurements is given by

$$\underline{Y} = \{y_1, y_2, \ldots y_m\},$$

and $y_1 \leq y_2 \leq \cdots \leq y_n$. In addition the enzymes are applied together to produce a list of the so-called double digest fragments. In the double digest the DNA is cut at all occurrences of pattern X or pattern Y. The set of double digest length measurements is given by

$$\underline{Z} = \{z_1, z_2, \ldots z_\ell\},$$

where as above $z_i \leq z_j$ when $i \leq j$. If all cuts are at distinct points, $\ell = n + m - 1$. Since the measurements are assumed exact,

$$\Sigma_i x_i = \Sigma_i y_i = \Sigma_i z_i = L.$$

The double digest problem(DDP) is to find permutations of $\underline{X}$ and $\underline{Y}$ such that the set of double digest fragments implied by these permutations is $\underline{Z}$. The problem can be stated more precisely as follows. For permutations $\sigma \in (123\ldots n)$, $\mu \in (12\ldots m)$ call (σ, μ) a configuration. By ordering $\underline{X}$ and $\underline{Y}$ according to σ and μ, respectively, the set of locations of cut sites are obtained

$$S = \{s : s = \sum_{1 \leq j \leq r} x_{\sigma(j)} \text{ or } s = \sum_{1 \leq j \leq t} y_{\mu(j)}; \ 0 \leq r \leq n, 0 \leq t \leq m\}.$$

The set S is not allowed repetitions, that is, S is not a multiset. Now label the elements of S such that

$$S = \{s_j : 0 \leq j \leq \ell\},$$

with $s_i \leq s_j$ for $i \leq j$. The double digest implied by the configuration (σ, μ) can be defined by

$$\underline{Z}(\sigma, \mu) = \{z_i(\sigma, \mu) : z_i(\sigma, \mu) = s_j - s_{j-1} \text{ for some } 1 \leq j \leq \ell\},$$

where it is assumed as usual that the set is ordered in the index i. The problem then is to find a configuration (σ, μ) such that $\underline{Z} = \underline{Z}(\sigma, \mu)$. As discussed next, this problem lies in the class of NP complete problems conjectured to have no polynomial time solution.

It should be pointed out that there is only one biologically correct solution, other than the obvious possibility of reversal of all digests. This distinguishes the problem from many problems in science where it is satisfactory to come close to the optimum.

3. DDP is NP complete.
A restriction map has been defined as permutations of the various digest fragments. Problem DDP requires that a pair of optimal permutations be found. It should be no surprise that the double digest problem is, along with the traveling salesman problem, NP complete. See [6] for definitions.

THEOREM 3.1. *The double digest problem DDP is NP complete.*

Proof. It is clear that the double digest problem DDP described above is in the class NP, since a nondeterministic algorithm need only guess a configuration (σ, μ) and check in polynomial time if $Z(\sigma, \mu) = \underline{Z}$. The number of steps to check this is in fact linear. To show that DDP is NP complete the partition problem is transformed to DDP.

In the partition problem, known to be NP complete [6], a finite set Q, say $|Q| = n$ is given along with a positive integer $s(q)$ for each $q \in Q$ and wish to determine whether there exists a subset $Q' \subset Q$ such that

$$\sum_{q \in Q'} s(q) = \sum_{q \in Q-Q'} s(q).$$

If $\sum_{q \in Q} s(q) = J$ is not divisible by two, there can be no such subset Q'. Otherwise input to problem DDP the data

$$\underline{X} = \{s(q) : q \in Q\}$$
$$\underline{Y} = \{J/2, J/2\}$$

and

$$\underline{Z} = \{s(q) : q \in Q\}.$$

It is clear that any solution to problem DDP with this data yields a solution to the partition problem through the order of the implied digest $\underline{Z}$. Therefore DDP is NP complete. □

This proof suggests that algorithms for the partition problem might be adapted to solve DDP. For example an algorithm might solve the problem of which subsets of $\underline{Z}$ sum to the elements of $\underline{X}$ and which subsets of $\underline{Z}$ sum to the element of $\underline{Y}$. Overlap between $\underline{X}$ and $\underline{Y}$ fragments is then determined by non-empty intersection of the associated subsets of $\underline{Z}$. In [4] the authors take this idea further and attempt to assign $\underline{Z}$ fragments to $\underline{X}$ and $\underline{Y}$ in a consistent way. Their algorithm does not yet seem to be effective for many examples encountered in practice but more research along these lines might result in a useful algorithm for construction restriction maps. Another approach to map construction is described in the section 5. Next non-uniqueness is discussed.

4. DDP has exponentially many solutions.

In many instances, the solution to the double digest problem is not unique. Consider for example

$$\underline{X} = \{1, 3, 3, 12\},$$
$$\underline{Y} = \{1, 2, 3, 3, 4, 6, \},$$

and

$$\underline{Z} = \{1, 1, 1, 1, 2, 2, 2, 3, 6\}.$$

This problem of size $4!6!/2! = 4320$; all 4320 permutations of $\underline{X}$ and $\underline{Y}$ have only 208 distinct double digests $\underline{Z}$. The $\underline{X}, \underline{Y}$, and $\underline{Z}$ given above was a test for the algorithm of section 5 and the large number of solutions was a surprise. Next it is demonstrated that this phenomenon is to be expected.

Below, the Kingman subadditive ergodic theorem is used to prove that the number of solutions to the double digest problem increases exponentially as a function of length under the probability model stated below.

For reference, a version of the subadditive ergodic theorem is given [10]. This remarkable theorem due to Kingman deserves to be more widely known and is very useful in some combinatorial situations. For s, t non–negative integers with $0 \leq s \leq t$ let $U_{s,t}$ be a collection of random variable which satisfy

(i) Whenever $s < t < u, U_{s,u} \leq U_{s,t} + U_{t,u}$,
(ii) The joint distribution of $\{U_{s,t}\}$ is the same as that of $\{U_{s+1,t+1}\}$,
(iii) The expectation $g_t = E[U_{0,t}]$ exists and satisfies $g_t \geq -Kt$ for some constant K and all $t > 1$.

Then the finite $lim_{t\to\infty} U_{0,t}/t$ exists and is a finite constant with probability one and in the mean.

To more fully examine this theorem, a version of the well–known strong law of large numbers is derived [5]. In this setting we are interested in the limit of the average $n^{-1} \sum_{1 \leq i \leq n} W_i$ where W_i are independent and identically distributed (iid) with expectation $E(W_i) = \mu$. Set

$$U_{s,t} = \sum_{s+1 \leq i \leq t} W_i.$$

It is easy to see that (i) is satisfied:

$$U_{s,u} = \sum_{s+1 \leq i \leq t} W_i + \sum_{t+1 \leq i \leq u} W_i.$$
$$= U_{s,t} + U_{t,u}.$$

Since W_i are iid (ii) is evidently true. Finally $g(t) = E(U_{0,t}) = t\mu$, so that (iii) holds with $\mu = -K$. Therefore the limit

$$\lim_{t \to \infty} \sum_{1 \le i \le t} W_i$$

exists and is constant with probability 1. Notice that this set up does not allow us to conclude that the limit is μ. This is a price of relaxing the assumption of additivity. Now we turn to the problem of the multiplicity of solutions to DDP.

THEOREM 4.1. *Assume the sites for two restriction enzymes are independently distributed with probability of cut p_1, p_2 respectively and $p_i \in (0,1)$. Let $V_{s,t}$ be the number of solutions between the s-th and the t-th coincident cut sites. Then there is a constant $\lambda > 0$ such that*

$$\lim_{t \to \infty} \frac{\log (V_{0,t})}{t} = \lambda.$$

Proof. Let a coincidence be defined to be the event that a site is cut by both restriction enzymes; such an event occurs at each site independently with probability $p_1 p_2 > 0$, and at site 0 by convention. At sites 1, 2, 3, ..., there will be an infinite number of such events. For $s, u = 0, 1, 2, \ldots, 0 \le s \le u$, consider the double digest problem for only that segment located between be s^{th} and u^{th} coincidence. Let $V_{s,u}$ denote the number of solutions to the double digest problem for this segment.

Let $s < t < u$. A solution for the segment between the sth and t^{th} coincidence and a solution for the segment between the tth and u^{th} coincidence can be combined for a solution for the segment between the sth and u^{th} coincidence. Thus

$$V_{s,u} \ge V_{s,t} V_{t,u}.$$

Note that the inequality may be strict as $V_{s,u}$ counts solutions given by orderings where fragments initially between, say, the sth and t^{th} coincidence now appear in the solution between the tth and u^{th} coincidence. Letting

$$U_{s,t} = -\log V_{s,t},$$

then $s \le t \le u$ implies $U_{s,u} \le U_{s,t} + U_{t,u}$.

Some additional technical details can be established to complete the proof of the theorem. The key to the proof is to set things up so that subadditivity holds. □

5. Stochastic annealing. Since the restriction mapping problem is solved by ordering the restriction fragments, it is natural to consider adapting algorithms for traveling salesman problem to restriction mapping. Recently [1] proposed applying the Metropolis algorithm, decreasing the temperature (a parameter) as the algorithm proceeds. The new algorithm is known as simulated or stochastic annealing.

Simulated annealing was first proposed by Metropolis et al[13] and has its origins in statistical mechanics. Suppose S is a finite set and $f : S \to R$. The problem is to find s^* satisfying $f(s^*) = \min_{s \in S} f(s)$. Think of S as the ensemble of configurations possible for some physical system and of $f(s)$ as the energy of the system in configuration s. Our analogy in statistical mechanics is with Gibbs distributions. For temperature T the Gibbs distribution is a probability distribution π_T with state space S defined by

$$\pi_T(s) = c\exp(-f(s)/T),$$

where c is chosen so that $\sum_s \pi_T(s) = 1$. When T is large (high temperature), π_T is nearly uniform over S, since $\exp(-f(s)/T) \approx 1$. For small T, π_T has its mass concentrated near the minima of f. It should be emphasized that the mathematical problem of locating minima is not a physical problem and that discussing the connection with statistical mechanics is only to motivate the annealing algorithm. The goal of this approach is to sample from π_T for small T. We can consider the notion of starting at large T and "cooling" or decreasing T in hopes that $\lim_{T \to 0} \pi_T = \pi_0$. Geman and Geman[1] show that this obtains if cooling is through a sequence $T_n \downarrow 0$ with $T_n \geqq c/\log(n)$, where $c = c(f)$. It is necessary to go into little more detail to carefully explain these ideas.

The strategy is to simulate a Markov chain X_n, $n \geqq 0$, with state space S which has π_T as its equilibrium or stationary distribution. To specify the transitions of the chain, for each $s \in S$, let the neighborhood N_s be a subset of S which satisfies for all $s \in S$, $s \in N_t$ if and only if $t \in N_s$, and $|N_s| = |N_t|$. Now define transition probabilities $p_T(s,t) = p_T(X_{n+1} = s | X_n = t)$ by

$$p_T(s,t) = 0, \quad \text{if} \quad s \in N_t^c,$$
$$p_T(s,t) = \exp\{-(f(s) - f(t))^+/T\}/|N_t|, \text{ if } s \in N_t \text{ and } s \neq t.$$

The quantity $p_T(t,t)$ is set so that $\sum_s p_T(s,t) = 1$. The fact that

$$p_T(s,t)\pi_T(t) = p_T(t,s)\pi_T(s)$$

implies that π_T is the unique stationary distribution of the Markov chain [5].

This method can be applied to our mapping problem as soon as we take care of three important specifications: (1) the set S, (2) the function f to be minimized and (3) the neighborhood structure. The set $S = \{(\sigma, \mu) : \sigma \text{ is a permutation of } (12\ldots n)$ and μ is a permutation of $(12\ldots m)\}$. Recall that σ denotes the permutation of the X digest and μ denotes the permutation of the Y digest. The energy function is taken to be the chi-square like function

$$f(\sigma, \mu) = \sum_{1 \leqq i \leqq n_{1,2}} (z_i(\sigma, \mu) - z_i)^2/z_i;$$

If all measurements are free of error then f attains its global minimum of zero for at least one choice (σ, μ).

Following [8], define the set of neighbors of a configuration (σ, μ) by

$$N(\sigma, \mu) = \{(\tau, \mu) : \tau \in N(\sigma)\} \cup \{(\sigma, \nu) : \nu \in N(\mu)\},$$

where $N(\rho)$ is defined next. The approach is to utilize the work of [1] on the traveling salesman problem where a neighborhood $N(\rho)$ is defined for a permutation ρ. There $\tau \in N(\rho)$ if $\rho = (i_1 i_2 \ldots i_\ell)$ and $\tau = (i_1 i_2 \ldots i_{j-1} i_k i_{k-1} \ldots i_{j+1} i_j i_{k+1} \ldots i_\ell)$. That is, a segment of the original permutation is reversed.

With these ingredients, the algorithm was tested on exact, known data from the bacteriophage lambda (of 48,502 nucleotides) with restriction enzymes BamHI and EcoRI, yielding a problem size of $n!m! = 6!6! = 518,400$. See Daniels et al. [3] for complete sequence and map information about Lambda. Temperature was not lowered at the rate $c/\log(n)$ as suggested by the theorem in Geman and Geman [7], but for computational reasons was instead lowered by the schedule $T_n = 1/n$. In three trials using various annealing schedules the solution was located after 29,702, 6895, and 3670 iterations from random initial configurations. While the map of lambda is known from its DNA sequence, it was used as a test case because it is a realistic problem of a size relevant to biology. We comment on the relevance and effect of assuming the length data is without errors in the next section.

6. Conclusion. An important aspect of this problem is that biologists want the map that is consistent with the corresponding DNA sequence of the organism. In this experimental science there is appropriately a strong emphasis on correct inference from the data. Maps that fit the data well according to some criterion such as our chi–square like function are of no value unless they are also biologically correct.

Currently most restriction mapping is done by hand. The scientist is usually unaware of the double hazards of mapping, one of an NP hard problem and the other of an explosion of the number of solutions. The scientist frequently has developed a good sense of how to construct as well as when to believe a map. A common approach to DNA regions that appear to be difficult to map is to change enzymes.

Much remains to be done in this area. The most important job for mathematical scientists is to devise an algorithm to solve problems of practical interest. It should be possible to include the strategy of introducing new enzymes described above. We have not discussed one additional difficulty: Experimental data of course does not consist of exact measurements. Instead the measurements are approximately lognormally distributed. The stochastic annealing method described above performs poorly on such data; we have little of interest to report other than the general failure of the algorithm.

REFERENCES

[1] E. BONOMI AND J-L LUTTON, *The N–city travelling salesman problem: Statistical mechanics and the Metropolis algorithm*, SIAM Rev. 26 (1984), 551–568.

[2] D. BOTSTEIN, R.L. WHITE, M. SKOLNICK, AND R. DAVIS, *Construction of a genetic linkage map in man using restriction fragment length polymorphisms*, Am. J. Hum Genet. 32 (1980), 314–331.

[3] D. DANIELS, J. SCHROEDER, W. SZYBALSKI, F. SANGER, A. COULSON, G. HONG, D. HILL, G. PESSON AND F. BLATTNER, *Complete annotated lambda sequence in "Lambda II,"* (R.W. Hedrix, J.W. Roberts, and F.W. Weisberg, Eds.), Cold Spring Harbor Laboratory, 1983.

[4] W.M. FITCH, T.F. SMITH, AND W.W. RALPH, *Mapping the order of DNA restriction fragments*, Gene, 22 (1983), 19–29.

[5] W. FELLER, *An Introduction to Probability and Its Applications*, third edition, John Wiley and Sons, New York, 1950.

[6] M.R. GAREY AND D.S. JOHNSON, *Computers and Intractability: A Guide to the Theory of NP–Completeness*, Freeman, San Francisco, 1979.

[7] S. GEMAN AND D. GEMAN, *Stochastic relaxation*, Gibbs distribution, and the Bayesian restoration of images, IEEE Trans Pattern Anal. Mach. Intell. 6 (1984), 721–741.

[8] L. GOLDSTEIN AND M.S. WATERMAN, *Mapping DNA by stochastic relaxation*, Adv. Appl. Math. 8 (1987), 194–207.

[9] K. ISONO, Y. KOHARA, AND K. AKIYAMA, *The physical map of the whole E. coli. chromosome: application of a new strategy for rapid analysis and sorting of a large genomic library*, Cell, Vol. 50 (1987), 495–508.

[10] J.F.C. KINGMAN, *Subadditive ergodic theory*, Ann. Probab., 1 (1973), 883–909.

[11] E.S. LANDER AND P. GREEN, *Construction of multilinkage maps in humans*, Proc. Natl. Acad. Sci. USA, 84 (1987), 2363–2367.

[12] B. LEWIN, *Genes III*, Second Edition, John Wiley & Sons New York, 1987.

[13] N. METROPOLIS, A ROSENBLUTH, M. ROSENBLUTH, A. TELLER AND E. TELLER, *Equations of state calculations by fast computing machines*, J. Chem. Phys. 21 (1953), 1087–1092.

[14] D. NATHANS AND H.O. SMITH, *Restriction endonucleases in the analysis and restructuring of DNA molecules*, Ann. Rev. Biochem., 44 (1975), 273–293.

[15] M.S. WATERMAN AND J.R. GRIGGS, *Interval graphs and maps of DNA*, Bull. Math. Biol., 48 (1986), 189–195.